The Intended and Unintended Effects of U.S. Agricultural and Biotechnology Policies

A National Bureau of
Economic Research
Conference Report

The Intended and Unintended Effects of U.S. Agricultural and Biotechnology Policies

Edited by **Joshua S. Graff Zivin and Jeffrey M. Perloff**

The University of Chicago Press

Chicago and London

JOSHUA S. GRAFF ZIVIN is associate professor of economics at the University of California, San Diego, where he holds faculty positions in the School of International Relations and Pacific Studies and the Department of Economics. He is research director for international environmental and health studies at the Institute on Global Conflict and Cooperation and a research associate of the National Bureau of Economic Research. JEFFREY M. PERLOFF is professor in the Department of Agricultural and Resource Economics at the University of California, Berkeley, and a former member of the National Bureau of Economic Research's Board of Directors.

The University of Chicago Press, Chicago 60637
The University of Chicago Press, Ltd., London
© 2012 by the National Bureau of Economic Research
All rights reserved. Published 2012.
Printed in the United States of America

21 20 19 18 17 16 15 14 13 12   1 2 3 4 5
ISBN-13: 978-0-226-98803-0 (cloth)
ISBN-10: 0-226-98803-1 (cloth)

Library of Congress Cataloging-in-Publication Data

The intended and unintended effects of U.S. agricultural and
    biotechnology policies / edited by Joshua S. Graff Zivin and
    Jeffrey M. Perloff.
        p. cm.—(National Bureau of Economic Research conference
    report)
    Includes bibliographical references and index.
    ISBN-13: 978-0-226-98803-0 (cloth : alk. paper)
    ISBN-10: 0-226-98803-1 (cloth : alk. paper) 1. Agriculture and
    state—United States. 2. Agricultural subsidies—Government
    policy—United States. 3. Agriculture and energy—Government
    policy—United States. 4. Agricultural biotechnology—Govern-
    ment policy—United States. 5. Crop insurance—United States.
    6. Biomass energy—Government policy—United States. I. Graff
    Zivin, Joshua S. II. Perloff, Jeffrey M. III. Series: National Bureau
    of Economic Research conference report.
    HD1761.I56 2012
    338.1'873—dc23

                                                                2011020188

⊗ This paper meets the requirements of ANSI/NISO Z39.48-1992
(Permanence of Paper).

Relation of the Directors to the
Work and Publications of the
National Bureau of Economic Research

1. The object of the NBER is to ascertain and present to the economics profession, and to the public more generally, important economic facts and their interpretation in a scientific manner without policy recommendations. The Board of Directors is charged with the responsibility of ensuring that the work of the NBER is carried on in strict conformity with this object.

2. The President shall establish an internal review process to ensure that book manuscripts proposed for publication DO NOT contain policy recommendations. This shall apply both to the proceedings of conferences and to manuscripts by a single author or by one or more co-authors but shall not apply to authors of comments at NBER conferences who are not NBER affiliates.

3. No book manuscript reporting research shall be published by the NBER until the President has sent to each member of the Board a notice that a manuscript is recommended for publication and that in the President's opinion it is suitable for publication in accordance with the above principles of the NBER. Such notification will include a table of contents and an abstract or summary of the manuscript's content, a list of contributors if applicable, and a response form for use by Directors who desire a copy of the manuscript for review. Each manuscript shall contain a summary drawing attention to the nature and treatment of the problem studied and the main conclusions reached.

4. No volume shall be published until forty-five days have elapsed from the above notification of intention to publish it. During this period a copy shall be sent to any Director requesting it, and if any Director objects to publication on the grounds that the manuscript contains policy recommendations, the objection will be presented to the author(s) or editor(s). In case of dispute, all members of the Board shall be notified, and the President shall appoint an ad hoc committee of the Board to decide the matter; thirty days additional shall be granted for this purpose.

5. The President shall present annually to the Board a report describing the internal manuscript review process, any objections made by Directors before publication or by anyone after publication, any disputes about such matters, and how they were handled.

6. Publications of the NBER issued for informational purposes concerning the work of the Bureau, or issued to inform the public of the activities at the Bureau, including but not limited to the NBER Digest and Reporter, shall be consistent with the object stated in paragraph 1. They shall contain a specific disclaimer noting that they have not passed through the review procedures required in this resolution. The Executive Committee of the Board is charged with the review of all such publications from time to time.

7. NBER working papers and manuscripts distributed on the Bureau's web site are not deemed to be publications for the purpose of this resolution, but they shall be consistent with the object stated in paragraph 1. Working papers shall contain a specific disclaimer noting that they have not passed through the review procedures required in this resolution. The NBER's web site shall contain a similar disclaimer. The President shall establish an internal review process to ensure that the working papers and the web site do not contain policy recommendations, and shall report annually to the Board on this process and any concerns raised in connection with it.

8. Unless otherwise determined by the Board or exempted by the terms of paragraphs 6 and 7, a copy of this resolution shall be printed in each NBER publication as described in paragraph 2 above.

Contents

Acknowledgments

We are very grateful to the National Bureau of Economic Research, the Giannini Foundation, and the Agricultural and Applied Economics Association Foundation for funding this conference. We thank Ernst Berndt, Martin Feldstein, Donald Fullerton, GianCarlo Moschini, James Poterba, and David Zilberman for extensive help in organizing this conference.

An Overview of the Intended and Unintended Effects of U.S. Agricultural and Biotechnology Policies

Joshua S. Graff Zivin and Jeffrey M. Perloff

The eight chapters in this volume were presented at the National Bureau of Economic Research Agricultural Economics Conference, March 4–5, 2010. The conference focused on two central issues confronting agricultural economists and policymakers today. Five of the chapters examine the efficiency and distributional effects of crop subsidy programs. The other three chapters examine the effects of biofuels and other biotechnology policies and innovations on commodity prices and social welfare. All the chapters bring new empirical methods and new data to analyze issues that are at the forefront of debates in agricultural policy circles. The importance of institutional details and unintended consequences are common themes that run throughout.

Agricultural Policy

Most introductory economics textbooks blithely describe agricultural markets as unregulated and perfectly competitive. Nonetheless, virtually every aspect of agricultural markets is regulated or subject to other government interventions. The U.S. government regulates farm production, prices, land use, environmental quality, organic foods, and food safety, among many other aspects of agriculture. The government also provides food or food stamps to poor consumers and trade assistance and insurance to farmers. Despite peri-

Joshua S. Graff Zivin is associate professor of economics at the University of California, San Diego, where he holds faculty positions in the School of International Relations and Pacific Studies and the Department of Economics. He is research director for international environmental and health studies at the Institute on Global Conflict and Cooperation and a research associate of the National Bureau of Economic Research. Jeffrey M. Perloff is professor in the Department of Agricultural and Resource Economics at the University of California, Berkeley, and a former member of the National Bureau of Economic Research's Board of Directors.

odic debates in Congress about deregulating these markets and some reduction in direct transfers to agricultural producers, the types of government intervention in food and agriculture has, if anything, increased over the years.

Shortly after the end of World War I, the U.S. government started aggressively intervening in agricultural markets. A complex set of regulations evolved that provided support for growers of major field crops (barley, corn, upland cotton, oats, rice, sorghum, and wheat). From 1973 to 1995, farmers almost always received deficiency payments: the difference between the market price and a target price times a quantity determined by a farmer's "base" acreage and crop yield.

Although it is not well known to most economists, agricultural price supports in the past several decades have been part of what are now known as "crop insurance" programs. Farmers used a specified quantity of their crop as collateral for a nonrecourse loan equal to that quantity times a price called the commodity loan rate. The loan rate serves as a support price. In most versions of the program, after the crop was harvested, farmers repaid the loans if the market price exceeded the support price; otherwise, they essentially sold the crop to the government by forfeiting the actual physical product.

In recent years, Congress passed an omnibus "Farm Bill" package of legislation that modifies the federal programs roughly every five years. The most recent federal agricultural legislation, the 2008 Farm Bill, allocates $288 billion to farmers (e.g., income support and insurance), to consumers (e.g., food stamps), and for biofuel research and production from 2008 through 2012. Of this total, $43.3 billion goes to the commodity loan program.

The traditional agricultural price support subsidies—as typically described in intermediate microeconomics textbooks—are no longer used. The 1996 Federal Agriculture Improvement and Reform (FAIR) Act, also known as the Freedom to Farm Act and as the Agricultural Market Transition Act, made major changes in agricultural policy. In theory, it was supposed to wean farmers from government support. A farmer's eligibility for support was no longer based on historical production; farmers could plant whatever crops they desired (including soybeans, which were not traditionally supported), and farmers received loan deficiency payments based on the difference between the market price and a target support price (the loan rate). These payments were relatively infrequent in some years for some crops.

To aid in the transition from agricultural price supports, FAIR provided a seven-year period of income support through direct payments that were (allegedly) independent of production and were known in advance. This "temporary" direct income support is now well entrenched—indeed, it was expanded in two subsequent farm bills—and costs $5 billion per year. The federal government also provides countercyclical payments that are triggered by low prices but not tied to current production; loan deficiency payments and marketing loans, which use loan rates to support market prices;

and a subsidized crop insurance program. According to proponents, this significantly modified and expanded insurance program would help farmers manage financial risk and reduce the likelihood that Congress would have to pass supplemental ad hoc disaster assistance. Insurance is provided for more than 350 commodities in all states and Puerto Rico, and more than 80 percent of eligible acres are now insured under the program.

All five of the chapters on agricultural policy focus on aspects of price supports or federal crop insurance. Two chapters examine who captures agricultural support, taking account of vertical relations. Barry K. Goodwin, Ashok K. Mishra, and François Ortalo-Magné estimate the distribution of agricultural subsidies in upstream land markets, while Rachael E. Goodhue and Carlo Russo investigate whether downstream processors capture a large share of the subsidies. These chapters examine many of the most important current policy questions.

The other three of these chapters focus on the newly important crop insurance program. Bruce A. Babcock examines the politics and the efficiency of the U.S. Crop Insurance Program. Ethan Ligon looks at the supply effects of the new specialty crop insurance program. Jeffrey LaFrance, Rulon Pope, and Jesse Tack examine how crop farmers respond to changes in risk more generally.

The Distribution of Agricultural Policy Rents and Market Power in Vertical Markets

The record clearly shows that, throughout the history of the commodity support program, Congress intended that the beneficiaries be family farmers, the salt of the earth. However, it is well known that most of the support funds go to relatively large farms owned by corporations. Although some of these are family farm corporations, many others are owned by Fortune 500 companies, members of Congress, and by other wealthy people with no obvious connection to farming. What is not as well known is that a large share of these transfers goes to owners of firms in other related markets.

Traditionally, most analyses of the effects of agricultural programs have focused—not surprisingly—on the directly regulated agricultural crop market. However, as these markets are intimately connected to other markets, it is important to examine the spillover effects. Two of the chapters in this volume examine how agricultural transfers are shared with other vertically related markets. One chapter looks downstream at processors, while the other looks upstream at landlords.

The two chapters individually and collectively make a significant contribution to our understanding of agricultural support programs by showing that much of the support payments go to people other than Congress's intended recipients, farmers. In addition, the chapters indicate new efficiency harms from the commodity support program that have not been previously identified.

Barry K. Goodwin, Ashok K. Mishra, and François Ortalo-Magné (GMOM) focus on how the commodity support program benefits are distributed. Because these subsidies have been provided for over three-quarters of a century and are expected to persist, GMOM argue that farm land values should capture the subsidy benefits. They estimate the extent of this capitalization and conclude that landowners are the major beneficiaries of the farm programs. New owners of land pay prices that reflect expected future benefits, so the ultimate recipients of the benefits are the land owners at the time the benefits are instituted. Moreover, as nearly half (45 percent) of U.S. farmland is owned by nonfarmers, and 57 percent of agricultural landlords are nonfarm corporations or individuals who work in or are retired from nonfarm-related activities, much of the benefits go to nonfarmers. Goodwin, Mishra, and Ortalo-Magné carefully examine the role of leasing and show that lease rates capture much of the agricultural subsidy payments.

Although a number of previous studies examined the capitalization of price subsidies into farm values, GMOM are the first to resolve a number of very vexing data problems. More important, they have data on lease arrangements and rates, so they can estimate how landlords and farmers split the government transfers. Also, for the first time in this literature, they determine how the market values the insurance feature of farm programs. They show that programs that provide insurance and lower the risk from uncertain farm earnings cause the share lease rate premium to fall substantially.

They examine a variety of farm support programs, which ranged from price-support payments based on current market conditions to other programs that were, in principle, not connected to market conditions or production. They find that the effects on land values were smallest for programs that were decoupled from output and were supposed to be transitory, moderate for programs whose payments may have signaled future benefits, and largest for price support programs.

In "Modeling Processor Market Power and the Incidence of Agricultural Policy: A Nonparametric Approach," Rachael E. Goodhue and Carlo Russo investigate the link between downstream, processor market power and U.S. agricultural policies. They examine marketing loan and pre-1986 deficiency payment programs in which payments are made to farmers only if the market price was less than a target price.

This chapter makes an important contribution to the literature by examining how federal programs affect firms' ability to exercise market power and how some of the rents intended for farmers are captured by downstream firms. Although earlier studies indicate that firms in many relevant markets have substantial oligopoly or oligopsony power, traditionally most analyses of federal subsidy programs have assumed that these markets are competitive. Given the assumption of competition, most previous studies of the effects of agricultural price support policies have examined the size of the transfer to farmers relative to the deadweight loss. However, if, for example, downstream processors possess oligopsony market power, they can extract some of the policy rents.

Goodhue and Russo examine the wheat market, which is highly competitive (with over 160,000 farms) and the vertically related milling market (with a four-firm concentration ratio of 57 percent), which they argue is not perfectly competitive. They focus on how the policy affects the margin between the price of flour and the price of wheat, taking account of the possibility that the policy may change millers' behavior. The premise of the chapter is that if, all else is the same, the oligopsony margin increases when farmers receive deficiency payments, then by comparing periods with and without payments, millers' oligopsony power can be detected. Their approach uses the policy as an experiment, exploiting the changes in millers' strategic pricing behavior as an identification device for oligopsony power.

They use an innovative nonparametric model. Hopefully, by using a nonparametric model, they are less vulnerable to specification bias than in the traditional structural approaches to estimating market power. Unlike the now standard structural or reduced-form approaches to estimating market power, they compare the outcomes of using constrained and unconstrained sliced inverse regressions to identify which factors affect millers' pricing behavior.

Constraining the factors in the sliced inverse regression to generate coefficients that are easily interpreted using economic theory does not affect the results. Their results are consistent with the story that a change in policy regime triggers a change in pricing behavior. For years when the policy is binding, millers appear to absorb as least as large of a share of a marginal cost increase as they do in years when the policy is not binding.

They conclude that wheat millers were able to extract an additional 23¢ or 24¢ per hundred weight of flour by increasing their marketing margins in years farmers received program payments. This increase was approximately 10 percent of the estimated marketing margin in years farmers received program payments. However, the amount of the increase depends on market conditions: the difference in the expected margin between the two policy regimes is smaller when farmer supply is low, suggesting that millers' extraction of policy rents has a nonlinear relationship with the market price.

The chapter concludes that U.S. wheat millers are able to extract some of the transfer payments intended for farmers. Thus, the analysis suggests that the general assumption that competitive models may be a good approximation for imperfectly competitive agricultural markets does not necessarily hold, particularly if distribution, as well as efficiency, is a concern.

Insurance

Bruce Babcock's chapter analyzes whether the crop insurance program provides farmers with an efficient risk transfer mechanism that the private sector cannot provide. Historically, private crop insurance markets have failed. A key question is whether this failure was due to the systemic nature of crop losses, which prevents pooling, or a lack of demand. As Babcock notes, if the private sector cannot provide adequate crop insurance, then a government risk transfer market could improve welfare. However, if the private insurance markets do not exist because of a lack of demand, then the government's crop insurance program could result in large welfare losses. Thus, he focuses on whether there is unmet demand for crop insurance.

Proponents' justification for heavily subsidizing insurance rates in recent years was that virtually no farmers were buying insurance. Babcock examines how farmers' demand for insurance responds to a change in the actuarial fairness of the insurance. His task is difficult because the government's rate making methods are not consistent with actuarial fairness. A significant contribution of his chapter is the construction of a data set that allows him to estimate the degree of actuarial fairness of corn, soybean, and wheat contracts.

In years in which insurance premiums exceed indemnities, crop insurance companies get to keep a portion of the difference, which is called the net underwriting gain. For example, in 2004, premiums exceeded claims by $979 million, and underwriters kept $848 million, including $692 in underwriting gains. From 1998 through 2009, underwriters had gains in all but one year.

The government provides reinsurance to underwriters. Babcock concludes that program costs would be substantially lower if the federal government took all the risk from the crop insurance program rather than buying an overpriced insurance policy from the crop insurance companies.

Farmers' demand for crop insurance depends on expected returns and risk reduction. Farmers pay only about 41 percent of the amount needed to cover insured losses. This large subsidy means that most farmers will get substantially more back from the program than they pay it, but there is wide variance in actuarial fairness over crops and over time. Babcock develops a method to isolate the demand for risk reduction from the demand for expected returns. He can estimate the demand for actuarially fair insurance contracts because before the recent increase in the size of the subsidies, rates were less than actuarially fair, whereas after the change, they were more than actuarially fair.

The vast majority of corn, soybean, and wheat farmers obtained less than the 85 percent maximum available coverage in 2009. Babcock finds that farmers who are offered more actuarially fair insurance would increase their insurance coverage. The reason that very few farmers bought insurance in the past was because the rates were much less than actuarially fair. He finds that more than half of the acreage that was able to be insured at higher coverage levels at actuarially fair incremental premiums was insured. That is, a large number of producers find that the risk reductions offered by revenue insurance generate significant value. Thus, he concludes that it is not necessary to heavily subsidize the rates to make them more than actuarially fair for many farmers to buy insurance.

Of course, a plausible explanation for the large premium subsidies is effective lobbying by interest groups. Consequently, he examines the interest groups who capture rents from the program to illuminate the political forces that have led to an expansion of these subsidies. This explanation is consistent with Becker's theory of legislation as a reflection of economic payoffs from the application of pressure by affected interest groups. The effectiveness of interest group lobbying is illustrated by the congressional ban on price competition between crop insurance agents striving for farmers' business. Babcock concludes that the chief beneficiaries of this lobbying activity are agents and farmers in the Great Plains.

The justification offered for providing such insurance is to help reduce risk of producers and thereby induce them to produce more to benefit consumers. We know from Babcock's careful study of the actuarial fairness of this insurance that the government is subsidizing the insurance so that taxpayers' outlays exceed the gains to producers. Thus, if insurance is to raise overall welfare, there needs to be a substantial benefit to consumers. Ethan Ligon asks whether insurance against low yields on fruits and vegetables has a substantial effect on quantities and prices and, hence, whether producers and consumers benefit from the provision of government insurance.

Insurance policies for wheat date back to 1938, and policies for other program crops were introduced many years ago. However, coverage for specialty crops—particularly fruits and vegetables—is a relatively recent event. In 1981, 28 crop-county insurance contracts were offered for fruits and vegetables in California. That number rose to 500 in 1989, and large increases followed in 1990 and 1995 so that now about 2,300 crop-county contracts are available.

Farmers face greater risk for specialty crops, particularly fruits and vegetables, than for cereal crops because specialty crops have greater price variation than do storable commodities. However, although that greater risk might increase the demand for insurance relative to storable commodities, most fruits and vegetables in California are marketed using vertical intermediaries, partially to help manage risk, which lowers demand for insurance all else the same. In addition, specialty crops are grown in small geographical areas so that a single weather or other shock tends to have a larger effect on aggregate supply than for crops grown over larger areas. Bad weather could reduce supply and dramatically increase price, given inelastic demand curves, so that the amount of harm (if any) to an individual producer is unclear. Consequently, specialty farmers' demand for yield insurance might be low. These factors may explain why only a quarter of eligible California acreage was insured in 1988. More recently, expansion of crops covered, provision of quasi-mandatory insurance, and subsidies have increased usage of insurance.

Ligon is able to identify the effects of insurance because coverage for particular fruits and vegetables in particular counties was introduced slowly over time. Expansion of coverage results from a government agency's bureaucratic decision, which Ligon shows is unlikely to be based on maximizing farmers' profits.

Thus, Ligon exploits variation in the timing of the introduction of crop insurance policies across crops and counties to estimate the effects of crop insurance on output and price, using instruments to control for the timing of the introduction of insurance. He finds that the introduction of insurance for a given crop statistically significantly increases production by 164 percent for tree crops but does not have a statistically significant effect on nontree crops. This difference across crops may be a consequence of the much larger investment at risk in perennial crops than in annual crops. Moreover, although he clearly shows that insurance leads to greater production of tree crops, he notes that some of this expansion may reflect substitution away from other crops so that total production may not rise in proportion.

Ligon then estimates a corresponding average demand curve and finds that it is extremely flat. Thus, from Ligon's careful study, we can conclude that expansion of coverage of insurance to more specialty crops has very small effects on consumer welfare even given substantial output effects.

Jeffrey LaFrance, Rulon Pope, and Jesse Tack develop a new approach to analyzing the impact of the government's insurance programs and other policies on farmers' risk response using aggregate data. Farmers have to choose how many inputs to use before output prices are revealed. When analyzing supply responses, economists have to model the process by which farmers form expectations, which is a particularly difficult challenge with aggregate data. LaFrance, Pope, and Tack, in a tour de force, develop a comprehensive structural econometric model of variable input use; crop mix and acreage choices; investment and asset management decisions; and consumption, savings, and wealth accumulation in a stochastic dynamic programming model of farm-level decision making over time. LaFrance, Pope, and Tack develop necessary and sufficient condition on cost and technology to allow variable input demand equations to be specified as functions of input prices, quasi-fixed inputs, and total variable cost. They develop and estimate a flexible, exactly aggregable, and economically regular model of ex ante variable input demands.

They develop a new class of variable input demand systems in a multi-product production setting. All of the models in this class can be estimated with observable data, are exactly aggregable, are consistent with economic theory for any von Neumann-Morgenstern expected utility function, and can be used to nest and test exact aggregation, economic regularity, functional form, and flexibility. Implications of monotonicity, concavity in prices, and convexity in outputs and quasi-fixed inputs are developed for a specific subset of this class of models.

LaFrance, Pope, and Tack use a coherent framework to estimate this model while dealing with the major questions that arise in estimating production functions: choice of functional form; degree of flexibility; conditions required for and regions of economic regularity; consistency with aggregation from micro- to macro-level data; and how best to handle simultaneous equations bias, errors in variables, and latent variables in a structural econometric model.

They use their new model to analyze acreage and supply decisions under risk for ten crops that represent roughly 95 percent of total farm revenue from crop production and nearly all crop acreage: soybeans, corn, cotton, hay, potatoes, rice, sugar beets, sugarcane, tobacco, and wheat. They estimate Euler equations for the excess return to investing in agriculture, personal consumption expenditures, and the rate of return to stocks as measured by the Standard & Poor's (S&P) 500 index. Because they are careful to include government payments in crop revenues, their estimated model can be used to examine how behavior changes in response to those payments. They take account of risk aversion in agricultural production and investment decisions.

Using the LaFrance, Pope, and Tack model, one can analyze the effects of any policy that alters the distribution of agricultural crop income on the choices that restore equilibrium. A major implication is that agricultural policy may affect other markets by changing nonagricultural investment and consumption. The estimated social value of public crop insurance is likely to be reduced as more margins for adjustment (arbitrage conditions) are included in the analysis. Unlike in traditional, static studies (that essentially estimate long-run effects), this model can be used to distinguish short-run and long-run effects. In particular, public subsidies that raise the return to insurance have larger long-run than short-run effects.

Biofuels and Biotechnology

While the processing of foodstuffs into fuel has been around for at least a century, interests in large-scale biofuel production in the United States dates back to the oil crises of the 1970s, which ultimately led to the creation of the National Renewable Energy Laboratory. Continued concerns about energy security as well as increasing awareness about climate change led to the promulgation of the Energy Policy Act of 1994, whose goal was to promote the production and use of renewable fuels. Nearly a decade later, the Energy Policy Act of 2005 was enacted, which created the first national renewable fuel standard (RFS). This RFS called for annual biofuel usage of 7.5 billion U.S. gallons by 2012 and established blending requirements for refiners and importers of gasoline.

Within the first year of the enactment of the Energy Policy Act of 2005, the United States surpassed Brazil as the largest producer of ethanol in the world, with nearly five billion gallons of mostly corn-based ethanol production in 2006. By the end of the following year, annual federal support for the ethanol industry had topped \$3.25 billion.[1] In 2008, the production of U.S. corn ethanol had nearly doubled, consuming one-third of U.S. corn production and raising the price of corn.[2] Concerns about the effects of these requirements on agricultural commodity prices led to a revision of the National Renewable Fuel Standard in 2009. The new requirements establish volumetric standards for cellulosic and advanced biofuels that would use nonedible feedstocks in the production process.

Although a growth industry has sprung up of academics studying these issues, most of these studies take place in isolation, typically focusing on one particular aspect of the problem rather than the linkages between them. The remaining three chapters in this volume examine the complex relationship between biotechnology policy and agriculture. Two chapters look at the link between policies that encourage biofuel production and food prices. Some commentators have speculated that biofuel demand contributed significantly to the worldwide food price crisis in 2008. A third chapter investigates whether genetic engineering can help keep food prices low in the face of the increased demand due, in part, to higher levels of biofuels production.

The immediate future of the U.S. ethanol market will be shaped by two policies. The U.S. RFS described in the preceding and the U.S. Environmental Protection Agency's ethanol "blend wall," which dictates the maximum amount of ethanol that can be mixed with petroleum in gasoline used by conventional automobiles. The blend wall was set at 10 percent. It has subsequently (October 2010) been raised to 15 percent, but only for recent model automobiles. Future increases are likely as long as the blend wall prevents ethanol from reaching the RFS level. Hertel and Beckman develop a theoretical framework to analyze the linkages between energy and agricultural markets under these policies. When neither policy is binding, consumers are able to respond to all realizations of oil prices by changing their biofuel mix. Thus, the transmission of energy price volatility to commodity price volatility is high and corn price volatility is low in response to traditional supply-side shocks. When either policy binds, the biofuel portion of demand for corn becomes more inelastic (on average). In this case, the agricultural commodity price impacts from energy price volatility are smaller, but the impacts from corn supply volatility are magnified.

The authors use an applied general equilibrium analysis to assess the magnitude of these effects. They begin by validating their empirical model against historical data and then use it to project the impacts in 2015 when the RFS takes effect. They find that when no policies bind, energy price volatility accounts for 0.53 of the total variation in corn prices. When the RFS is just binding, the share of energy price volatility is cut in half; when

1. "Federal Financial Interventions and Subsidies in Energy Markets in 2007," Energy Information Administration, http://www.eia.doe.gov/oiaf/servicerpt/subsidy2/pdf/execsum.pdf.
2. See http://www.nass.usda.gov/.

the blend wall is binding, it accounts for an even smaller share of total corn price variation. Where both policies are on the verge of binding—as will be the case if the blend wall is raised until it reaches fifteen billion gallons in total—the relationship between energy and corn price volatility is almost entirely eliminated. In contrast, the impact of corn supply shocks on prices is 57 percent larger than under the nonbinding case, boosting world price volatility by 25 percent. These numbers are large and suggest that, at least for biofuel feedstocks, energy policy may become a more important source of agricultural price uncertainty than traditional agricultural policies.

Concerns about energy security and the environment have led to a range of policies to promote the production of biofuels. The impact of this production on food prices and land use has led to additional interventions designed to shift policy incentives toward advanced biofuels made from non-food-based feedstocks. Xiaoguang Chen, Haixiao Huang, Madhu Khanna, and Hayri Önal (CHKO) develop a dynamic, multimarket equilibrium model that analyzes the markets for fuel, food, and livestock. After validating the model with historical data, CHKO use the estimated model to simulate the impacts of two U.S. policies: the biofuel mandates under the U.S. RFS and these mandates accompanied by targeted subsidies designed to make advanced biofuels more competitive with gasoline for fuel blending.

Simulations under the biofuel mandate alone suggest that 50 percent of the cumulative biofuel production from 2007 to 2022 under the RFS will be met through corn ethanol production. As a result, corn prices will be significantly higher and gasoline prices will be lower than they would be without the mandate. On net, this policy will result in a $122 billion increase in the present value of social welfare compared to a no-biofuel policy scenario. Adding subsidies to the mandate reduces the share of corn ethanol to 10 percent of biofuel totals, leading to lower food prices (due to assumed technological improvements in corn production and reduced demand for corn for ethanol) as well as lower fuel prices. Despite these gains by consumers, the policy results in a net welfare loss of roughly $79 billion as the cost to tax payers and agricultural producers are quite substantial. Thus, in order for the current configuration of biofuels policy to improve social welfare, the incremental environmental and energy security benefits resulting from these policies must exceed roughly $200 billion over the 2007 to 2022 period.

Increased global demand for biofuels is placing increased pressure on agricultural systems at a time when traditional sources of yield improvements have been mostly exhausted, generating concerns about the future of food prices. Steven Sexton and David Zilberman estimate the impact of global adoption of genetically engineered (GE) seeds on food supply. The conceptual model illustrates two key points. First, because GE crops are typically designed to minimize crop damages, they will be most valuable in locations facing high pest pressure. Second, because GE crops increase damage abatement, they increase the value of the marginal product of other production inputs and thus help to increase yield beyond the typical "gene effect" estimated in previous literature.

Sexton and Zilberman's estimation model exploits spatial and temporal variation in the adoption of GE crops to identify the average yield effect due to GE technologies among adopters—the average treatment effect on the treated. The yield gains range from 65 percent for GE cotton to 12.4 percent for soybeans. Separating the analysis by developed and developing countries, yield increases appear to be three- to fivefold higher in the developing world, consistent with the notion that they are experiencing higher pest pressure. They also find modest evidence in support of learning-by-doing as farmers gain experience with these new technologies. Applying these figures to the 2008 food crisis suggests that corn, soybean, wheat, and rapeseed prices would have been between 27 and 43 percent higher absent GE crop plantings. Stated another way, achieving the 2008 harvest without GE technologies would have required an additional twenty million hectares of land planted using traditional seeds. Genetically engineered crops appear to play an important role in arbitrating tensions between energy production, environmental protection, and global food supplies.

This book has two main themes. First, to accurately analyze agricultural and biotechnology policies, one must pay attention to complex institutional rules: the devil is in the details. Second, these policies have many unintended consequences—fostering inefficiency and the redistribution of income intended for producers—in large part because of complex linkages within and across markets.

Chapters 1 (Goodwin, Mishra, and Ortalo-Magné), 2 (Goodhue and Russo), 3 (Babcock), and 4 (Ligon) find that much of the rent from agricultural subsidies designed to help agricultural producers is extracted by processors, landowners, and insurers or lost due to inefficiency. Chapter 5 (LaFrance, Pope, and Tack) argues that current risk policies affect income and wealth over time, thereby affecting future production choices so that the long-run responses to subsidized crop insurance are likely to be larger than the short-run effects. The three chapters on biotechnology emphasize the increasing connections between agricultural and energy markets. Chapter 6 (Hertel and Beckman) shows that ethanol policy is a significant driver of agricultural commodity price levels and volatility. Chapter 7 (Chen, Huang, Khanna, and Önal) emphasizes that policies to manage future food price impacts by moving to cellulosic feedstocks in the production of biofuels will come at high social costs. Chapter 8 (Sexton and Zilberman) finds that the adoption of genetically engineered seeds has played an important role in mitigating recent food price effects—in part due to biofuel subsidies—although its role going forward is less clear. Thus, together, these chapters demonstrate that institutional details matter and that even evaluations of fairly narrow policies should be viewed through a broad lens.

I

Agricultural Policy

The Buck Stops Where?
The Distribution of
Agricultural Subsidies

Barry K. Goodwin, Ashok K. Mishra,
and François Ortalo-Magné

1.1 Introduction

A 2002 news report posed the following question. What do former basketball star Scottie Pippen, publisher Larry Flynt, and stockbroker Charles Schwab all have in common? The surprising answer is that all are recipients of farm program subsidies.[1] Other notable payment recipients include nine U.S. Members of Congress; David Rockefeller, former chairman of Chase Manhattan and grandson of oil tycoon John D. Rockefeller, who received ninety-nine times more in subsidies than the median farmer; Ted Turner, the twenty-fifth wealthiest man in America, who received thirty-eight times more subsidies than the median farmer; and the late Kenneth Lay, the ousted Enron CEO and multimillionaire (Reidl 2004). Several Fortune 500 companies have also received substantial farm program payments, including John Hancock Mutual Insurance ($2.5 million in 2002), the Chevron corporation, and the Caterpillar corporation.

In arguing for program reforms, U.S. Senator Amy Klobuchar (D-MN)

Barry K. Goodwin is the William Neal Reynolds Professor in the Departments of Economics and Agricultural and Resource Economics at North Carolina State University. Ashok K. Mishra is professor in the Department of Agricultural Economics and Agribusiness at Louisiana State University AgCenter. François Ortalo-Magné is the Robert E. Wangard Professor in Real Estate at the University of Wisconsin.

We thank our discussant, Gilbert Metcalf, for his insightful comments. We are also grateful to Erzo G. J. Luttmer, Jim Shilling, Fred Sterbenz, and seminar participants at Michigan State University and the Universidade Federal de Viçosa, Brazil, for useful discussions and comments on an earlier draft.

1. "Farm Subsidies Help Those Who Help Themselves," a Fox News report by William LaJeunesse, July 15, 2002. This article is available from http://www.foxnews.com/story/0,2933,57602,00.html. These statistics are all drawn from the Environmental Working Group's farm subsidy database (www.ewg.org).

stated that "$3.1 million in farm payments went to the District of Columbia, $4.2 million has gone to people living in Manhattan, and $1 billion of taxpayer money for farm payments has gone to Beverly Hills 90210."[2] The fact that support for U.S. "farmers" is often directed to individuals and corporations that seem to be some distance from the farm has been the topic of considerable debate in recent years, in particular because congressional support for U.S. agriculture continues to expand. The 2008 Farm Bill (P.L. 110-246) will provide in excess of $284 billion in financial support to U.S. agriculture over the 2008 to 2012 period. Commodity program payments account for $43.3 billion of this total.

To the extent that eligibility for government benefits is tied to the ownership or operation of certain assets, the market values of these assets will reflect expected future benefits. Such is the case with farmland. Considerable variation exists in agricultural land values across the United States (see figure 1.1). U.S. Department of Agriculture (USDA) statistics indicate that 45.3 percent of U.S. farmland is operated by someone other than the owner (U.S. Department of Agriculture, National Agricultural Statistics Service [USDA-NASS] 1999). Mishra et al. (2002) report that, contrary to conventional wisdom, most agricultural landlords (57 percent) are nonfarm corporations or individuals that work in or are retired from nonfarm-related activities. In light of these facts, a fundamental question arises regarding the distribution of farm support programs and the extent to which those who operate the farms actually receive the benefits. This is a critical issue, not only for policymakers but also for farm operators who might benefit from a better understanding of the implications of the various programs they tend to support.

The relevant question is, of course, who are the policies intended to benefit? The capture of agricultural benefits by farmland values is problematic if the policies aim to support farmers, and these farmers do not own their land when the policies are announced. To the extent that (young) expanding farmers are paying for the expected policy benefits in the farm assets they acquire, the present value of future benefits is captured by the (old) sellers. New owners only benefit from surprise increases in public transfers. Given the large share of U.S. farm land that is operated by tenant farmers, the extent to which lease rates capture program benefits is also important to the distribution of these benefits.

The concern with the capture of agricultural policy benefits by the initial land owners is not new. A number of papers have attempted to estimate the capitalization of aggregate agricultural transfers into farmland values.[3]

---

2. Quote is from the December 12, 2007 Senate floor statement of Senator Amy Klobuchar.

3. See Barnard et al. (1997), Goodwin and Ortalo-Magné (1992), Ryan et al. (2001), Shertz and Johnston (1997), Shoemaker, Anderson, and Hrubovcak (1990), and Weersink et al. (1999). These papers only examine aggregate policy effects on land values, thus ignoring the

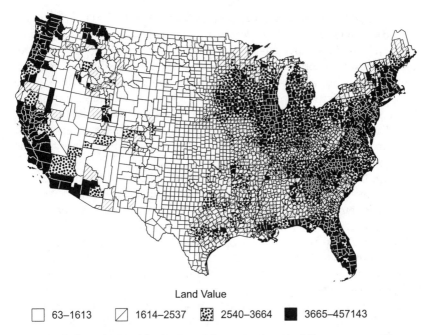

Land Value

☐ 63–1613   ▨ 1614–2537   ▩ 2540–3664   ■ 3665–457143

Fig. 1.1   U.S. agricultural land values ($/acre), land and buildings as reported in 2007 Agricultural Census

These papers suffer from a number of shortcomings that we are able to address here through an empirical analysis of a unique set of farm-level data. We contribute to the understanding of the distribution of farm subsidies in several ways. First, we are able to investigate the differential impact of the principal farm programs because we are able to observe the breakdown of government payments at both the farm and the county level. Second, because we know the location of each farm, we are able to control for non-agricultural pressures on the land and determine how they affect its value. Third, we observe not only land values but also the terms of lease arrangements and rates. This puts us in a unique position to be able to assess directly the extent to which owners and farmer operators share the benefits of various agricultural programs, a useful complement to the indirect assessment we obtain from investigating land values. Finally, variations in the difference between cash lease rates and share lease rates enables us to investigate the extent to which the market values the insurance features built into some farm programs, a feature ignored by the literature.

Our analysis makes use of a data set drawn from an annual survey of

_____

myriad effects of different programs. In addition, the extraction of policy benefits through lease arrangements has not been widely investigated. Important exceptions exist in the recent studies of Kirwan (2009) and Patton et al. (2008), which we discuss in the following.

approximately 10,000 farms per year over the 1998 to 2005 period. This period was characterized by a variety of different farm programs, including some that were not connected in any way to market conditions or production, at least in theory. At the other extreme are output price-support payments that are intimately tied to contemporaneous market conditions. We find that payments that are decoupled from output and are supposed to be transitory yield the smallest effects on land values. Payments that may signal future benefits, even in cases where they are not a permanent part of farm legislation, have stronger effects. Price-support payments have the strongest effects.

U.S. farm legislation typically intends benefits to be "shared" between the owner and operator of a farm. Under cash lease arrangements, the entire subsidy is sent to the operator. However, the law does not regulate lease rates; they are set by the market. Our empirical analysis indicates that owners extract a large proportion of farm benefits from tenants through the lease rates. From the study of lease rates, we also find that programs with strong insurance objectives, such as output price-support payments, significantly affect the gap between cash and share lease rates. In particular, the share rate premium is significantly diminished by programs that serve to lower the risk associated with uncertain farm earnings. This provides direct evidence of the insurance component of agricultural policy in the land market.

Accounting for the benefits of decreased earnings volatility raises two issues with the traditional approach to the assessment of the contribution of agricultural policy to farm land values. First, the insurance feature of several governmental programs raises questions about the traditional implicit assumption that a dollar of transfer today conveys the same information about future transfers, regardless of market conditions and local agricultural output characteristics. Instead, a low price support payment this year may be due to high market prices and, thus, in no way indicates a decrease in the expected stream of long-run benefits from the price-support program. Second, those government transfers whose level are negatively correlated with farm earnings from the market decrease the volatility of farm land returns. They must, therefore, decrease the discount rate required to hold farm land and, thus, the discount rate applied to earnings from the market. Hence, regression estimates of the contribution of market earnings to the value of land depend on the policy environment. In particular, it is wrong to assume that such estimates would not change to reflect a more volatile environment if price-support programs were to be dismantled.

We have noted that the empirical literature has largely been focused on policy effects on land values and the incidence of policy benefits in rental arrangements—which is an increasingly prominent feature of U.S. agriculture—has not received the same level of scrutiny. Two important exceptions lie in the recent analysis of Kirwan (2009) and Patton et al. (2008), who both evaluated the effects of policy benefits on land rental rates. Kirwan (2009) used farm-level panel data taken from the 1992 and 1997 Agricultural Censuses to evaluate the incidence of policy benefits in farmland rental arrangements. However, the census data only included rents for land leased on a cash or free basis, and he, therefore, largely ignored the potentially important role of share lease contracts. In the sample of farms evaluated in this analysis, 63.6 percent of the farms reported renting land and, of those that rented land, 36.3 percent reported leasing land on a share basis. As Kirwan points out, to the extent that a significant share of rented land is leased under share arrangements, this may raise an errors-in-variables problem that results in biases if the measurement error is correlated with policy benefits. Kirwan undertook an analysis intended to demonstrate that the biases raised by ignoring share lease arrangements were modest. To do so, he used a single year of data from a related survey, the Agricultural Economics Land Owners Survey (AELOS) in 1999, to investigate the extent to which the measurement error in rental rates arising from the omission of share rents resulted in biases in his estimates of benefit incidence. On the basis of results for this single year, he concluded that the biases were small and generally positive. While these arguments are persuasive, the reliance upon a single year of data in a case where policy benefits are very dependent upon market conditions in any given year may make it hard to generalize his results. Further, as we argue in the following, it may be important to segregate benefits across different policy types because the effects on land values and rents may vary substantially for different types of policies—a point demonstrated by Goodwin, Mishra, and Ortalo-Magné (2003b).

An important point of relevance is the significant variety of agricultural programs used by policymakers to convey support to the farm sector (see appendix table 1A). Kirwan (2009) argues that policy benefits after the 1996 Farm Bill were exogenously determined by underlying program parameters. As we discuss in greater detail in the following, this is not entirely the case because a wide range of policies are used to convey benefits to agricultural producers. Although certain payments were exogenously determined by congressional mandate and were known with certainty over the life of the legislation, other significant benefits, including price supports, disaster payments, and market loss payments were not exogenously known prior to the year in which they were received.

Patton et al. (2008) draw a careful distinction between payments that are "coupled" and "decoupled." Although disagreement exists over what constitutes coupling of payments, a formal definition is afforded by Annex 5 of the 1996 World Trade Organization (WTO) Uruguay Agreement on Agriculture (URAA), which defines a decoupled payment as one that is not dependent upon production or price in the year in which it is made. Patton et al. (2008) adopt an instrumental variables (IV) approach to recognize the fact that payments are not known with certainty at the time rental contracts are determined and thereby represent expected values of policy benefits by using instruments. We follow a similar approach in certain portions of our analysis. Their results indicate that different types of agricultural policy benefits have different effects on rental rates, thereby confirming the earlier assertions of Goodwin, Mishra, and Ortalo-Magné (2003a) that predicted such differential effects. Patton et al.'s (2008) results also raise important questions regarding the validity of the assumed operation of agricultural programs and modeling of benefit incidence presented by Kirwan (2009).

The remainder of our chapter is organized as follows. Section 1.2 gives a brief overview of the history and nature of U.S. farm programs. We are particularly concerned with providing a careful description of the different mechanisms commonly used to convey policy benefits to U.S. agricultural producers. Section 1.3 discusses issues pertaining to model specification, estimation, and measurement of the relevant variables. Section 1.4 presents the results of our empirical analysis and discusses their implication for the distribution of agricultural policy benefits. Section 1.5 offers some concluding remarks.

1.2   A Brief Overview of U.S. Farm Policy

Most U.S. farm programs have their origins in the New Deal legislation of the Great Depression. A variety of price- and income-support programs have been used over time to increase and stabilize farm earnings. These programs are revised approximately every five years by an omnibus "Farm Bill" package of legislation. In addition to this major package of farm programs, support is provided through a number of other legislative channels. This is the case with farm programs such as crop insurance and conservation measures. On a regular basis, agriculture also benefits from ad hoc support (though emergency bills) that is not a part of any budgeted legislation.

Over most of its history, U.S. agricultural policy has used price supports coupled with production controls, with the declared objective to provide income support to the farm sector. Some support was made on the basis of a need for "parity" with the high relative agricultural prices of 1910.[4] In more recent times, support was provided only to program crops (corn, wheat, cotton, rice, grain sorghum, rye, barley, and oats). Deficiency payments, determined by the difference between market and target prices, were paid to producers on the basis of their "base" acreage and yields. This base reflected historical production (in most cases, acreage and yields during the 1980s). The fact that price supports were tied to historical production patterns implied a lack of planting flexibility for producers. In addition, soybeans, a major U.S. crop, was largely omitted in provisions for support

---

4. Though any link with market and production conditions in 1910 would seem difficult to make, arguments in favor of such "parity" pricing are still heard on occasion in farm policy debates.

due to the fact that it was not an important crop when most farm programs began.

In 1996, Congress agreed to what was intended to be a major overhaul of U.S. farm policy—the Federal Agriculture Improvement and Reform (FAIR) Act. This act is also known as the Agricultural Market Transition Act (AMTA). The nomenclature "Reform" and "Market Transition" was meant to indicate a major shift in policy away from government involvement and toward market-oriented policies. Eligibility for price support was no longer based upon historical production—producers were free to plant whatever crops they desired, and prices were supported at a legislatively determined loan rate. Soybeans were made eligible for price supports, which were now provided through the Loan Deficiency Payment (LDP) program. Loan Deficiency Payment payments were made on the basis of the difference between market and support prices (the loan rates). The rhetoric accompanying the act implied, in principle at least, that the legislation signaled a transition to an environment with limited government support. A program of direct payments to those producers with base acreage (historical rights to program benefits) was instituted to compensate producers over this transition, at least in theory. These payments were known as AMTA or Production Flexibility Contract (PFC) payments. By design, AMTA payments were completely decoupled from the market—the only requirement for receiving AMTA payments was that the producer (or landowner) had to have base acreage. Eligibility for such payments in no way depended upon current production patterns. In some cases, payments were made on land no longer in production. The AMTA payments were set to decline each year until the FAIR Act expired in 2002. Of course, the extent to which such payments were perceived to be temporary is a subject of debate, especially because the payments were continued (and even increased) in the 2002 and 2008 Farm Bills. Further, the 2002 Farm Bill allowed landowners the option to update their base acreage using production and yields over the 1998 to 2001 period. Many critics of U.S. farm programs have argued that this updating made it much harder to characterize the payments as decoupled because farmers and landowners may factor such updating possibilities into their future production decisions.

Over its history, U.S. farm policy has provided benefits through three general channels—price supports (sometimes tied to acreage restrictions) that are tied to production (i.e., benefits are provided on a per-unit basis); decoupled income support, which has no production requirements; and disaster or market assistance payments, which provide benefits intended to offset poor production or bad market conditions. Since the 1996 Farm Bill, U.S. agricultural policy has been characterized by three specific program mechanisms, together with a large collection of various minor programs. These mechanisms include the aforementioned direct payments (PFC and fixed, direct payments); market loss assistance and countercyclical payments

(payments that are triggered by low prices but are not tied to current production); and loan deficiency payments and marketing loans, which use loan rates to support market prices. Each of these policies functions in unique ways to provide support.

Direct payments were introduced in 1996 and were specified for the subsequent seven years. Payment recipients knew in advance exactly what their payments would be because they were determined exogenously. However, other major components of farm program benefits are not known in advance. Market loss assistance and its successor—countercyclical payments—are triggered by low market prices. The market loss assistance program that was introduced in 1999 was entirely ad hoc and was determined outside of the farm bills. Its successor, countercyclical payments, formally brought these price supporting payments into the farm legislation. In both cases, these programs are triggered by market prices falling beneath a legislatively defined target price. Because market prices are not realized until after harvest, agents do not know what payments will be in advance.

Figure 1.2 illustrates the evolution of these three types of payment programs over the last twenty years. Note that coupled price supports and countercyclical payments are very volatile from year to year. This is because they are based on market prices.[5] The fixed, decoupled payments, which were known in advance over the life of the legislation, began in 1996 and are much less variable by design.

The important point regarding these payment programs is that, contrary to arguments advanced in the literature (e.g., Kirwan 2009), the bulk of farm program payment benefits is not predetermined by legislation, and payments are not known in advance because they are triggered by market conditions. Such arguments simply mischaracterize the basic operation of farm programs. Agents' actions and the effects of policy on asset values and rental agreements will, therefore, be based upon *expectations* of such payments—a point well noted by Goodwin, Mishra, and Ortalo-Magné (2003b) and Patton et al. (2008). Further, the level of support varies substantially from year to year, and, thus, any analysis that focuses on one or two years (e.g., the 1992 and 1997 Census years or the 1999 AELOS survey year) is faulty because benefits will most certainly reflect market conditions in those two years, which are volatile over time but highly systemic in nature and, therefore, highly correlated in the cross-section.

Ad hoc disaster assistance has been a fixture in U.S. agricultural policy for many years. Periods of drought or poor market conditions frequently trigger ad hoc assistance labeled as disaster payments. Under provisions of other farm legislation (the Crop Insurance Reform Act of 1994), Congress stated an intention to make subsidized insurance the only mechanism for

---

5. This degree of volatility increases substantially when one considers individual commodities and support at lower levels of aggregation (i.e., the state or county).

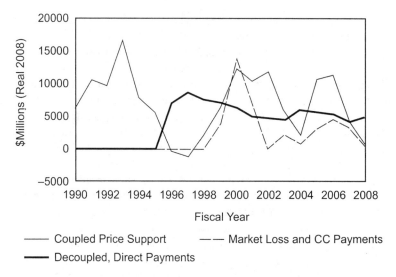

Fig. 1.2   U.S. farm program payments by category (real $2008)

providing disaster relief.[6] However, localized droughts and low market prices led Congress to rapidly retreat from this position and conclude that the support provided to farmers under the FAIR Act was not sufficient. Ad hoc assistance, in the form of yield compensations and the aforementioned payments for low market prices (market loss assistance), were then instituted. Again, such support cannot be perfectly anticipated because it is based upon random production and market conditions.

A number of other programs have been important to agricultural policy. For example, a considerable amount of farm land (approximately thirty-five million acres) has been removed from production through the Conservation Reserve Program (CRP). The CRP pays producers annual rents to place their land in reserve under a ten-year lease agreement. In order to be eligible for the CRP program, land must be "erodible" and environmentally fragile. Such land is typically of a lower value in terms of crop production.

In spite of rhetoric to the contrary, congressional support for U.S. agriculture continues to expand. President Bush signed an omnibus package of farm program support on May 13, 2002 that was scored at $190 billion. The Food, Conservation, and Energy Act of 2008 (P.L. 110-246) was enacted into law on June 18, 2008. These two packages of farm programs did not make substantial changes to farm policy. Notable is the fact that direct,

6. As an aside, an interesting policy situation exists for crop insurance, which recently has returned about $2.00 in indemnity payments for every dollar of premiums paid by farmers. This program, also in existence since the 1930s, runs hand in hand with ad hoc disaster assistance—a form of free insurance. Note that disaster assistance is an obvious impediment to a well-functioning insurance program.

decoupled payments were maintained in both farm bills, thereby eliminating any doubt regarding the extent to which these payments were transitory. One important exception to this general lack of change in programs occurred in the introduction of countercyclical payments (CCPs) in the 2002 Farm Bill. These payments formally brought the ad hoc market loss assistance support that characterized the late 1990s into farm legislation. The CCP program established target prices for program commodities. If market prices fall beneath a target, payments are made on the basis of the price deficiency and the base yield and acreage. The CCP program was continued with little modification in the 2008 Farm Bill.[7]

Congressional debate over the 2008 farm legislation and the generous level of support that emerged from these deliberations have made clear Congress's intent to continue taxpayer support for agriculture. The most recent policy debate centered on means testing for payment eligibility and limits on the amount of payments any individual could receive. Under the 2002 Farm Bill, individuals with an adjusted gross income over $2.5 million were ineligible for payments unless more than 75 percent of this total came from agriculture. Payments to an individual farm were limited to $360,000 although price-support payments made on actual production were essentially unlimited due to program loopholes. The 2008 Farm Bill essentially removed payment limits on coupled support and provided limited income limitations on some payments.[8]

In all, support for agriculture remains strong in the U.S. Congress. A wide variety of programs are used to convey significant benefits to the farm sector. The latest omnibus farm bill is projected to cost U.S. taxpayers nearly $300 billion to provide agricultural and nutritional support.

1.3   Modeling Framework

1.3.1   The Income Approach to Farm Land Valuation

All government transfers help the farmers in at least one of two ways: by raising the returns to farming and by decreasing the volatility of these returns. The LDP and DP programs have major insurance components. The AMTA payments are lump-sum transfers determined by farmers' activities prior to their implementation. The same is true with CRP payments; they

7. The 2008 Farm Bill did introduce an optional alternative to the CCP program—the Average Crop Revenue Election (ACRE) program. Enrolling farmers agreed to cuts in some program benefits and the elimination of CCP payments in order to obtain a crop revenue guarantee. Only about 12.8 percent of eligible acreage was enrolled in the ACRE program.

8. In particular, a person or legal entity with adjusted farm gross income of over $750,000 is not eligible for direct (decoupled) payments. A person or entity with average adjusted gross nonfarm income in excess of $500,000 is not eligible for any program payments. However, the legislation allowed a husband and wife to allocate income as if they had filed separate returns, essentially doubling these limits.

are lump-sum additions to the return of farming that are uncorrelated with present or future earnings from the market. In addition to all these transfers, farm land also gives the farmer the opportunity to generate nonagricultural earnings. The jackpot is to own land in an area under strong urban pressure with friendly zoning authorities, hence providing the opportunity to realize substantial capital gains by converting the land to residential or commercial use.

The value of a parcel of land is the present discounted value of expected cash flows from agricultural activities plus the value of the option to convert the land to nonagricultural use.

$$(1) \quad V_0 = E_0 \left[\sum_{t=1}^{\infty} \frac{MKT_t + LDP_t + DP_t + AMTA_t + CRP_t}{(1 + r)^t} \right] + CONV_0,$$

where MKT and DP denote earnings from the market and from disaster payments, CONV is the value of the conversion option, and r is the discount factor. The discount factor reflects the risk of the overall portfolio of individual streams of cash flow. This risk is not simply the sum of the individual risks because of the nonzero covariance, by design, between MKT payments, LDP and DP.

As mentioned earlier, AMTA and CRP are, for the most part, lump-sum transfers whose levels are independent of current and future earnings from MKT, LDP and DP, and from each other. We can, therefore, rewrite equation (1) as

$$(2) \quad V_0 = E_0 \left[\sum_{t=1}^{\infty} \frac{MKT_t + LDP_t + DP_t}{(1 + r_1)^t} + \frac{AMTA_t}{(1 + r_2)^t} + \frac{CRP_t}{(1 + r_3)^t} \right] + CONV_0,$$

where r_1, r_2, and r_3 denotes the discount factors for output-related earnings, AMTA payments, and CRP payments, respectively.

Implicit in equation (2) is the assumption of a constant discount rate. If we are willing to assume that farm land buyers and sellers expect the various earnings to grow at a constant rate, then the regression coefficients we will obtain will be the inverse of the capitalization rates, or cap rates. The valuation formula can indeed be rearranged as

$$(3) \quad V_0 = E_0 \left[\frac{MKT_1}{\kappa_1} + \frac{LDP_1}{\kappa_2} + \frac{DP_1}{\kappa_3} + \frac{AMTA_1}{\kappa_4} + \frac{CRP_1}{\kappa_5} \right] + CONV_0,$$

where the cap rates are denoted by κ. It is easy to check that if a stream of cash flows is expected to grow at the constant rate g and is discounted at the constant rate r, then the cap rate κ satisfies $\kappa = r - g$.

To estimate the contribution of each source of value in equation (3), we need estimates of expected next-period cash flows for each source of agricultural earnings. This raises a serious measurement issue. As mentioned in the preceding, it has been largely ignored in the literature, which tends to rely on current payments as an indicator of future payments. This is the issue to which we now turn.

1.3.2   Measuring Expected Cash Flows

Let us suppose that agents correctly assess the true determinants of land values, but the econometrician, working with actual realizations of policy outcomes from year to year, is unable to observe these determinants. Instead, the econometrician relates the observable annual realizations of market and policy outcomes to land prices. In this case, the econometrician is confronted with the classical errors in the explanatory variables problem. Errors-in-variables results in an attenuation bias that forces coefficients toward zero and, thus, yields inconsistent estimates.[9] This problem is compounded by the fact that the government operates more than one program of payments, hence suggesting that traditional empirical approaches suffer from multiple explanatory variables observed with error.

A complicating factor arises in that the errors applying to observed policy benefits may be correlated in a typical sample. This correlation may assume two different forms—correlation of the errors across different programs (for a given farm) and correlation of errors across different farms in a sample. Both circumstances are likely to exist when one considers a pooled cross-section of farms (as is the case in our empirical analysis). Consider a case of two programs—price supports and market loss assistance payments. The extent of support provided from the government is likely to vary considerably from year to year according to market conditions. Low price years realize larger payments for both programs. Thus, the errors associated with using realized benefits are likely to be highly correlated across the programs. The correlation could also be negative. Consider the case of yield disaster relief and price supports. In low yield years, market prices are likely to be high and, thus, price-support payments will be low, though disaster benefits will be higher to compensate for the production shortfalls.

Another form of correlation is likely to be relevant when a pooled sample of individual farms is considered. Because realized program benefits are dependent upon aggregate market conditions, the errors are likely to be highly correlated across observational units (farms) in a given year. In a sample consisting of only a few years of data, the correlation across farms increases the estimation error and may further exaggerate the bias; year-to-year shocks may not average out when only a few years are observed.

Furthermore, if realizations are highly correlated across units within a year, parameter estimates may shift considerably from year to year. If only a few years are observed, the estimates from a pooled sample may be sensi-

---

9. This problem is analogous to the standard omitted variable problem, where the omitted factor is the difference between what is observed and the true, latent value.

tive to events in the years observed and, thus, may vary substantially across years and be more variable in a pooled analysis.[10]

The standard approach to addressing this problem is to obtain instruments or proxy variables for those latent variables that are measured with error. An instrument should be correlated with the variable of interest but uncorrelated with the error pertaining to the observation. We represent the expected payment benefits by constructing average values of each relevant policy variable over the preceding four years. This approach raises one complicating factor. As we discuss in detail in the following, our data set is not a true panel in that a different set of farms is sampled each year—meaning that repeated observations for an individual farm are not available.

To represent expected payment benefits, we utilize the four-year average value of real payments per farm acre *in the county* where the individual farm is located.[11] We argue that this is a superior measure of long-run expected benefits as compared to realized payments because values in the county more closely represent the long-run potential benefits associated with agricultural policy. Payments on an individual farm, in contrast, may reflect individual policy choices and characteristics of the farm operation. Transfer of the land to a new operator may result in different subsidy realizations (for example, because of a different crop mix), which are better represented using county-level averages.[12]

We adopt a number of different historical averages to represent expected policy benefits. We use a four-year average of county-level total payments in our aggregate policy models. In contrast, because LDPs were not the main instrument for providing price support prior to the 1996 Farm Bill, we use a two-year average for LDP payments at the county level. We should note that this errors-in-variables problem does not apply to all sources of government subsidies. Subsidies provided through AMTA payments and rents earned on land enrolled in the CRP program are known with certainty a priori. It is only those payments that are triggered by market and production events (price supports and disaster payments) that must be proxied.

1.3.3   Data

The primary source of our farm-level data is a data set collected from a large sample of farms through the USDA's National Agricultural Statistics Service (NASS) Agricultural Resource Management Survey (ARMS)

10. See Goodwin, Mishra, and Ortalo-Magné (2003b) for a quantitative assessment of this issue in the farm land valuation context.

11. A standard IV estimation approach is also feasible, though the fact that payment realizations in any given year may be very weakly tied to long-run expected benefits makes the utility of such an approach limited. This problem is exacerbated in a short sample when realizations are highly correlated in the cross section, as is true in our application.

12. Observations for an individual farmer in a particular year might reflect crop rotation patterns. We expect county-level acreage to be more reflective of the expected crop mix.

project. The ARMS survey is a large probability-weighted, stratified sample of about 8,000 to 20,000 U.S. farms each year. The survey collects detailed government payments information for individual farm program benefits as well as extensive farm and operator characteristics. We focus on data collected over the years 1998 to 2005. Thus, our empirical analysis focuses on these years. All monetary values in our sample were adjusted to 2005 equivalent real values by deflating by the consumer price index. Given the relatively short nature of our sample, such deflation had only minor effects on the results.

Besides detailed farm earnings, the survey also reports farm land value. Farm operators are asked to estimate the market value as of December 31 of the preceding year of their land, dwellings, and other farm buildings and structures. We restrict our attention to the value of land only (excluding trees and orchards).[13]

In order to eliminate hobby and retirement type farms and to focus on commercial agricultural operations, we eliminated any farm of less than fifty total acres. We also excluded farms located in counties with less than 100 total farm acres, thereby excluding urban counties that have no production agriculture. We excluded farms for which incomplete data were available. This left us with a small number of extreme outlier observations (land values less than $200 per acre or those exceeding $20,000 per acre). Such extreme observations represent nontypical agricultural properties, such as vineyards and properties with characteristics (e.g., riverside properties) not recorded in the survey. Only a very small number of observations (less than 1 percent) were excluded on this basis. In the portion of our analysis that addresses rental markets, we excluded any observation for which a share or cash lease rate in excess of $1,000 per acre was reported. Again, such outliers occurred in only a very small number of cases.

A variety of sources were used to collect county-level observations on crop acreage and state-level prices (unpublished USDA-NASS statistics) and data relevant to county population and trends (unpublished U.S. Census data). Aggregate (county-level) agricultural market performance (total sales and production costs) and population statistics were taken from the Bureau of Labor Statistics' (BLS) Regional Economic Information System (REIS). Total farm acres for each county were taken from the 1997, 2002, and 2007 Agricultural Censuses. We used linear splines to interpolate between census years. In that these values evolve slowly over time but vary significantly in the cross-section, such interpolation should provide valid estimates in non-census years. Unpublished data on calendar-year total program payments at the county level for individual farm programs were collected from the Farm Service Agency (FSA) of the USDA.

---

13. Confidentiality of responses is maintained, and farmers do not have any incentive to distort their response to the survey.

Our empirical approach involves consideration of farm-level observations of land values, cash rental rates, and share rental rates. We use explanatory variables that are measured at the farm level as well as at a more aggregate county level. It is important to note that this is not analogous to analysis at the county level as the left-hand-side variables in our empirical models are measured at the farm level. Further, in our models of actual realized payment benefits, the right-hand-side explanatory factors are also measured at the farm level. In the case of models that utilize aggregate averages of payment benefits, right-hand-side explanatory factors are constant across all farms within a county, while the dependent variables of interest vary within and across counties. It is relevant to note that, because the ARMS is a national survey, it is uncommon for a large number of farms to be sampled in a single county. In our estimation sample, each county had an average of 1.15 farms each year, and the number of farms sampled per county ranged from one to seven.

Summary statistics and definitions for the key variables of our analysis are presented in table 1.1. To the variables aimed at capturing expected cash flows from farming, we added three factors intended to represent the additional value of land in areas facing nonagricultural pressures. First, to represent nonagricultural demand pressures, we included the population growth rate in each county. We also include a series of discrete indicator variables (obtained from the USDA) that represent the extent of urbanization for each county. The ordinal ranking ranges from 1 = rural to 4 = urban. Finally, we considered the ratio of total population to total farm acres to again capture the effects of residential and nonagricultural commercial demands for farm land.

We are interested in evaluating the differential effects of benefits provided by the government versus those returns generated by the market. Of course, a risk-neutral farmer will not care where a dollar comes from, though alternative sources of revenue may have different levels of risk, thus affecting the preferences of a risk-averse farmer. We acknowledge at the outset that any representation of market earnings should not be interpreted as a measure of the market returns that would be generated in the *absence* of farm policy. Returns in such a situation are difficult to assess, especially in light of the long history of government involvement in U.S. agriculture. Likewise, the relevance of such a consideration is limited—it is unlikely, to the authors at least, that the U.S. government will completely remove policies that currently support agriculture. Having acknowledged these limitations, we construct a measure of net returns from the market using county-level averages of the difference in total agricultural sales receipts (exclusive of government payments) and total production costs (dollars per acre of farm land). We considered using measures of market returns from individual farm records. However, farm-level financial records are highly volatile across individual years and individual farms due to any number

Table 1.1   Variable definitions and summary statistics

Variable	Definition	Mean	Standard deviation
Total payments	Total government payments operator and landlord ($/farm acre)	29.4142	49.1946
Average total payments	County average payments over previous 5 years ($/farm acre)	23.1093	20.6402
Land value	Land value for owned land ($/acre)	2,258.4700	2,485.6500
Cash rent	Cash rental rate ($/acre)	76.7523	89.1673
Share rent	Share rental rate, including payments to landlords ($/acre)	99.8997	109.1717
Share-cash difference	Share-cash differential ($/acre) on farms with both lease types	11.8615	41.5512
LDP payments	Loan deficiency payments ($/acre)	6.5507	17.8325
Direct payments	Direct payments (including countercyclical payments) ($/acre)	10.9385	22.8224
Disaster payments	Disaster payments (including market loss assistance) ($/acre)	5.0161	25.7740
Other payments	All other federal and state government payments ($/acre)	11.0282	29.1607
Average LDP payments	County average loan deficiency payments ($/acre)	7.1304	8.4403
Average direct payments	County average direct payments (includes countercyclical and market loss payments) ($/acre)	13.6405	14.0972
Average disaster payments	County average disaster payments ($/acre)	2.3527	2.8248
Average other payments	County average of all other payments ($/acre)	3.4808	4.4062
Average market returns	County average cash crop and livestock sales less costs ($/acre)	36.3074	145.0764
Aggregate market returns	Current year county average sales less costs ($/acre)	37.6490	168.9909
Population growth	Annual population growth rate (percentage change year $t-1$ to t)	0.4069	1.5293
Urban$_1$	Urban indicator variable 1 (most rural counties)	0.6660	0.4717
Urban$_2$	Urban indicator variable 2	0.1043	0.3057
Urban$_3$	Urban indicator variable 3	0.0902	0.2865
Urban$_4$	Urban indicator variable 4 (most urban counties)	0.1395	0.3465
Population/Farm acres	Ratio of county population to farm acres	0.6648	13.3475

of idiosyncratic factors, and, therefore, we use the countywide average to represent market returns.

A few final aspects of the construction of our sample merit discussion. The ARMS survey is conducted annually from a stratified random sample of farms. Strata are defined by farm size, sales class, and area. While it is possible that individual farms may be sampled in multiple years, the identity of any individual farm is unknown (at least to the researcher), though we do know the county in which the farm was located. Thus, it is impossible to track an individual farm across time and, even if such identification were possible, it is likely that farms would be sampled infrequently and without regularity. This fact complicates inferences in that unobserved heterogeneity concerns and endogeneity of key variables may be difficult to address using standard econometric practices. Our use of county-level aggregates, which should be exogenous to individual farm observations, alleviates these concerns in many cases.[14]

A second point of relevance pertains to the timing of production decisions, including rental agreements, and the administration of the survey. In most cases, rental agreements are set prior to planting and, in some cases, may be long-term agreements that extend across multiple years. Such agreements are, therefore, clearly based upon expected values of returns and policy benefits. A subtle difference exists in the case of land values. Farm operators are asked to assess the value of their land holding as of December 31 of the previous year. Such an assessment would be made with full knowledge of realized returns and policy benefits. However, in that returns and program payments are very time dependent, observed returns and payments may not accurately reflect the long-run expected values that influence land values and rental rates—a point demonstrated by Goodwin, Mishra, and Ortalo-Magné (2003a). We, therefore, use an average of the preceding five years for individual program payments and county-level market returns. The extent to which a five-year historical average accurately represents long-run expected values is debatable, but such a measure should control for year-specific effects that may move realized benefits in any given year far from expected values.[15]

1.4   Empirical Results

Our empirical analysis utilizes three distinct approaches to modeling policy effects on land values and rental rates. The first simply considers the effect of farm-level, realized payments on farm-level land values and rental rates. This approach is analogous to that adopted in many studies (see, for example, Kirwan 2009) and ignores the fact that payment benefits are largely unknown at the time asset values and rental arrangements are determined. A second approach constructs explicit measures of expected policy benefits by considering an average of historical county-level aggregates. A third

14. Approaches to directly addressing endogeneity and unobserved heterogeneity remain important topics for future research.

15. Consider, for example, basic price supports. These programs (e.g., deficiency payments) support prices by making a payment any time market prices fall beneath a target support level. In a year of strong prices, no payments may be made. However, in light of the considerable volatility of basic commodity prices, a subsequent year may realize substantial payments due to low prices.

approach adopts an IV model in which the aggregate measures of policy benefits are used to form instruments that represent expected payments in a generalized method of moments (GMM) context. In the case of standard regression models, we also considered clustered and robust standard error estimation techniques. We allowed for clustering among regions, states, counties, and population weights. Controlling for clustering generally produced larger standard errors but did not alter the overall conclusions of the analysis. We present conventional standard errors in the results contained in the following.

1.4.1   Land Values

We first consider the relationship between land values and agricultural policy benefits. As we have noted, our individual farm-level data are collected using a complex survey design. The individual strata used in collecting the data are not identifiable, again reflecting confidentiality considerations. This precludes efficiency gains that could be achieved from incorporating information about the construction of strata. However, we can observe population weights for each farm and, thus, have pursued both weighted and unweighted regression methods. In every case, the weighted and unweighted results were quite similar, and, thus, we only present unweighted results.[16] However, the unweighted results which follow should be interpreted as applying to this sample of farms only and should not be directly extended to the entire population.

Our analysis of the determinants of land values is conducted in three segments. In the first, we consider models that aggregate all program payments into a single category. Such a model is useful in that it provides a summary of the impacts of additional federal subsidy dollars on land values at the margin. This analysis also permits a straightforward comparison to the large literature on this topic. Two versions of this model are considered. The first uses actual, observed payments for each farm. The second uses county-level historical averages to assess the total, expected per-acre receipts from farm program payments. Note again that expected payments are represented using the county average over the preceding five-year period. The results are presented in table 1.2.

The model using actual observed farm-level payments (Model 1) indicates that $1 of farm payments tends to add $2.93 per-acre to the value of farm land. The effect, though highly statistically significant, is unreasonably low and suggests a very high rate of discounting payment benefits (approximately a 30 percent rate of discounting). Such a high rate of discounting would necessarily imply that land market agents either anticipate the elimi-

---

16. Because strata are defined using size and sales class, dropping very small farms from our sample mitigates bias concerns resulting from the nonrandom sampling, at least to a degree. Weighted regression results are available from the authors on request.

Table 1.2   **Aggregated policy models of land value determinants: Parameter estimates and summary statistics**

Variable	Model 1		Model 2	
	Estimate	t-ratio	Estimate	t-ratio
Intercept	2,995.9138	131.69	2,679.4802	107.68
	(22.7502)		(24.8831)	
Total payments	2.9304	17.62	13.1309	32.19
	(0.1664)		(0.4080)	
Market returns	3.1442	64.00	3.4549	58.92
	(0.0491)		(0.0586)	
Population growth	385.1069	70.40	408.7445	73.49
	(5.4701)		(5.5623)	
Urban$_1$	−1,395.2725	−58.38	−1,290.0735	−53.57
	(23.8995)		(24.0821)	
Urban$_2$	−931.7608	−28.70	−879.6743	−27.00
	(32.4653)		(32.5752)	
Urban$_3$	−667.5923	−19.97	−626.3230	−18.69
	(33.4375)		(33.5112)	
Population/Farm acres	0.4298	0.76	15.9674	11.62
	(0.5657)		(1.3747)	
No. of observations	83,936		83,790	
R^2	0.1766		0.1758	

Notes: Numbers in parentheses are standard errors. Model 1 uses current year realized values for payments and market returns. Model 2 uses the historical average values of payments and market returns over the preceding five-year period to represent expected values.

nation of such benefits or that considerable uncertainty exists regarding the future of agricultural programs. Neither explanation seems persuasive in light of the previous seventy years of generous support for U.S. agriculture.

It is interesting to compare the effects of government payments on land values to the effects of market returns. The results indicate that an additional $1 obtained from the market would raise land values by $3.14, a figure very similar to what is implied for subsidy payments. The results reflect the expected influences of urban pressures on land values, with more highly populated and less rural areas having higher land values. Although these urban effects are interesting in their own right, it is important that they be accounted for (a step that has generally been neglected in previous analyses) in order to obtain accurate measures of the policy effects on land values.[17] Land in the most rural areas tends to be valued at $1,395 per acre less than

17. Hardie, Naryan, and Gardner (2001) estimate the effects of urban pressure on agricultural land. They are not concerned, however, with the contribution of agricultural policy to the returns from land.

that in the most urban areas, other things constant. Population growth and a greater population relative to agricultural land in a county also both positively contribute to land values.

We have argued that the use of observed payments may result in an attenuation bias that forces the implied capitalization rates toward zero. As an alternative, we have argued that a measure of expected payments may be preferred. Model 2 replaces the total realized payment measure with the five-year average measure noted in the preceding. As expected, the results suggest much larger and more reasonable effects of agricultural policy benefits on agricultural land values. An additional $1 of government payments raises land values by $13.13 per acre. Such a finding implies a much more reasonable capitalization rate of policy benefits into agricultural land values. The effect of historical average market returns is quite similar across the alternative models, with an additional $1 of net market returns corresponding to an increase of $3.45 in land values. The fact that market returns appear to be much more heavily discounted than is the case for government payments may seem puzzling at first glance. However, an examination of the historical patterns of returns and payments may help to explain this finding. Figure 1.3 illustrates the patterns of net returns (given by total marketings less total production costs) and total government payments over the 1970 to 2006 period. Three different levels of aggregation are presented—the entire United States, Iowa (a major agricultural state), and Kossuth County, Iowa (a major agricultural county in Iowa). In each case, the diagram illustrates the fact that real net farm market returns have been falling over time and that market returns are much more volatile than government payments. In many cases, net returns from the market are actually negative. Aggregation conceals much of the volatility in returns that is actually present at the farm level. This is demonstrated by the increased level of volatility across the successively less aggregated statistics. At the individual farm level, at least to the extent that individual risks are not perfectly correlated across farms, one would expect an even higher level of variation in net market returns.

In light of the observed behavior of market returns over time, a high degree of discounting by risk averse agents is not unexpected. Of course, one cannot fully decouple market returns from government payments because most agricultural programs are intended to provide countercyclical benefits intended to offset decreases in market earnings. Such countercyclical behavior is evident in the diagrams in that benefits are higher when market returns fall. It is important to again emphasize that it is not our intention to interpret the full or average impact of payments and thereby to make inferences about the total impact of payments on land values. Such inferences may be impossible given the fact that payments are so deeply embedded in asset markets and are so closely tied to market swings. Rather, our intent is to examine marginal impacts of changes in payments and market returns on land values—the effects that are represented by our model coefficients.

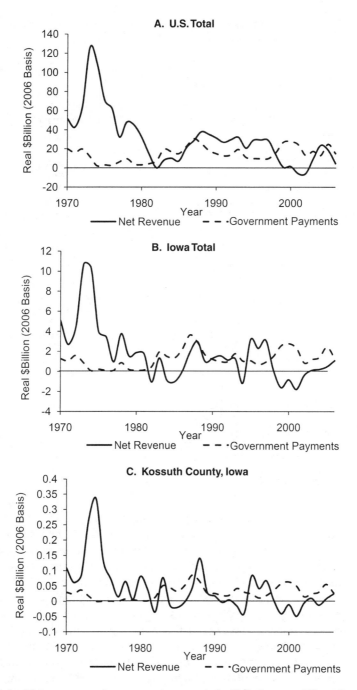

Fig. 1.3   Net revenues and government payments for U.S., Iowa, and Kossuth County, Iowa

Table 1.3   **Disaggregate policy models of land value determinants: Parameter estimates and summary statistics**

Variable	Model 3 Estimate	t-ratio	Model 4 Estimate	t-ratio
Intercept	2,911.0958	106.31	2,649.6005	104.40
	(27.3840)		(25.3794)	
Other payments	3.1634	8.44	−0.9048	−0.49
	(0.3747)		(1.8495)	
LDP payments	2.3818	3.96	21.0658	11.58
	(0.6015)		(1.8189)	
Direct payments	2.0045	3.96	5.3529	4.71
	(0.5056)		(1.1370)	
Disaster payments	5.0215	13.94	31.1035	10.36
	(0.3602)		(3.0022)	
Aggregate market returns	2.7420	42.88	3.3864	56.48
	(0.0639)		(0.0600)	
Population growth	350.4934	53.12	403.5037	72.22
	(6.5988)		(5.5875)	
$Urban_1$	−1,414.6055	−49.38	−1,242.8038	−50.91
	(28.6454)		(24.4121)	
$Urban_2$	−936.4743	−24.04	−849.7788	−26.01
	(38.9550)		(32.6728)	
$Urban_3$	−631.2550	−15.84	−607.3020	−18.07
	(39.8520)		(33.6108)	
Population/Farm acres	−0.2463	−0.43	36.9117	20.16
	(0.5744)		(1.8312)	
No. of observations	53,542		83,135	
R^2	0.1645		0.1786	

Numbers in parentheses are standard errors. Model 3 uses current year realized values for payments and market returns. Model 4 uses the historical average values of payments and market returns over the preceding five-year period to represent expected values. LDP = loan deficiency payment program.

A second segment of the analysis breaks out the overall government payments into their individual components, generated from different programs. We have argued that it is likely that different policies, which operate through widely varied support mechanisms, may have different effects on land values. Models 3 and 4 use actual payment receipts and our measure of expected payments (historical averages), respectively. We segment payments into four different components (see table 1.3). The first consists of LDP payments, which include marketing loan gains and deficiency payments in years prior to 1996. These payments are directly tied to production and are intended to support the price of actual production of commodities. A second component of payments is direct payments. This comprises payments that are not tied to production but rather are based upon historical "base" acreage and yield, which was largely established in the early

1980s.[18] These payments include direct, decoupled payments, market loss assistance, and countercyclical payments. Although these payments are all based upon historical base production and are not tied to current production and acreage, the market loss and countercyclical payments are triggered by market prices and, thus, may not be known in advance.[19] Ad hoc disaster payments are also identified separately and are included in the disaggregated regression. A distinction between the aggregate and disaggregate segmentation of payment data should be noted at this point. The ARMS survey collected market loss assistance data jointly with disaster payments, whereas the aggregate FSA data groups countercyclical payments (the successor to the market loss assistance program) together with fixed, direct payments. In light of this fact, we group together direct and countercyclical payments for the disaggregated data and direct payments, market loss assistance, and countercyclical payments for the aggregate data. Finally, we have a category of all other payments, which consists of conservation payments, state and local government benefits, and any other miscellaneous government subsidies.

The model of observed payments suggests that an additional dollar of LDP payments (direct price supports) will increase land values by only $2.38 per acre. When realized payments are replaced by the five-year average at the county level, the LDP effect rises to $21.07, again perhaps reflecting the attenuation biases inherent in using observed payments in any given year on an individual farm. The significantly higher value of an additional dollar of price support is consistent with expectations and suggests a reasonable discounting rate close to 5 percent.

Disaster payments tend to exhibit a large effect on agricultural land values, especially in the model using aggregate data. These payments are largely ad hoc by design and encompass a wide range of congressional objectives. Expectations regarding the impacts of aggregated disaster payments are unclear because so many different programs are of such an ad hoc nature and are included in this category (see appendix table 1A.1). However, direct monetary subsidies may certainly be expected to raise the returns to ownership of an asset and, thus, should increase land values. On the other hand, disaster relief is often targeted to higher risk, marginal areas. Thus, it would not be unexpected to see disaster payments being correlated with lower land values. Our results suggest that additional support in the form of disaster payments does indeed increase asset values in agriculture. An additional $1 of disaster relief raises land values by $5.02 in the case of the realized payments model (Model 3) and by $31.10 in the model based upon long-run average values of disaster payments. The 1990s were a period that experi-

---

18. The 2002 Farm Bill gave landowners the option of updating their base using 1998 to 2001 production and acreage records.

19. This particular grouping of payments was dictated by the available FSA data.

enced significant ad hoc disaster relief and, thus, may certainly have had significant impacts on farmland values.[20]

Direct payments also exert a significant effect on agricultural land values. An additional dollar in payments raises land values by $2.00 per acre in the model using realized payments and $5.35 per acre in the model based upon long-run average values. If land market agents truly believed that these payments were transitory, as the 1996 legislation seemed to imply, these impacts would seem to be larger than expected.[21] It is likely that these payments were a signal of future benefits to be paid on a decoupled basis. Indeed, in its generosity, Congress not only continued these payments under the 2002 and 2008 farm legislation, but also expanded and enhanced the benefits. More important, the new Farm Bill made soybean acreage eligible for direct payments in 2002. Thus, our results suggest that agents anticipated such legislative actions—any implicit threats to terminate this avenue of support with the expiration of the 1996 legislation were heavily discounted.

Similar values of the impacts of market returns and nonagricultural demands for farmland are revealed in the disaggregate policy models. A larger impact of market returns is exhibited in the model using the long-run average value of historical returns than occurs in the case of realized returns in the year of the survey. Urban pressures again play an important role in determining agricultural land values. In every case, the effects are statistically significant.

In all, the results confirm that government payments exert a significant effect on land values. The (marginal) rates of capitalization suggest that in the current policy context, a dollar in benefits typically raises land values by $13 to $30 per acre, with the response differing substantially across different types of policies. This response certainly suggests that agents expect these benefits to be sustained for some time. In terms of the implications for the distribution of farm program benefits, our results confirm that a substantial share of the benefits is captured by landowners. Recall that, in many cases, landowners may be a very different entity than farmers. Farmers wishing to expand or enter production will realize much smaller benefits than the policy rhetoric tries to substantiate. An important finding is that market returns, which are much more volatile than government payment benefits, tend to have a much lower influence on land values. Such returns have often been negative over the last several years, and the degree of volatility increases at less aggregate levels of measurement.

---

20. A 2006 report from the Environmental Working Group (EWG; 2006) reports that U.S. taxpayers provided nearly $26 billion in emergency agricultural disaster aid to more than two million farm and ranch operations over the 1985 to 2005 period, with payouts exceeding one billion dollars in eleven of the twenty-one years.

21. Cynics often note that, beyond naive academics, few farm policy observers believed these so-called transition payments were temporary. The empirical evidence has confirmed such suspicions.

The results on farm land values provide evidence that land captures policy benefits as land values are enhanced by the subsidies. When the farm operator owns the land, the transfers go to him. Likewise, in the absence of effective limits on payments (there are none), the larger a farm is, the greater will be total payment benefits.[22] However, as we have noted in the preceding, about 45 percent of U.S. farmland is operated by someone other than the owner.

1.4.2   Land Rents

Our findings so far raise the important question—how do the generous provisions for support of agriculture affect the significant share of farmers that rent the land used in production? Likewise, how much of the support goes to landowners? Again, the stated intent of the legislation is a "fair and equitable" sharing of program payments, with an owner that shares no risk (i.e., rents under cash lease arrangements) receiving none of the benefits. The real answer to this question lies in an evaluation of the terms of lease arrangements—do lease rates reflect policy benefits? If, as we have demonstrated, the value of land is increased by policy transfers, given that value of land is a present discounted value of expected cash flows plus an option to convert, one would expect that lease rates reflect payments from the government. Lease rates provide direct evidence on the proportion of farm payments passed on to landlords, something much more difficult to assess from land values.

For those farmers in our sample that were engaged in renting land, we were able to obtain the rental rates paid per acre for land rented under both share and cash arrangements. This is an important distinction because both types of rental arrangements are common. In our sample, 65.6 percent of farm operators reported renting some land and of those that rented, 84.6 percent used cash leases, 37.8 percent used share leases, and 22.4 percent used both cash and share leases.

It is likely that some frictions exist in lease arrangements for farm land because these arrangements may not be negotiated every year. In this light, it may take some time for lease markets to respond to increases or decreases in the level of support provided to producers, in particular for cash leases. On the other hand, we should expect share lease payments to reflect the ex post contribution of every single source of agricultural earnings. Share rents are indeed paid at the end of the season, once all uncertainty has been realized. Share lease payments are supposed to reflect the agreed proportion of

22. The extent to which farm program payments should be limited was an important point of considerable debate in recent farm bill deliberations. Any support based on production (such as LDP payments) will naturally favor larger producers. In the end, any limits on benefits tied directly to production were eliminated in 2008, and very loose income means tests were imposed. Goodwin (2008) investigated the likely impacts of binding payment limits and found that, for the vast majority of producers, limits have no impact on production.

cash flows from all sources of earnings related to the farming of the land, including government payments, though again share arrangements may be subject to the terms set through individual negotiations. In both share and cash leases, the terms of the lease are set in advance of the realization of farm earnings and program benefits, at least in most cases. Thus, it is again the case that rental rates will be based upon expectations of returns and further that the terms of the lease are set prior to the realization of these returns.

A subtle distinction exists in the role of expectations in our analysis of land values and rental rates. The data are collected early in the year following the survey year. At this point, survey respondents have full knowledge regarding realized payment benefits and market returns and are free to factor such knowledge into their assessment of land values. However, as we have noted, it is not realized payments in the preceding year but rather long-run payment expectations that will influence land values. In the case of rental rates, realized returns and policy benefits are not known at the time lease terms are determined. In the end, the distinction does not alter the fundamental analysis in that a measure of expectations of payments and market returns is necessary. To this end, we again utilize the historical five-year average value of the payment variables and of market returns. In addition, we adopt an approach similar to that used by Patton et al. (2008) and utilize the generalized method of moments approach of Hansen and Singleton (1982) and utilize IV to model expectations. We choose instruments available in agents' information sets at the time rental terms are defined.

We considered regressions of cash and share rents, respectively, against the factors expected to be relevant to land values and rents, including the indicators of expected payments. The results are presented in tables 1.4 to 1.7. We consider both aggregate policy benefits (for all programs at the farm level) and the alternative specification that distinguishes benefits from different types of policies. Table 1.4 presents estimates from a regression of farm-level cash rental rates on aggregated, historical payment benefits (Model 5) and the GMM estimates that are based upon instruments that include lagged payment variables, preplanting futures prices, annual fixed effects, and lagged county-level market returns (Model 6).

The key question is the extent to which higher government payments are reflected in higher cash rental rates. Kirwan (2009) found that the incidence of subsidy benefits fell mainly upon tenants, who received about $0.79 of each $1 of total payments. Put differently, cash rents tended to increase by $0.21 for each $1 of payments. However, as we have noted in the preceding, these estimates may be subject to measurement error biases due to the ignorance of share leases and the assumption that payment benefits are predetermined by exogenous policy parameters. It is again important to note that payments are delivered to farmers through many different mechanisms and, in most cases, are unknown until after harvest.

Our analysis reveals a substantially higher share of payments being dis-

Table 1.4   **Aggregate models of cash rental rate determinants: Parameter estimates and summary statistics**

	Model 5		Model 6	
Variable	Estimate	*t*-ratio	Estimate	*t*-ratio
Intercept	74.7981	68.02	56.4471	47.51
	(1.0997)		(1.1882)	
Total payments	0.3207	39.76	1.0137	52.89
	(0.0081)		(0.0192)	
Aggregate market returns	0.0785	30.98	0.1159	38.42
	(0.0025)		(0.0030)	
Population growth	3.6276	13.31	5.5842	20.31
	(0.2726)		(0.2750)	
Urban$_1$	−15.3141	−13.22	−12.9287	−11.28
	(1.1581)		(1.1460)	
Urban$_2$	−14.7203	−9.4	−13.3325	−8.61
	(1.5652)		(1.5482)	
Urban$_3$	−17.2977	−10.78	−15.6342	−9.85
	(1.6046)		(1.5869)	
Population/Farm acres	0.7389	8.96	0.8131	8.87
	(0.0825)		(0.0917)	
No. of observations	50,611		50,571	
R^2	0.0601		0.0806	

Notes: Numbers in parentheses are standard errors. Model 5 uses current year realized values for payments and market returns. Model 6 uses the historical average values of payments and market returns over the preceding five-year period to represent expected values.

tributed to landlords engaged in cash rental arrangements with farmer tenants. For each $1 of aggregated payments (across all program types), landlords claim $0.32 in benefits, other things constant. When actual payment receipts are used within an IV-GMM context (Model 6), this amount rises to $1. This does seem unreasonably high, but both results are indicative of a situation somewhat counter to the results of Kirwan (2009) in which landlords are effective at extracting payment benefits through higher cash leases. Table 1.5 presents results for disaggregated programs. The results again indicate that landlords are effective in extracting a large share of payment benefits through higher cash rental rates. In the model using historical average payments (Model 7), cash rents rise anywhere from $0.73 to $1.64 for each $1 of payments received, depending on the program mechanism used to deliver these payments. Direct payments, which are not tied to production and which, at least in part, were known with certainty over the period of study, raise rents by $0.73 for each additional $1 of payments. Disaster payments are actually correlated with lower cash rents, a result that is not consistent with our earlier findings regarding land values. However, disaster payments are, by definition, directed toward more marginal areas of production and, therefore, may be correlated with lower productivity and

Table 1.5   Disaggregate models of cash rental rate determinants: Parameter estimates and summary statistics

Variable	Model 7		Model 8	
	Estimate	*t*-ratio	Estimate	*t*-ratio
Intercept	58.4650	48.8	54.9059	17.21
	(1.1981)		(3.1901)	
Other payments	0.9007	10.72	2.2508	7.27
	(0.0840)		(0.3098)	
LDP payments	1.6367	20.93	2.9856	8.84
	(0.0782)		(0.3379)	
Direct payments	0.7295	14.73	0.6020	3.22
	(0.0495)		(0.1867)	
Disaster payments	−2.1341	−15.28	−4.2835	−8.74
	(0.1397)		(0.4904)	
Market returns	0.1287	42.17	0.1574	21.53
	(0.0031)		(0.0073)	
Population growth	6.0780	22.14	4.3409	7.22
	(0.2745)		(0.6013)	
Urban$_1$	−12.2037	−10.6	−12.9420	−5.1
	(1.1510)		(2.5388)	
Urban$_2$	−13.0745	−8.48	−12.1397	−3.58
	(1.5424)		(3.3888)	
Urban$_3$	−15.9286	−10.07	−18.5700	−5.52
	(1.5814)		(3.3651)	
Population/Farm acres	0.9721	11.99	1.2248	6.87
	(0.0811)		(0.1783)	
No. of observations	50,115		32,526	
R^2	0.0962		0.0336	

Notes: Numbers in parentheses are standard errors. Model 7 uses the historical average values of payments and market returns over the preceding five-year period to represent expected values. Model 8 uses instrumental variables–generalized method of moments estimation methods to incorporate expectations of current period values. LDP = loan deficiency payment program.

lower rents. The results again indicate a relatively low incidence of market returns on cash rental rates. The GMM estimates (Model 8) imply even larger impacts of payments on cash rents, though the general implications of the analysis are the same—landlords are effective at extracting policy benefits through higher cash rental rates.

Table 1.6 repeats the analysis for share rental rates. An important point regarding the construction of share rental rates should be noted. These rental rates include payments going directly to the landlord. This allows a direct comparison with cash rental rates. If the landlord's direct share of payments were removed from the calculation of rental rates, one would expect coefficients to be zero if the landlords were unable to extract additional benefits through higher share rates. The results are largely similar to those

Table 1.6   Aggregate models of share rental rate determinants: Parameter estimates and summary statistics

Variable	Model 9 Estimate	t-ratio	Model 10 Estimate	t-ratio
Intercept	98.0877	48.34	81.5600	37.65
	(2.0291)		(2.1665)	
Total payments	0.4978	31.64	1.1635	35.94
	(0.0157)		(0.0324)	
Aggregate market returns	0.0527	7.25	0.1421	16.26
	(0.0073)		(0.0087)	
Population growth	5.4354	11.14	−22.6754	−10.66
	(0.4878)		(2.1265)	
Urban$_1$	−23.7259	−11.09	−8.5924	−3.07
	(2.1400)		(2.8032)	
Urban$_2$	−7.8767	−2.79	−11.2415	−3.81
	(2.8187)		(2.9505)	
Urban$_3$	−12.9986	−4.38	7.0523	14.44
	(2.9671)		(0.4884)	
Population/Farm acres	−0.3355	−1.53	−0.2611	−1.20
	(0.2189)		(0.2177)	
No. of observations	23,627		23,601	
R^2	0.0594		0.0716	

Notes: Numbers in parentheses are standard errors. Model 9 uses current year realized values for payments and market returns. Model 10 uses the historical average values of payments and market returns over the preceding five-year period to represent expected values.

for cash rental rates, with an additional $1 of total payments raising share rental rates by $0.50 to $1.16 per acre. This again indicates that landlords are likely able to extract additional policy benefits beyond those received directly, though if the rental agreements are on a 50-50 share basis, the lower estimate would suggest no additional benefits for landlords over those that they receive directly.[23] Again, significant differences in rental impacts of policies are apparent across different policy types (see table 1.7). Disaster payments tend to lower share rental rates, though the effect is statistically significant only in the case of the GMM estimates (Model 12). This is consistent with expectations in that share lease rates are usually considered to carry a risk premium over cash rental arrangements. To the extent that disaster payments lower risk as they are designed to do, they should lower share rental rates. Each additional $1 in direct payments raise share rental rates by $0.33 to $0.70, again indicating a significant benefit for landlords.

23. Legislation mandates a "fair and equitable" allocation of policy benefits, which in share leases typically corresponds to the overall terms of the share lease.

Table 1.7   Disaggregate models of share rental rate determinants: Parameter
                   estimates and summary statistics

	Model 11		Model 12	
Variable	Estimate	t-ratio	Estimate	t-ratio
Intercept	84.3849	37.18	65.8711	9.77
	(2.2697)		(6.7400)	
Other payments	−0.2209	−1.04	6.8322	6.35
	(0.2124)		(1.0758)	
LDP payments	2.4906	21.08	0.0008	0.00
	(0.1182)		(0.6939)	
Direct payments	0.3302	4.45	0.6957	2.57
	(0.0741)		(0.2703)	
Disaster payments	−0.2583	−0.85	−4.5738	−5.55
	(0.3030)		(0.8248)	
Market returns	0.1390	15.88	0.1772	7.48
	(0.0088)		(0.0237)	
Population growth	6.6198	13.58	3.4059	3.41
	(0.4876)		(0.9993)	
Urban$_1$	−21.6807	−10.14	−26.8882	−6.51
	(2.1379)		(4.1282)	
Urban$_2$	−8.5882	−3.08	−10.1961	−1.87
	(2.7916)		(5.4412)	
Urban$_3$	−11.7639	−4.00	−13.5427	−2.32
	(2.9416)		(5.8366)	
Population/Farm acres	−0.2124	−0.98	−0.0287	−0.05
	(0.2167)		(0.6288)	
No. of observations	23,466		15,143	
R^2	0.0852		0.0352	

Notes: Numbers in parentheses are standard errors. Model 11 uses the historical average values of payments and market returns over the preceding five-year period to represent expected values. Model 12 uses instrumental variables–generalized method of moments estimation methods to incorporate expectations of current period values. LDP = loan deficiency payment program.

1.4.3   The Insurance Component of Agricultural Support

The typical approach to the assessment of the total contribution of agricultural policy to land values relies on the coefficient from the land value regressions. This is problematic for two reasons. The first one, usually mentioned in the literature, is due to the fact that regressions yield the effects of the marginal dollar for each type of policy. The literature has, however, overlooked the second reason. If we think about land as a portfolio of securities each delivering its stream of cash flow, it is obvious that the risk of the portfolio depends on the covariance of the various underlying securities. Therefore, eliminating one or more of the underlying securities will affect the risk of the portfolio. In terms of the analysis of the policy contribution, this implies that eliminating a policy that provides an insurance benefit will not only decrease expected returns, but it will also increase the volatility of the remaining (market) returns. In other words, we should expect the coefficient on market earnings to decrease in response to an increase in uncertainty. The capitalization rate of earnings will be lower, reflecting the higher opportunity cost of capital for an asset with more volatile earnings.

This raises the following question: if there is a theoretical argument in favor of an insurance component to the contribution of agricultural policy to land, can we find evidence from the market that it matters quantitatively? Unfortunately, there are no counties targeted by the ARMS survey that exempt all farmers from the benefits of agricultural policy. However, as we have noted in the preceding, farm land is rented under both cash lease and share lease arrangement. Cash lease rate are set ex ante, while the share payment depends upon the actual earning of the parcel, thus implying a risk sharing arrangement.

The main programs designed to reduce the variability of farm earnings and insure the cash flow to farmers are price supports and disaster payment programs. If the insurance component matters, we should find that higher payments should be correlated with a lower risk premium on rental arrangements. By committing to an ex ante fixed payment, the farmer provides insurance to his landlord for which we should expect him to be rewarded (unless we observe cash rents only when the farmer is not risk averse).

To evaluate this risk premium, for the subset of 11,227 farms that have both cash and share rental agreements in place, we consider the impact of different policies on the share-cash rental rate differential. These results are presented in table 1.8. We find that disaster payments do indeed tend to exhibit an insurance benefit effect in that they lower the share-cash rental rate differential. In contrast, LDP payments tend to increase the differential. The insurance properties of disaster payments are obvious, but the reason for the positive relationship between LDP payments and the risk premiums is less clear. Because LDP payments tend to be higher for crops and regions that experience more price volatility, this may reflect the greater price risk associated with such crops and regions. The category of "other payments" appears to lower the share-cash differential, likely reflecting the insurance benefits provided by this large grouping of payments, which includes conservation program payments.

1.5   Concluding Remarks

Policy rhetoric often justifies Farm Bill expenditures with the argument that impoverished farmers are in need of governmental support to remain in business. This view is pervasive outside of Washington. For example, consider the annual Farm Aid events intended to draw attention to the plight of the American farmer. Our analysis challenges this view. We demonstrate that land owners capture substantial benefits from agricultural policy. However,

Table 1.8   **Analysis of share-cash rental rate differentials: Parameter estimates and summary statistics**

	Model 13		Model 14	
Variable	Estimate	t-ratio	Estimate	t-ratio
Intercept	13.5978	10.38	26.9354	4.50
	(1.3106)		(5.9900)	
Other payments	−0.1328	−1.04	−2.7560	−2.29
	(0.1282)		(1.2040)	
LDP payments	0.0966	1.36	2.1743	3.00
	(0.0713)		(0.7236)	
Direct payments	0.0660	1.40	−0.1708	−0.94
	(0.0471)		(0.1827)	
Disaster payments	−0.7841	−3.95	−2.1911	−2.85
	(0.1985)		(0.7698)	
Market returns	−0.0214	−3.39	0.0184	0.93
	(0.0063)		(0.0197)	
Population growth	0.2551	0.88	0.0415	0.07
	(0.2908)		(0.6037)	
$Urban_1$	−2.9239	−2.35	−0.8763	−0.34
	(1.2449)		(2.5514)	
$Urban_2$	−3.4084	−2.12	−1.4991	−0.45
	(1.6064)		(3.3340)	
$Urban_3$	1.8332	1.11	1.4492	0.42
	(1.6534)		(3.4305)	
Population/Farm acres	0.0189	0.20	−0.5210	−1.21
	(0.0960)		(0.4293)	
No. of observations	11,227		7,514	
R^2	0.0069		0.0021	

Notes: Numbers in parentheses are standard errors. Model 13 uses the historical average values of payments and market returns over the preceding five-year period to represent expected values. Model 14 uses instrumental variables–generalized method of moments estimation methods to incorporate expectations of current period values. LDP = loan deficiency payment program.

in many cases, land owners are distinct from the farmers whose plight we are told we should be concerned with.

Of course, many farmers are also landowners and, thus, have an important stake in maintaining agricultural policy benefits. A farmer that purchased land that reflected the value of anticipated benefits would certainly suffer a damaging capital loss if such support were to be withdrawn. Furthermore, all farmers have a strong interest in congressional surprises, whvereby more transfers are allocated than anticipated by the land market. As owners, they benefit from the unexpected capital gains. The 2002 and 2008 Farm Bills, with their large increases in support expenditures, may have been such nice surprises.

Tenants also gain from positive surprises as long as lease rates do not adjust instantaneously. However, the 2002 Farm Bill seems to have shut down this avenue for a temporary increase in the share of transfers captured by farm operators. One provision of the bill is that it offers to farmers the opportunity to update the factors that determine the level of some of the payments they receive. The power to decide whether to update has been given to the owners of the land, not the operator. Not surprisingly, tenant farmers complained that land owners used this opportunity to impose a renegotiation of the existing leases that did not foresee the generosity of the 2002 Farm Bill. No base updating provisions were included in the 2008 legislation. However, the precedent for such updating has been established, and agents most certainly have some expectation, however much it is discounted, that such future opportunities will again be presented.

Appendix

Table 1A.1   **U.S. Department of Agriculture program payments by category: Outlays and number of recipients (1990–2005)**

Program	Total	No. of recipients
Coupled payments		
Acreage Grazing Payments	11,475,210	9,386
Barley Assessment Deficiency	37,303,627	89,793
Cotton Deficiency	3,434,395,526	694,440
Crop Special Grade Rice LDP	4,719,159	285
Feed Grain Deficiency	15,328,664,623	6,072,369
LDP, Noncontract PFC Growers	85,305,152	58,536
Loan Deficiency	29,732,547,354	5,192,213
Market Gains	4,476,129,696	633,196
Rice Deficiency	3,338,380,074	218,421
Rice Marketing	34,014,757	20,274
Wheat Deficiency	7,923,366,487	3,223,695
Winter Wheat Deficiency	682,864,667	248,373
Direct payments		
Amlap–Apportioned	94,934,998	7,647
Apple Market Loss Assistance	166,373,534	13,160
Dairy Market Loss Assistance	968,612,817	187,732
Direct and Countercyclical	25,068,153,272	4,218,971
Lamb Meat Adjustment Assist	86,071,348	71,706
Marketing Loss Assistance	18,260,407,458	5,366,287
Oilseed Program	950,113,825	1,184,806
Peanut Marketing Assistance	119,010,211	27,094
Peanut Marketing Assistance Program III	53,924,599	17,277
Production Flexibility	35,210,684,603	9,667,805
Supplemental Oilseed Payment Program	418,811,924	586,572
Supplemental Tobacco Loss	127,461,626	335,871
Tobacco Loss Assistance	346,044,295	361,113

(*continued*)

Program	Total	No. of recipients
WAMLAP II–Apportioned	18,637,475	20,985
WAMLAP III–Apportioned	16,730,874	20,974
Wool and Mohair Market Loss Assistance	10,228,857	18,629
Disaster payments		
01-02 Crop Disaster Assistance	2,547,849,688	389,516
2000 Florida Nursery Losses	29,437	3
AILFP–Apportioned	6,480,878	1,180
American Indian–Livestock Feed	12,458,007	2,389
Apple and Potato Quality Loss	34,199,943	1,681
Avian Influenza Indemnity Program	52,980,294	163
Cattle Feed Program	136,401,954	49,580
Citrus Losses in California	2,154,433	987
Crop Disaster Program	3,060,477,581	555,263
Crop Loss Disaster Assistance	1,857,480,163	249,555
Dairy Disaster Assistance	7,495,444	1,161
Dairy Indemnity	2,450,691	456
Disaster	5,532,181,025	1,504,547
Nonprogram Crops	42,215	29
Program Crops	–112,369	74
Disaster Reserve Assistance	145,110,728	85,247
Emergency Conservation	312,905,164	124,459
Emergency Conservation Program	70,106,623	26,205
Emergency Feed	–1,029,779	1,303
Flood Compensation Program	706,144	38
Idaho Oust Program	4,888,638	71
Karnal Bunt Fungus Payment	38,897,325	912
LIP–Contract Growers	1,031,180	1,229
Livestock Compensation Program	1,096,133,267	578,840
Livestock Emergency Assistance	1,550,736,935	781,983
Livestock Indemnity Program	305,696	164
NAP–Supplemental Appropriation	3,917,572	1,379
Noninsured Assistance Program	672,291,473	170,099
Nursery Losses–Florida	7,316,930	195
Pasture Flood Compensation	20,387,735	12,252
Pasture Recovery Program	52,971,866	35,093
Poultry Enteritis Syndrome	1,768,271	136
Quality Losses Program	148,615,562	35,246
Sugar Beet Disaster Program	45,636,494	2,745
Tobacco Disaster Assistance	2,696,981	343
Other payments		
Additional Interest	56,214	279
Agricultural Conservation	1,132,520,907	739,873
Agricultural Management Assistance	5,752,517	796
Animal Waste Management	256,368	26
Arkansas Beaver Lake	2,464,632	477
Auto Agricultural Conservation Program— Environment Long Term	402,632	109
Auto Agricultural Conservation Program— Environment Annual	1,163	1
Auto Ana–Conservation Annual	1,875	2

Table 1A.1 (continued)

Program	Total	No. of recipients
Auto CRP–Cost Shares	353,698,363	143,683
Auto EQIP	173,468,007	37,592
Auto LTA–Conservation Long Term	704,059	164
Clean Lakes	9,999	1
Colorado River Salinity	31,832,222	1,992
Cotton Diversion	−15,095	22
CRP Annual Rental	24,695,070,732	5,701,530
CRP Cost–Shares	840,994,086	429,665
CRP Incentives	483,637,540	219,732
Dairy Termination	237,026,377	19,893
Environment Quality Incentives	477,768,620	126,317
Extended Farm Storage	171,409,332	72,879
Extended Warehouse Storage	44,481,468	18,694
Feed Grain Diversion	−395,250	4,040
Finality Rule	1,007,752	1,403
Forestry Incentive–Annual	51,322,552	25,100
Forestry Incentive–Long Term	12,254,173	6,120
Fresh Market Peaches Program	783,991	126
Grasslands Reserve Program	9,275	4
Hard White Winter Wheat	3,517,590	3,301
Interest on CCC	1,624	38
Interest on NAP Payment	4,678	184
Interest Payments	29,003,888	1,046,365
Klamath Basin Water Program	−4,299	4
Milk Diversion	30,576	20
Milk Income Loss Contract Transition	547,209,081	73,836
Milk Income Loss Contract	1,403,354,665	247,585
Milk Marketing Fee	265,896,171	249,035
National Wool Act	895,921,293	442,720
NRCS EQIP	283,707,027	32,930
Options Pilot Program	39,762,496	4,128
Payment Limitation Refund	−6,983,394	2,411
Peanut Quota Buyout Program	1,220,640,857	80,080
Potato Diversion Program	20,263,929	1,222
Rice Diversion	−12,567	11
Rural Clean Water	3,126,831	618
Small Hog Operation Program	121,376,613	57,952
Soil/Water Conservation Assistance	10,358,605	2,383
Sugar PIK Diversion Program	180,690,205	15,126
Tobacco Payment Program	50,887,278	297,921
Water Bank–Annual	43,879,235	30,317
Water Bank–Practice Cost/Share	11,046,258	7,682
Wetlands Reserve	34,315,395	1,830
Wheat Diversion	−2,237	85

Notes: LDP = Loan Deficiency Payment Program; PFC = Production Flexibility Contract; WAMLAP = Wool and Mohair Market Loss Assistance Program; AILFP = American Indian Livestock Feed Program; LIP = Livestock Indemnity Program; NAP = Noninsured Crop Disaster Assistance Program; CRP = Conservation Reserve Program; EQIP = Environmental Quality Incentives Program; LTA = Loan Term Agreement; CCC = Commodity Credit Corporation; NRCS = National Resources Conservation Service; PIK = Payment in Kind.

References

Barnard, C. H., G. Whittaker, D. Westenbarger, and M. Ahearn. 1997. "Evidence of Capitalization of Direct Government Payments into U.S. Cropland Values." *American Journal of Agricultural Economics* 79:1642–50.

Environmental Working Group (EWG). 2006. *After Two Decades of Agricultural Disaster Aid a Chronic Dependency Takes Root.* EWG Research Report, September 25. Washington, D.C.: Environmental Working Group. http://www.ewg.org/book/export/html/22592.

Goodwin, B. K. 2008. "The Incidence and Implications of Binding Farm Program Payment Limits." *Review of Agricultural Economics* 30:554–71.

Goodwin, B. K., A. K. Mishra, and F. Ortalo-Magné. 2003a. "Explaining Regional Differences in the Capitalization of Policy Benefits into Agricultural Land Values." In *Government Policy and Farmland Markets: Implications of the New Economy,* edited by C. Moss and A. Schmitz, 744–52. Ames, IA: Iowa State University Press.

———. 2003b. "What's Wrong with Our Models of Agricultural Land Values?" *American Journal of Agricultural Economics* 85:744–52.

Goodwin, B. K., and F. Ortalo-Magné. 1992. "The Capitalization of Wheat Subsidies into Agricultural Land Values." *Canadian Journal of Agricultural Economics* 40:37–54.

Hansen, L. P., and K. J. Singleton. 1982. "Generalized Instrumental Variables Estimation of Nonlinear Rational Expectations Models." *Econometrica* 50:1269–86.

Hardie, I. W., T. A. Naryan, and B. L. Gardner. 2001. "The Joint Influence of Agricultural and Nonfarm Factors or Real Estate Values: An Application to the Mid-Atlantic Region." *American Journal of Agricultural Economics* 83:120–32.

Kirwan, B. 2009. "The Incidence of U.S. Agricultural Subsidies on Farmland Rental Rates." *Journal of Political Economy* 117:138–64.

Mishra, A. K., H. S. El-Osta, M. J. Morehart, J. D. Johnson, and J. W. Hopkins. 2002. *Income, Wealth, and the Economic Well-Being of Farm Households.* Agricultural Economic Report no. 812. Washington, D.C.: Economic Research Service, U.S. Department of Agriculture.

Patton, M., P. Kostov, S. McErlean, and J. Moss. 2008. "Assessing the Influence of Direct Payments on the Rental Value of Agricultural Land." *Food Policy* 33 (5): 397–405.

Reidl, B. 2004. "Another Year at the Federal Trough: Farm Subsidies for the Rich, Famous, and Elected Jumped Again in 2002." Heritage Foundation Backgrounder no. 1763. Washington, D.C.: The Heritage Foundation.

Ryan, J., C. Barnard, R. Collender, and K. Erickson. 2001. "Government Payments to Farmers Contribute for Rising Land Values." *Agricultural Outlook* June–July:22–6.

Shertz, L., and W. Johnston. 1997. "Managing Farm Resources in the Era of the 1996 Farm Bill." Staff Paper no. AGES 9711. Washington, D.C.: Economic Research Service, U.S. Department of Agriculture.

Shoemaker, R., M. Anderson, and J. Hrubovcak. 1990. "U.S. Farm Programs and Agricultural Resources." Agriculture Information Bulletin no. 1990. Washington, D.C.: Economic Research Service, U.S. Department of Agriculture.

U.S. Department of Agriculture, National Agricultural Statistics Service (USDA-NASS). 1999. *1999 Agricultural Economics and Land Ownership Survey (AELOS).* Washington, D.C.: U.S. Department of Agriculture.

Weersink, A., S. Clark, C. G. Turvey, and R. Sarker. 1999. "The Effect of Agricultural Policy on Farmland Values." *Land Economics* 75:425–39.

2

Modeling Processor Market Power and the Incidence of Agricultural Policy
A Nonparametric Approach

Rachael E. Goodhue and Carlo Russo

Johnson (1979) identified six justifications for government policy, including agricultural policy. One of these was the provision of a stable minimum level of income commensurate with that of other groups in society. Equally classic analyses of the incidence of agricultural subsidies have focused on comparing the deadweight loss across policies, given the amount of income transferred to farmers (Wallace 1962; Gardner 1983). For the most part, these policy analyses have assumed that agricultural markets are competitive enough that any market power on the part of processors can be safely assumed away and the market treated as perfectly competitive (Rude and Meilke 2004). Given this assumption, the focus of these analyses is efficiency, and the assessment of the economic cost of a particular support policy depends only on its deadweight loss and the size of the transfer to farmers. Russo, Goodhue, and Sexton (forthcoming), however, demonstrate theoretically that even small degrees of market power can enable processors to extract considerable rents from taxpayers, affecting the distribution of the benefits (and the costs) of the policy. Our analysis focuses on distribution, rather

Rachael E. Goodhue is professor of agricultural and resource economics at the University of California, Davis. Carlo Russo is associate professor of agricultural economics at the University of Cassino.

The authors thank Jeffrey Williams for useful conversations. Philippe Botems, Anna-Célia Disdier, Julie Holland Mortimer, Jeffrey Perloff, and conference participants at the National Bureau of Economic Research Agricultural Economics Conference and the 8th Institut National de la Recherche Agronomique-Institut d'Economie Industrielle (INRA-IDEI) Conference on Industrial Organization and the Food Processing Industry provided helpful suggestions. Special thanks to Richard Sexton for his detailed and thoughtful comments on the paper and for his many insights developed in our joint research that have enriched this analysis. Goodhue is a member of the Giannini Foundation of Agricultural Economics. Authors are listed alphabetically.

than efficiency. Any distribution of policy rents to nonfarmer agents may make support policies less attractive politically.

The reliance on the perfectly competitive framework for analyzing policy incidence is interesting from a historical perspective. Reducing the exercise of market power in order to increase economic efficiency was among Johnson's other justifications for agricultural support policies. The economic history of agriculture suggests that it would be appropriate to address processor market power when analyzing government support policies. Farmer protests against the exercise of market power by other parties predate the major agricultural support programs developed in the 1930s; for example, the Grange and Populist movements in the nineteenth century were driven in part by farmers' protests regarding their perceptions of the exercise of market power against them in transportation and procurement (Stewart 2008). This earlier movement resulted in the Interstate Commerce Act of 1887. The Capper-Volstead Act of 1922, which exempted farmer cooperatives from antitrust regulations, was designed to enable farmers to organize collectively in order to exercise countervailing market power against buyers. In this chapter, we do not assess whether the current U.S. policy is effective in alleviating the consequences of market power on farmers. Instead, we focus on the unintended implications for welfare distribution of the interaction between middlemen's market power and government intervention.

This analysis examines the interactions between market power and agricultural policy in the U.S. wheat flour milling industry. It has two main objectives: to assess if the payments trigger a change in the underlying economic behavior of the milling industry, and to estimate if the spread between the price of wheat and the price of wheat flour is affected by the policy regime, holding everything else constant. Results indicate that wheat millers alter their pricing behavior when the program is making payments, and they are able to extract a rent from government intervention. These findings are consistent with a static model of oligopsony power. Theory suggests that deficiency payments reduce the elasticity of farmers' supply (e.g., Wallace 1962; Alston and James 2001). Consequently, the expectation is that, holding everything else constant, the oligopsony markdown is larger when the policy results in payments to farmers than otherwise. In this context, deficiency payments can be used as a natural experiment for identifying millers' oligopsony power, similar to other policy measures (Ashenfelter and Sullivan 1987).

Previous literature has tested for market power in the U.S. wheat flour milling industry. Brester and Goodwin (1993) found that the degree of cointegration of the price time series over space and across the vertical wheat chain was negatively correlated with the concentration of the largest four firms (CR4) index and argued that the increase in concentration was lessening competition. On the other hand, the price series exhibited a high degree of cointegration, consistent with the possibility that the industry remained competitive. Because the use of cointegration analysis may lead to ambiguous conclusions, as it did in this instance, later studies have relied on structural models. Kim et al. (2001) used a Poisson regression model to investigate changes in industry structure and found evidence of oligopoly with price leadership. Stiegert (2002) tested for upstream and downstream market power in the U.S. hard wheat milling industry and found that the null hypothesis of perfect competition could not be rejected. These analyses did not take into account the possibility of interactions between government support policies and the exercise of market power in the wheat market. Russo, Goodhue, and Sexton (2009) did so using a standard New Empirical Industrial Organization (NEIO) approach (Applebaum 1982; Bresnahan 1989). This approach relies on shifts in supply, demand, or policy to identify the exercise of market power and identify its magnitude.

The test for market power in these structural models is, implicitly, a joint test regarding market power and the functional forms specified in the empirical model (Genesove and Mullin 1998). Consequently, the estimation is vulnerable to misspecification of cost, supply, and demand relationships (Perloff, Karp, and Golan 2007). Furthermore, the standard NEIO analysis leaves many questions regarding industry behavior and its impacts unanswered. When economic agents are strategic players, are market power parameters sufficient for describing their behavior? Theory suggests that this is not necessarily the case. Although the so-called agnostic interpretation of the NEIO market power coefficient is an effort to avoid this criticism, misspecification of the economic game can still lead to biased estimation (Corts 1999). Strategies may be more complex than simple Cournot strategies or may vary over time, such as collusion-sustaining price wars in oligopolies or oligopsonies (Green and Porter 1984).

In the case of government intervention in agriculture, the unbiasedness of the NEIO estimator is conditional on specifying the agents' strategic reaction to the policy correctly. Because this modeling choice requires prior information regarding agents' economic behavior that is not available in the case at hand, nonparametric techniques are used to characterize the pricing behavior of the wheat milling industry without introducing assumptions about the nature of the economic game governing processors' conduct and without specifying functional forms. A change in strategic behavior may be postulated if processors react to exogenous shocks in different ways when a policy is in effect than when it is not. Moreover, if the price margin under the policy regime is larger, then one may conclude that the millers are acting strategically to extract a rent from the policy at taxpayers' expense, ceteris paribus.

2.1   Background: The U.S. Wheat Milling Industry

U.S. farmers harvested 2.1 billion bushels of wheat from fifty-one million acres in 2007. The total value of production including government payments was $13.7 billion (National Agricultural Statistics Service 2008). Wheat production is concentrated geographically; the three major production regions are the southern Great Plains (primarily Kansas and Oklahoma), the northern Great Plains (Montana and the Dakotas), and the Northwest (primarily southeastern Washington). The 2007 Census of Agriculture reported 160,818 farms classified as primarily wheat-producing farms (U.S. Department of Agriculture [USDA] 2007). The total number of farms producing wheat is even larger. Flour milling is the primary domestic use of wheat, although some is used for livestock feed and other purposes. The milling process generates both flour and by-products. By-products account for approximately 10 percent of the revenue from flour milling (Brester and Goodwin 1993).

The milling industry displays a number of characteristics that are consistent with an ability to exercise market power at the national level.[1] The four-firm concentration ratio in the flour milling industry is reasonably high and has increased over time. In 1974, the top four firms accounted for 34 percent of total milling capacity (Wilson 1995). In 1980, their share had increased slightly to 37 percent, further increasing to over 65 percent in 1991 (Brester and Goodwin 1993). Over that time period, consolidation was not limited to the firm level; between 1974 and 1990, the number of mills declined by a quarter, and the average plant capacity almost doubled (Wilson 1995). More recent data regarding concentration, the number of mills, or plant capacity are not available for the wheat flour industry alone; in 2007, the four-firm concentration ratio for the entire flour milling and malt manufacturing sector was 56.6 percent, and wheat flour milling accounted for 60 percent of the sector (IBISWorld 2007). Three of these large firms are large multicommodity agrofood firms; Archer Daniels Midland, ConAgra, and Cargill compete with each other across a number of markets, which potentially could strengthen their ability to collude. These firms increased their share of the number of wheat flour milling plants operated from 14 percent in 1974 to almost two-thirds in 1992 (Wilson 1995).

Between the mid-1970s and 1997, per capita wheat consumption increased, even though its share of total per capita grain consumption declined. There are a number of factors that may have contributed to this increase, including increased consumption of meals away from home, increased awareness of the health benefits of eating grain-based foods, and the promotion of wheat products by industry organizations (Vocke, Allen, and Ali 2005; Brester 1999). Since 1997, per capita wheat consumption has declined, due in part to a new technology for extended shelf life bread that has reduced the share of unsold bread and due in part to an increased interest in low-carbohydrate diets (Vocke, Allen, and Ali 2005). Another factor behind the continuing decline in wheat's share of total grain consumption has been increased con-

---

1. In theory, it is possible that any market power exercised by millers is due to local monopsony power.

sumer interest in eating a variety of grain products, driven in part by an increasingly diverse population (Putnam and Allshouse 1999).

Wheat is one of the major agricultural support program commodities, and government payments are a nonnegligible share of farm income for wheat producers. For farms characterized as primarily wheat producers, government payments were approximately 20 percent of average gross cash income in 2003. Government payments to other wheat-producing farms were about 8 percent of average gross cash income (Vocke, Allen, and Ali 2005). These numbers are quite dependent on the difference between the policy price set by the government and the market price; in 2007, average government payments equaled 5 percent of the market value of agricultural products sold for farms characterized as primarily wheat producers (USDA 2007).

U.S. farm policy is governed primarily by federal "farm bills" legislated every few years. Wheat producers were eligible for three basic types of program payments during the period of study (1974 to 2005), although implementation details differed. Direct payments are not linked to market conditions, while countercyclical payments depend on a season's average market price. Beginning with the 1985 Farm Bill, direct and countercyclical payments were restricted to a share of production defined by base acres and base yields. Federal commodity loan and marketing loan programs are the source of the third type of payment. Historically, these programs were intended to promote orderly marketing by providing farmers with income at harvest time that enabled them to repay operating loans without forcing them to sell their crops. Because farmers could wait to market their production, harvest-time prices would not be depressed by credit-driven sales. In addition to promoting orderly marketing, loan programs have become an important source of farm income support in years with low market prices.

Some variant of a commodity loan program has been available to farmers since the 1930s. Under a loan program, a farmer pledges a specified quantity of wheat as collateral for a loan valued at that quantity of wheat multiplied by the loan price. Farmers can choose to repay loans at the market price, rather than the loan price, when the market price is lower. Depending on the year, repayment could occur via forfeiting the actual physical product (a nonrecourse loan) or redeeming commodity certificates as well as through an exchange of funds. The resulting difference in price is referred to as a "marketing loan gain." Alternatively, for some years in the sample, the farmer could choose to receive a loan deficiency payment in lieu of an actual loan. The policy price on which loan deficiency payments and marketing loan gain payments are calculated is the loan rate. The relevant market price for loan repayments is the "posted county price" set daily by the government. It is intended to reflect market conditions in a county by adjusting major market prices for transportation costs and temporary cost differences. Farmers can lock in the loan rate as the price for their production by choosing to repay their loan at the posted county price rather than the loan rate, resulting in a marketing loan gain, or by requesting a loan deficiency payment in the amount of the difference between the two prices on a given day.

We focus on loan deficiency payments and marketing loan gains for three reasons. First, some variant of this program has been available to producers throughout the study period. Second, there has been no change in the share of production eligible for at least one of these payments. Finally, whether farmers receive payments is linked to the market price via the posted county price.

2.2   Oligopsony Power and Marketing Loan Rates

This analysis addresses the possibility that a marketing loan policy may enable millers with oligopsony power to increase their margins. Figures 2.1 to 2.3 illustrate the argument graphically by comparing the effects of a marketing loan policy under perfect competition and monopsony. Units are normalized so that one unit of wheat produces one unit of flour and normalize millers' production costs other than wheat to zero. Cases where the loan rate is the relevant price for farmers are referred to as cases where the marketing loan policy binds.

Under perfect competition, the market price $P*$ is determined by the intersection of supply and demand. If the loan rate is lower than the perfectly competitive expected price ($LR < P*$), then farmers will repay the loan, and the loan price will not affect the market outcome. If the loan rate is higher than the perfectly competitive expected price ($LR > P*$), as depicted in figure 2.1, then farmers will increase production because the loan rate is the effective price they receive. In this case, production $Q_s(LR)$ is independent of the market price. Millers will pay farmers the price $P_d[Q_s(LR)]$ defined by the demand curve.

Figures 2.2 and 2.3 address the effects of the marketing loan program when the miller is a monopsonist. If the loan rate is lower than the price the miller would pay farmers in the absence of the policy (P_o), then the market price determines output Q_o (figure 2.2). The margin received by the miller, defined as the difference between the price of flour and the procurement price of wheat at the quantity produced, equals $W_o - P_o$. If the loan rate is higher than P_o, then the policy is binding. Production is determined by the loan rate and is independent of the market price. The equilibrium quantity $Q_s(LR)$ is found by evaluating the supply function at the loan rate.

Given that supply is perfectly inelastic with respect to the market price when the policy binds, the price the miller pays farmers is indeterminate. Institutional factors suggest that millers are in a relatively strong position to extract policy rents by reducing the market price of wheat. Obviously, farmers have a weaker incentive to bargain for a higher price and greater

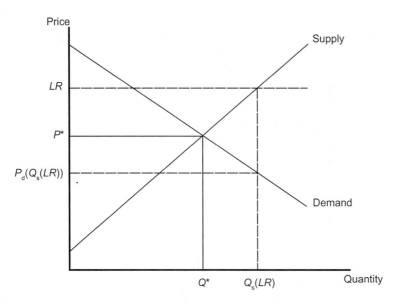

Fig. 2.1 Marketing loan program under perfect competition

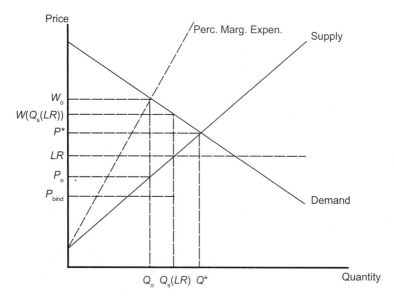

Fig. 2.2 Marketing loan program under oligopsony: Loan price below perfectly competitive price and above market price

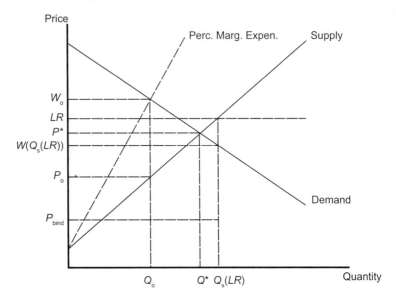

Fig. 2.3   **Marketing loan program under oligopsony: Loan price above perfectly competitive price**

share of the surplus under a marketing loan program than in an unregulated market. When offered a low price, they are less likely to continue to incur the cost of keeping their wheat in storage because any expected gain of selling the wheat later is (at least partially) offset by the loan rate-based government payment they receive.

Furthermore, as discussed earlier, the wheat milling industry is relatively concentrated, while there are a very large number of wheat-producing farms. Consequently, individual farmers have relatively little ability to negotiate effectively.[2] These factors are reflected in the organization of farmgate grain markets; generally prices are set by buyers and farmers choose whether to accept the take-it-or-leave-it offer.

There are two cases of binding policies defined by the relative magnitude of the loan rate and the perfectly competitive price. The first case of a binding policy is shown in figure 2.2. If the loan rate is higher than P_o and lower than P^*, then farmers produce more as a result of the policy. The miller maximizes profits by paying farmers a farmgate price no more than the loan

---

2. While some wheat farmers sell their output through marketing cooperatives, and the share of wheat marketed via cooperatives was nonnegligible during the sample period, the number of cooperatives is too large to suggest that there may be off-setting oligopoly power. In 1973, a total of 1,965 cooperatives as a group accounted for 29 percent of first-handler sales. In 1993, 1,243 cooperatives accounted for 38 percent of first-handler sales (Warman 1994).

rate LR. As can be seen in the figure, at LR, the loan marketing program reduces the miller's margin, all else equal. At farmgate prices no higher than P_{bind}, the miller's margin would increase.

The second case of a binding policy is shown in figure 2.3. If the loan rate is higher than P^*, then the farmgate price will be no higher than the price that will equate the quantity demanded with the quantity produced in response to the loan rate, or $W[Q_s(LR)]$. At that upper bound, the miller will have a zero margin. At prices no higher than P_{bind}, the miller's margin would increase.

The maximum price that the miller would be willing to pay farmers results in a lower margin when the policy binds than when the policy does not bind in both cases. Intuitively, the result is driven by the assumption that, if the policy is binding, farmers' supply is less elastic and the markdown of the farmgate price due to oligopsony power is expected to increase. Consequently, whether the marketing loan program allows millers to increase their margins is an empirical question. This analysis considers whether millers are able to drive the price sufficiently low when the policy is binding that they increase their margins.

2.3 Methodology

The structure of the empirical test regarding the millers' margin is simple. Define Y as the millers' margin calculated as the difference between the price of a hundredweight of wheat products and the price of the equivalent quantity of wheat, d as a dummy variable defining the policy regime (d = 1 if the policy's target price is above the procurement market price, and d = 0 otherwise), and X as a matrix of exogenous variables representing supply, demand and millers' marginal cost shifters. The null hypothesis

$$H_0: E(Y|X,d = 0) = E(Y|X,d = 1)$$

is tested versus the alternative hypothesis

$$H_1: E(Y|X,d = 0) < E(Y|X,d = 1),$$

where $E(Y|X,d) = f(X,d)$ and f is a function linking the exogenous variables and the policy regime to the conditional mean of Y. Rejection of the null hypothesis is statistical evidence that, holding everything else constant, the millers' margin increases if the policy is binding. Given our theoretical model, we interpret this result as a consequence of millers' oligopsony power.

Running the test is problematic because there is no clear and reliable a priori information about the linking function $f(X,d)$ and which variables are in the matrix X. Consequently, the test may be biased because of possible model misspecification or arbitrary exclusion restrictions. Much of the literature on competitive behavior addresses this information problem by introducing assumptions regarding the linking function and using information available a priori regarding the "most important" exogenous variables, such as marginal-cost components, demand shifters, or supply shifters. It is widely acknowledged that these studies are joint tests on the assumptions regarding the behavioral model, the link functions, and the exclusion restrictions (Genesove and Mullin 1998; Corts 1999). An alternative approach to the information problem is based on pairwise comparisons of alternative models or nested models (e.g., Gasmi, Laffont, and Vuong 1992; Karp and Perloff 1993). This strategy shares two major limitations with the first approach: it relies on specific assumptions regarding demand and cost functions, and it selects the alternative that fits the data best among those tested, which does not necessarily correspond to the true data-generating process.

Given the challenges involved in implementing either of these approaches satisfactorily, this section presents a nonparametric approach that is able to compare the conditional expectations in the policy regimes even in the absence of information about the link function and without imposing arbitrary exclusion restrictions in the matrix of the exogenous variables. Assume that the available information can be divided into two matrices: a $T \times S$ matrix of all observable exogenous variables (X) that may or may not affect millers' pricing behavior and a $T \times 1$ vector representing the millers' margin (Y). The goal is to calculate the conditional mean of Y without knowing which variables in X are relevant and without knowing the function linking Y to X. The nonparametric approach used here addresses the two problems separately using a two-step procedure (Russo 2008).

The first step uses a sliced inverse regression (SIR) to identify the linear combination of the exogenous variables (the SIR factors) that are the best predictors for the millers' margin (Li 1991). The use of this dimension-reduction technique eliminates the need to use arbitrary exclusion restrictions and specify functional forms in the estimation of the conditional expectation. This step allows us to run the test even in the absence of reliable a priori information regarding the relevant explanatory variables. We collect the largest possible matrix X and use SIR to collapse it into a small (and manageable) set of factors.

The second step uses the SIR factors as the independent variables in nonparametric Nadaraya-Watson regressions (NW) in order to compare how the millers' margin changes with changes in the independent variables for years in which the policy resulted in payments to farmers to those for years when it did not (Nadaraya 1964; Watson 1964). The use of kernel estimators does not require imposing assumptions about the unknown linking function. Consequently, it is possible to estimate the conditional means and variances of the millers' margins under the two policy regimes. A simple test on the equality of means allows us to establish if the two estimates are significantly different.

The logic motivating this approach is intuitive. The obvious methodological approach to estimating how the exogenous variables affect the margin without imposing specific function forms is to use nonparametric regression techniques. Yet if S, the number of possible exogenous regressors, is large, this approach is likely to suffer from the *curse of dimensionality:* adding extra dimensions to the regression space leads to an exponential increase in volume, which slows the rate of convergence of the estimator exponentially. In order to avoid this curse, the original variables are compressed into a smaller number of factors that are linear combinations of the variables using SIR.

Importantly, the use of SIR factors in the second-stage regression does not prevent linking the pricing behavior of the milling industry to the original S exogenous variables. The SIR factors are linear combinations of the original variables. The coefficients are estimated by decomposing the consistent estimator of M, the variance-covariance matrix of $E(X|Y)$. Accordingly, the coefficients for the original variables can be computed and their significance tested (Chen and Li 1998).

One drawback to the standard SIR approach is that the factors it identifies are not necessarily interpreted easily using economic theory, which can make it challenging to utilize the results to identify plausible behavioral models. In order to address this shortcoming, the SIR was reestimated with a set of constraints restricting the number and composition of dimension-reduced shifters to correspond to the predictions of economic theory. The results are not affected substantially by the constraints. Thus, while testing for difference in the millers' margins, we obtain additional information about the underlying economic model governing the industry's behavior.

The restricted model uses Naik and Tsai's (2005) constrained inverse regression approach (CIR), a special version of SIR, which enables the classification of the exogenous variables in the matrix X as possible shifters of demand, farmer supply, or processor nonwheat marginal costs ex ante, using economic theory. Formally, given q linear constraints of the form $A'\beta = 0$ (where A is the $S \times q$ constraint matrix), the constrained efficient dimension reduction directions are given by the principal eigenvector of $(I - P)\,\mathrm{cov}[E(z|y)]$, where $P = \tilde{A}(\tilde{A}'\tilde{A})^{-1}\tilde{A}$ and $\tilde{A} = \Sigma_{xx}^{-1/2}A$. The output of the CIR is dimension-reduced shifters (DRS) that are linear combinations of exogenous variables, summarizing—in the case at hand—the effects of demand, supply, and marginal cost shifters, respectively.

The link function F_0 is estimated by regressing Y non-parametrically on the L linear combinations of X instead of on the S original variables. Using the consistent estimates of the βs (instead of the true values) in a kernel regression does not affect the first-order asymptotic properties of the estimator, and the error term has the same order of magnitude (Chen and Smith 2010). The output from this step allows the examination of how shifts in the significant SIR and CIR factors affect the millers' margin in binding and nonbinding policy years.

Table 2.1   **Descriptive statistics: Real prices for wheat and wheat products by location, 1974–2005**

	Wheat price		Wheat products price		Price margin	
	Minneapolis	Kansas City	Minneapolis	Kansas City	Minneapolis	Kansas City
Mean	9.30	8.87	11.44	10.98	2.14	2.10
Standard deviation	1.57	1.51	1.64	1.43	0.49	0.24
No. of observations	32	32	32	32	32	32

Source: USDA Wheat Yearbook 2006, available at http://www.ers.usda.gov/data/wheat/wheatyearbook .aspx.

2.4 Data

The data set contains information on wheat prices, flour prices, and other variables for 1974 to 2005. Data have been deflated using the producer price index (base year 1982) provided by the Bureau of Labor Statistics. The prices of wheat and wheat flour are those reported in the USDA's *Wheat Yearbook* for two locations: Kansas City and Minneapolis.[3] These cities are traditional areas of geographic concentration for wheat milling because they are major markets near important wheat production regions (Wilson 1995). The price of wheat is reported in terms of the cost to produce a hundredweight of flour, and flour and by-product prices are reported directly. The price margin is defined as the difference between the price of a hundredweight of flour and by-products and the price of the wheat used to produce it.

Table 2.1 reports descriptive statistics for these price series by market. Average real prices in Minneapolis are higher, although the difference is not statistically significant at the 90 percent confidence level. Real price margins are similar in the two markets: the average was equal to $2.14/hundredweight of flour in Minneapolis and $2.10/hundredweight in Kansas City. Figure 2.4 illustrates the real price trends in the two markets.

Table 2.2 reports descriptive statistics for the other variables in the data set. The data sources are the USDA, the Bureau of Labor Statistics, the Census Bureau, the Energy Information Agency, the University of Michigan, and the World Bank. Increases in the cost of fertilizer per acre (FERT), agricultural fuel per acre (FUEL) and hired agricultural labor per hour (HLB) are predicted to shift farmer supply upward. The policy price (POL) is predicted to increase supply when the policy is binding. Increases in hourly manufacturing wages (RHW), the price of gas (GAS), the transportation price index (TPI), and the bank prime loan rate (IR) are predicted to shift processors' nonwheat marginal cost up. Demand is predicted to shift out as population (USPOP), per capita income (USINC), wheat weight (WGHT) and protein content (PRTN)—as proxies for quality, the share of the population that

---

3. Firm-level price data are not available publicly.

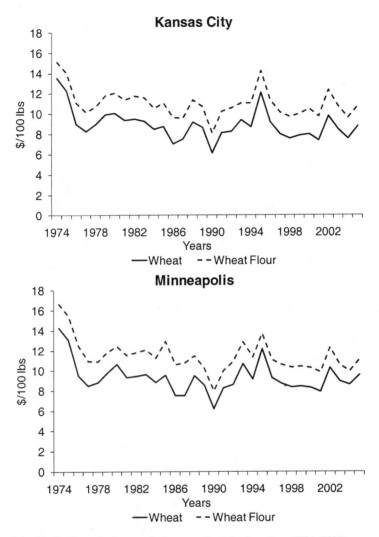

Fig. 2.4   Real prices of wheat and wheat products by location: 1974–2005

identifies as caucasian (CAUC), and per capita income in Japan (JINC), the largest importer of U.S. wheat during the sample period, increase.[4] Table 2.3 reports the pairwise correlation matrix of these variables. In addition, a

4. The analysis is limited to the market for wheat. Of course, in reality, farmers' wheat production is part of a larger acreage allocation decision. Depending on the region, barley, canola, corn, hay, rye, and other crops are substitutes in production for wheat. Including data regarding farmers' potential substitutes for wheat would introduce endogeneity concerns. Consequently, the analysis does not include data regarding the production, spot market prices, policy prices, or futures prices of substitute crops. Similarly, the Conservation Reserve Program (CRP) competes with wheat and other crops for acreage, although multiyear contracts limit its endogene-

Table 2.2   Descriptive statistics: Explanatory variables, 1974–2005

Variable	Definition	Mean	Min.	Max.	Standard deviation
FERT	Cost of fertilizer (real $/acre)	16.0	9.3	23.0	3.0
FUEL	Cost of agricultural fuel (real $/acre)	8.4	5.1	14.3	2.1
HLB	Cost of hired labor (real $/hour)	3.1	1.9	5.3	0.9
POL	Policy price (real $/hundredweight flour)	9.6	6.0	13.5	2.2
RHW	Industry wages (real $/hour)	15.3	14.7	16.3	0.5
GAS	Gas price (real $)	112.2	76.4	193.7	25.9
TPI	Transportation price index	114.3	45.8	173.9	35.5
IR	Bank prime loan rate (%)	9.0	4.1	18.9	3.2
USPOP	U.S. population (millions)	251.9	213.3	293.9	25.3
USINC	U.S. per capita income (real $)	4.1	1.1	9.5	2.6
WGHT	Wheat weight (pounds/bushel)	60.4	58.4	61.6	0.7
PRTN	Wheat protein content (%)	12.1	11.2	13.4	0.6
CAUC	Caucasian share of population (%)	0.8	0.8	0.9	0.0
JINC	Japan per capita income (real $)	90.6	56.7	103.8	13.3

Kansas City dummy variable (KANS) is included in order to allow for any location-dependent effects.[5]

The data set includes a dummy variable identifying the years when the policy is binding (BIN), that is, years in which the policy price is higher than the market price. Although the posted county prices are announced daily, data limitations require the use of less frequent observations.[6] Consequently,

---

ity to some extent. On the other hand, the criteria for acreage selection varied by enrollment round. Depending on the type of environmental protection targeted, the importance of CRP as a competitor for wheat varied considerably during the sample period. The analysis does not attempt to control for the farm crisis of the early 1980s because it was generated in part by low commodity prices.

5. The USDA time series for Kansas City prices is for No. 1 hard winter wheat, and the Minneapolis price series is for No. 1 dark northern spring wheat. Thus, the location dummy may include quality-related effects not captured by the weight and protein variables, as well as other factors that differ between the two locations.

6. Choosing the frequency of data was a difficult modeling decision. Annual data balance competing concerns regarding the unit of observation. Because wheat is storable, more frequent observations are more likely to be influenced by short-term storage decisions by farmers and millers. Farmers market their entire wheat crop within a year, except under very unusual conditions, and millers seldom hold flour more than one or two months (Brorsen et al. 1985). Inventories of wheat do extend across crop years; they are not addressed here due to the complications created by the presence of government-owned and exporter-owned stocks. On the other hand, as discussed in the preceding, the actual policy is implemented on a county-day basis. Incorporating this complexity into our analysis would be difficult, if not impossible, due not least to the increasing importance of storage as the frequency of observations increases. An additional practical difficulty is that some of the variables are provided on an annual basis, such as wheat quality. Specifying a time period that is less than a year would make it more difficult to collect information on exogenous variables, as would specifying a smaller unit of observation, such as a county or even a state. The National Agricultural Statistics Service reports that over 1,800 counties in forty-two states harvested wheat acreage in 2005 (National Agricultural Statistics Service 2010).

Table 2.3   Correlation matrix

	FERT	FUEL	HLB	POL	RHW	GAS	TPI	IR	USPOP	USINC	WGHT	PRTN	CAUC	JINC	Price margin
FERT	1.00														
FUEL	0.51	1.00													
HLB	-0.32	-0.08	1.00												
POL	-0.43	0.03	0.26	1.00											
RHW	0.34	0.18	-0.73	-0.53	1.00										
GAS	0.72	0.85	-0.32	-0.30	0.50	1.00									
TPI	0.68	0.48	0.02	-0.38	0.23	0.72	1.00								
IR	-0.21	0.24	-0.01	0.36	-0.26	-0.11	-0.45	1.00							
USPOP	0.68	0.36	-0.09	-0.54	0.39	0.69	0.97	-0.56	1.00						
USINC	-0.47	-0.12	-0.14	0.11	0.00	-0.35	-0.82	0.62	-0.79	1.00					
WGHT	-0.02	-0.07	-0.39	-0.33	0.31	-0.06	-0.34	0.15	-0.26	0.42	1.00				
PRTN	-0.01	0.16	0.25	-0.11	-0.03	0.14	0.26	0.04	0.23	-0.06	-0.37	1.00			
CAUC	-0.66	-0.32	0.09	0.49	-0.35	-0.62	-0.85	0.51	-0.88	0.67	0.21	-0.16	1.00		
JINC	0.41	0.32	0.33	-0.03	-0.16	0.42	0.87	-0.30	0.77	-0.87	-0.51	0.27	-0.63	1.00	
Price margin	-0.04	-0.03	-0.11	0.39	-0.13	-0.08	-0.17	0.03	-0.21	-0.09	-0.04	-0.26	0.26	-0.05	1.00

Note: See table 2.2 for explanation of abbreviations.

the years in which the policy was binding are defined using USDA yearly average data. A binding year (BIN = 1) is defined as one in which the average market price in that location is lower than the average "policy" price. The policy price is defined as the average yearly loan rate from 1996 on and is the maximum of the average yearly loan rate reported by the USDA and the target prices of deficiency payments prior to 1996 (before this date, all production was eligible for deficiency payments so the program provided the same incentives as the marketing loan program). Because both the policy and the market prices vary over the sample period, one does not expect, necessarily, that binding policy years correspond exactly to those years with lower market prices.

Figures 2.5 and 2.6 confirm that expectation. Figure 2.5 plots the policy price against the market price for the Kansas City market, distinguishing between binding and nonbinding years. Figure 2.6 plots the same information for the Minneapolis market. The figures are quite similar. While for the very highest market prices the policy is never binding, there is no clear pattern between the realized market price and whether the policy binds. The policy price appears to be the primary determinant. This pattern is consistent with the policymaking process. Prior to the 1985 Farm Bill, agricultural price support program parameters were set for the next few years in each farm bill and were not adjusted for market conditions (Love and Rausser 1997). Since 1985, national marketing loan rates have continued to be set as part of farm bills and do not respond to market conditions (USDA 2009).

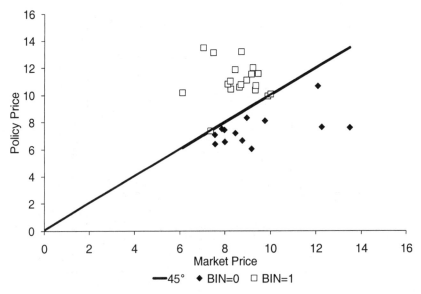

Fig. 2.5   Market and policy prices, binding and nonbinding policy years: 1974– 2005, Kansas City

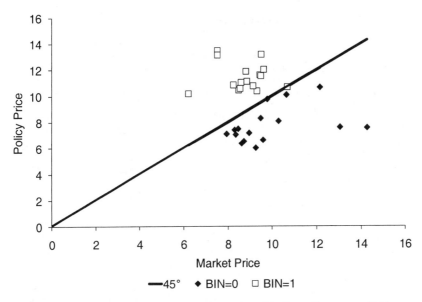

Fig. 2.6   Market and policy prices, binding and nonbinding policy years: 1974–2005, Minneapolis

2.5   Sliced Inverse Regression Results

Table 2.4 reports the results of the SIR estimation. SIR identifies two significant factors in the data. Factor one increases in a statistically significant fashion with the price of gasoline. It decreases when the following variables increase: wheat protein content, farmers' cost of hired labor, the price of fertilizer, the transportation price index, the policy price, the percentage of the population that identifies as caucasian, and the Kansas City dummy. The second factor increases significantly with wheat weight and the gasoline price and decreases with the price of fertilizer, the interest rate, the policy price, and the Kansas City dummy.

2.5.1   SIR Factors and the Policy Regime

Figures 2.7 and 2.8 plot realizations over time of the two SIR factors, distinguishing between years when the policy was binding and years when it was not. In figure 2.7, the first factor decreased steadily until the mid-1990s, then remained stable until about 2000, when it began declining again. There is no clear link between the level of the factor and whether the policy is binding. The second factor displays less of a trend over time, as shown in figure 2.8. It tended to have higher realizations in years when the policy was not binding. Figure 2.9 plots realizations of the second factor by realizations of the first factor, again distinguishing between binding and nonbinding policy years. As the figure demonstrates, there is no clear link between the relationship

Table 2.4   **Results: Sliced inverse regression**

| | Dimension-reduced factor | | | |
| | Factor 1 | | Factor 2 | |
	Coefficient	t-statistic	Coefficient	t-statistic
FERT	−4.65	−79.91**	−0.44	−6.55**
FUEL	0.47	2.66**	−0.22	−1.18
HLB	−3.73	−17.51**	0.10	0.42
POL	−10.58	−63.21**	−1.14	−6.37**
RHW	0.54	0.78	0.25	0.33
GAS	1.21	72.36**	0.87	48.69**
TPI	−5.84	−89.98**	−0.04	−0.60
IR	0.19	1.56	−0.96	−7.36**
USPOP	0.02	0.19	0.06	0.63
USINC	0.00	0.00	0.17	0.89
WGHT	−0.17	−0.79	0.86	3.79**
PRTN	−0.61	−3.24**	−0.27	−1.33
CAUC	−24.44	−2.39**	−2.94	−0.27
JINC	0.01	0.10	0.00	0.00
KANS	−0.37	−2.27**	−0.53	−3.09**

Notes: See table 2.2 for explanation of abbreviations. KANS = Kansas City dummy variable.

**Significant at 5 percent level.

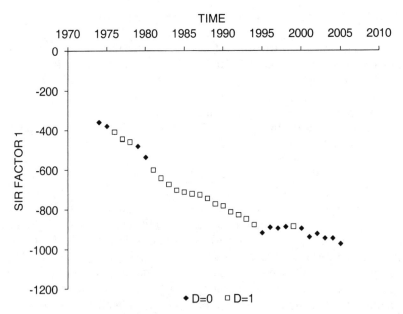

Fig. 2.7   Realizations of the sliced inverse regression (SIR) first factor: 1974–2005, binding and nonbinding policy years

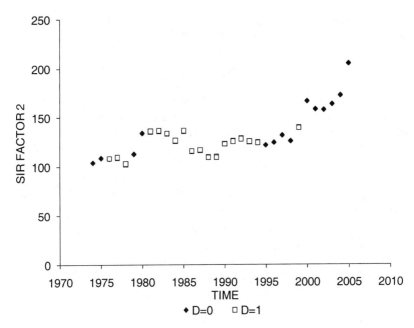

Fig. 2.8   Realizations of the sliced inverse regression (SIR) second factor: 1974–2005, binding and nonbinding policy years

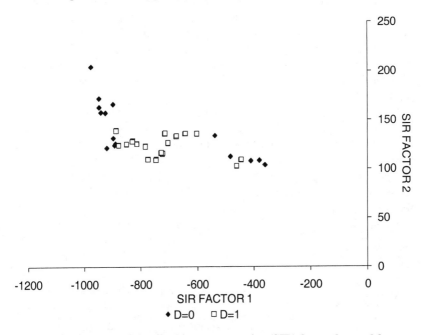

Fig. 2.9   Realizations of the sliced inverse regression (SIR) first and second factors: binding and nonbinding policy years

of the two SIR factors and whether the policy is binding even though the coefficient on the policy price is statistically significant for both factors.

2.6   Constrained Sliced Inverse Regression Results

Table 2.5 reports the CIR results. It includes the constrained efficient dimension reduction directions and the t-statistics for each coefficient on the exogenous variables in each DRS.[7] Overall, the CIR performs well. The signs of the coefficients match predictions. In the demand DRS, the U.S. population, Japanese per capita income, the share of the U.S. population identifying as caucasian, and wheat weight have statistically significant coefficients with the predicted signs. The Kansas City dummy has a statistically significant positive coefficient. In the farmer supply DRS, all three input costs have statistically significant coefficients with the predicted sign. In the miller marginal cost DRS, wheat weight and wheat protein content have statistically significant coefficients. The costs of nonwheat inputs have statistically significant, negative coefficients, as predicted. The Kansas City dummy has a statistically significant negative coefficient.

2.6.1   DRS and the Policy Regime

The CIR results allow us to examine the relationships between the three DRS and the policy regime. Figures 2.10 to 2.12 illustrate the distribution of the DRS over time, differentiating between binding and nonbinding policy years. The figures show that there is a concentration of binding years before the 1996 policy reform, when the policy target price was relatively high. The binding policy years are not associated with particularly low or high realizations of the demand or marginal cost DRS. Realizations of the supply DRS tended to be lower in nonbinding policy years.

Figures 2.10 to 2.12 each plot the realizations of a single DRS for binding and nonbinding policy years. Thus, they do not address the possibility that binding policy years are characterized by interactions between the realizations of the DRSs that lead to low prices. Figures 2.13 and 2.14 examine this possibility. Figure 2.13 plots the policy regime against the demand and farm supply DRS. To fix ideas, years in which the demand DRS has a large realization and the supply DRS has a small realization appear in the bottom right-hand quadrant of the graph. In a partial equilibrium graph of a market, these points would correspond to market outcomes with relatively high prices and low quantities. For a given realization of the demand DRS,

---

7. The signs of the coefficients in the farmer supply DRS are reversed relative to the conventional format of theoretical predictions in the table. That is, the positive signs on the cost of hired labor, fertilizer, and agricultural fuel indicate that an increase in any of these costs will shift supply inward. This is simply an artifact of the sliced inverse regression approach and does not affect the economic interpretation of the relationship between the exogenous and the endogenous variables.

Table 2.5   **Results: Constrained inverse regression**

	Dimension-reduced shifter					
	Demand		Supply		Marginal cost	
	Coefficient	*t*-statistic	Coefficient	*t*-statistic	Coefficient	*t*-statistic
FERT	0.00		0.22	3.81**	0.00	
FUEL	0.00		0.14	3.35**	0.00	
HLB	0.00		0.71	7.02**	0.00	
POL	0.00		−0.17	−0.59	0.00	
RHW	0.00		0.00		−0.88	−3.46**
GAS	0.00		0.00		−0.28	−3.28**
TPI	0.00		0.00		−0.79	−3.92**
IR	0.00		0.00		−0.48	−2.82**
USPOP	0.03	3.16**	0.00		0.00	
USINC	0.03	0.30	0.00		0.00	
WGHT	1.96	9.95**	0.00		0.85	5.16**
PRTN	−0.18	−0.74	0.00		−0.63	−3.82**
CAUC	167.78	12.32**	0.00		0.00	
JINC	0.19	20.49**	0.00		0.00	
KANS	0.95	3.79**	−0.43	−1.67	−1.04	−5.19**

Notes: See table 2.2 for explanation of abbreviations. KANS = Kansas City dummy variable.

**Significant at 5 percent level.

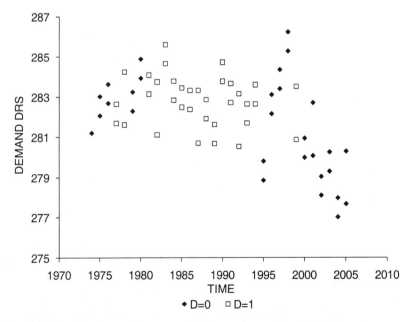

Fig. 2.10   Realizations of the constrained inverse regression (CIR) demand dimension-reduced shifter: 1974–2005, binding and nonbinding policy years

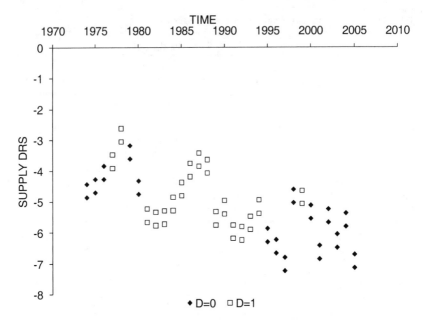

Fig. 2.11 **Realizations of the constrained inverse regression (CIR) wheat supply dimension-reduced shifter: 1974–2005, binding and nonbinding policy years**

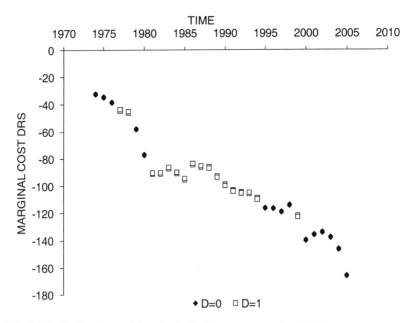

Fig. 2.12 **Realizations of the constrained inverse regression (CIR) processor nonwheat marginal cost dimension-reduced shifter: 1974–2005, binding and nonbinding policy years**

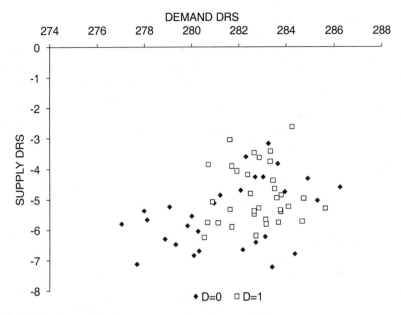

Fig. 2.13 Constrained inverse regression (CIR) demand and supply dimension-reduced shifters (DRS): binding and nonbinding policy years

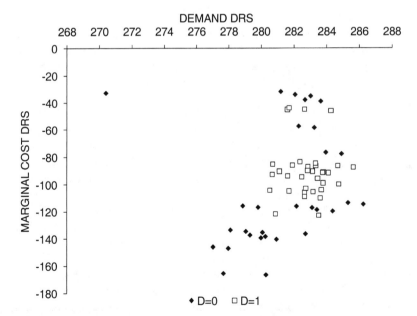

Fig. 2.14 Constrained inverse regression (CIR) demand and marginal cost dimension-reduced shifters (DRS) binding and nonbinding policy years

as the supply DRS realization increases in a partial equilibrium depiction of the market, the price will fall and the quantity produced and consumed will increase as the supply curve shifts out. If the target price was constant, then binding years should be associated with high realizations of the supply DRS for a given realization of the demand DRS. This pattern does not appear in figure 2.13. Figure 2.14 plots annual values of the demand and marginal cost DRS. This figure does not demonstrate any predictable pattern between the relationship between the two DRSs and whether the policy is binding. Consistent with figures 2.10 to 2.12, figures 2.13 and 2.14 indicate that high target prices are a more important determinant of the policy regime than market conditions are.

2.6.2   Comparison of SIR and CIR Results

There are differences in which variables have significant coefficients between the SIR and the CIR estimations. Two variables that were significant in the CIR demand DRS, U.S. population and Japanese per capita income, were not significant in either SIR factor. Manufacturing wage was not significant in either SIR factor, although it was significant in the CIR processing marginal cost DRS. The most important difference was that the policy price has an insignificant effect on the farmer supply DRS in the CIR results, although it had significant coefficients in both SIR factors.

2.7   NW Nonparametric Estimation Results

The second step of the procedure uses the SIR factors and the CIR DRS as regressors in a Nadaraya-Watson kernel estimator of the price margin with a cross-validation bandwidth. This step defines the link function and allows us to compute the conditional mean of the millers' margin.

2.7.1   SIR Factors

Figures 2.15 and 2.16 plot how each factor affects the flour-wheat price margin for binding and nonbinding policy years. In figure 2.15, the price margin is always higher when the policy is binding than when it is not, regardless of the realization of the first factor. The level of the margin is virtually constant for the nonbinding policy regime, regardless of the level of the factor. The level of the margin first increases, then decreases for the binding policy regime as the factor increases. There is no consistent change in the difference in the margins as a function of the level of the factor.

In figure 2.16, the price margin is always higher when the policy is binding than when it is not, except for very high realizations of the second factor. Because only observations for nonbinding policy years include very high values of the second factor, the behavior of the two regressions at the very end of the domain is not emphasized. The margin first increases, then decreases as the second factor increases for both the binding and nonbinding

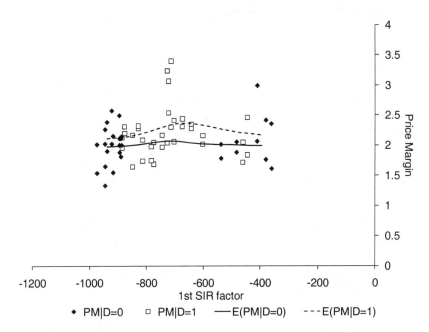

Fig. 2.15 Nadaraya-Watson (NW) nonparametric estimation of the relationship between the flour price-wheat price margin (PM) and the sliced inverse regression (SIR) first factor

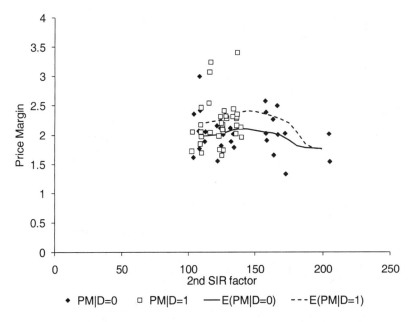

Fig. 2.16 Nadaraya-Watson (NW) nonparametric estimation of the relationship between the flour price-wheat price margin (PM) and the sliced inverse regression (SIR) second factor

policy regimes. The difference between the two margins remains relatively constant until the right-hand end of the domain.

The conditional mean of the millers' margin is \$2.07 per hundredweight of wheat when the policy is not binding and \$2.31 when it is binding. The \$0.24 per hundredweight difference is highly statistically significant, with a t-statistic of 3.398 obtained via bootstrapping.

2.7.2   CIR DRS

Figures 2.17 to 2.19 plot how the reduced-form demand, processor marginal cost, and wheat supply DRS affect the flour-wheat price margin. Figure 2.17 addresses demand. The estimated magnitude of the price margin depends on program status. For any given value of the demand DRS, the flour-wheat price margin is larger when the program is binding than when it is not. For both the binding and nonbinding policy regimes, the margin first increases with the demand DRS, then decreases. In both cases, the absolute values of the changes are small.

Figure 2.18 evaluates the effect of processors' marginal cost on the price margin. The observations for both regime types are clustered with respect to the realized values of the processor marginal cost DRS, with the nonbinding years at the extreme values of the DRS and the binding years in the middle. This pattern suggests caution when interpreting the results.

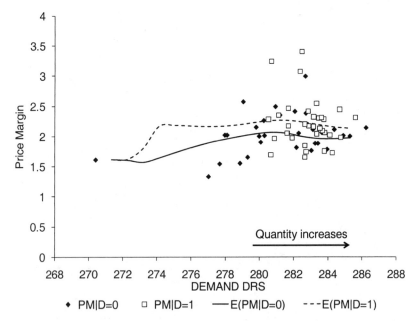

Fig. 2.17   **Nadaraya-Watson (NW) nonparametric estimation of the relationship between the flour price-wheat price margin (PM) and the dimension-reduced shifters (DRS) for demand**

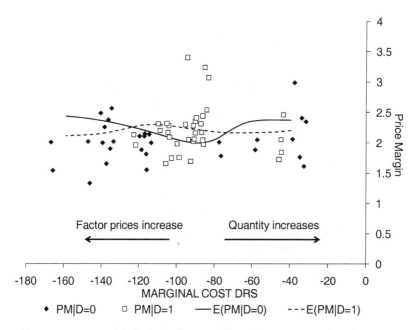

Fig. 2.18 Nadaraya-Watson (NW) nonparametric estimation of the relationship between the flour-wheat price margin (PM) and the dimension-reduced shifters (DRS) for processor marginal cost

There are differences between the binding policy and nonbinding policy regimes. In the middle of the range, the price margin is higher for a given realization of the marginal cost DRS when the policy is binding, but the opposite is true on the extremes. When the policy is binding, the price margin is virtually constant across values of the marginal cost DRS. When it is not binding, the price margin first declines as input prices decline and quantity increases, then increases. Thus, for low realizations of the marginal cost DRS when the policy is not binding, the result is consistent with Brorsen et al. (1985), who found that an increase in milling costs increases the flour-wheat price margin on a one-for-one basis. However, for high realizations of the marginal cost DRS when the policy is not binding and for all realizations when it is binding, the outcome is not consistent with Brorsen et al. (1985). Regarding the primary research question, the results are consistent with the possibility that a change in policy regime triggers a change in pricing behavior. For years when the policy is binding, millers appear to absorb as least as large of a share of a marginal cost increase as they do in years when the policy is not binding.

Figure 2.19 evaluates the effect of farmers' DRS of wheat supply on the price margin. As supply shifts out, the price margin first increases and then decreases in years when the policy is binding. In years when payments are not

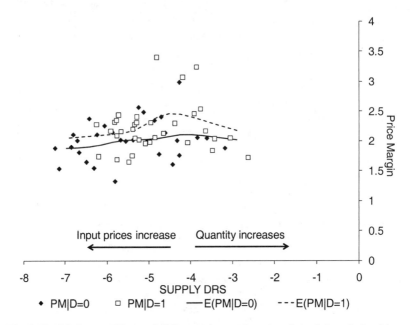

Fig. 2.19 **Nadaraya-Watson (NW) nonparametric estimation of the relationship between the flour price-wheat price margin (PM) and the dimension-reduced shifters (DRS) for wheat supply**

made, the price margin follows the same general pattern, although it is much less responsive to changes in the supply DRS. These policy-dependent relationships between supply and the price margin suggest that millers' strategies differ depending on whether the policy is binding.

The restricted model generates a conditional expectation of the millers' margin of \$2.02 per hundredweight of wheat when the policy is not binding and \$2.25 when it is binding. The \$0.23 per hundredweight difference is statistically significant, with a t-statistic of 2.5701 obtained via bootstrapping.

2.7.3   Implications

Overall, the analysis of the patterns obtained from the SIR-NW algorithm suggests that the data are consistent with a simple static model of market power. The figures suggest that millers are able to impose higher price margins in years in which the policy is binding. When payments are made, farmers respond to the target price and are less likely to store their grain and wait for a higher price to be offered by buyers. This circumstance allows millers to exploit market power and reduce the price of wheat relative to the price of flour.

The results of the two models differed in one important respect: the policy price was a significant explanatory variable for both SIR factors but was insignificant in the estimate of the supply DRS in the CIR model. Even though the two models varied in terms of which variables were significant, there was very little difference in the results regarding the research question of interest: whether the flour-wheat price margin is affected by the policy regime and, if so, by how much. The results of the two models were consistent regarding the difference in millers' marketing margins in binding and nonbinding policy years. The NW analyses based on both models indicate that the margins were higher in binding policy years regardless of the realizations of the shifters and DRS, respectively. That is, the higher margin found in binding policy years is not due to characteristics of demand, supply, or processor marginal costs in those years. In both models, margins were about 10 percent higher in binding years, and the hypothesis that the difference in the margins was not statistically significant was rejected.

2.8   Conclusion

As a sector, agriculture is subject to a great deal of government intervention. Although expenditures have declined substantially in the past decade due in part to international trade negotiations, in the last three years, Commodity Credit Corporation total net outlays for commodity programs have ranged between $9 and $13 billion, depending on economic conditions. For wheat alone, net outlays ranged between $0.7 and $1.2 billion (USDA 2010). Given the magnitude of these expenditures, there is an obvious public interest in efficient policy measures and the beneficiaries of these measures.

This analysis demonstrates that market power might redistribute the benefits of government intervention. It provides empirical evidence that U.S. wheat millers were able to increase their marketing margins, on average, by approximately 10 percent when farmers received payments through a marketing loan program. This expected increase in margins was computed controlling for the realizations of a broad set of supply, demand, and processor marginal costs shifters in those years. In turn, these findings suggest that millers are extracting a rent from the deficiency payment/marketing loan gain policy. Thus, the analysis suggests that the general assumption that competitive models may be a good approximation for imperfectly competitive agricultural markets does not necessarily hold, particularly if distribution, as well as efficiency, is a concern.

References

Alston, J., and J. James. 2001. "The Incidence of Agricultural Policy." In *Handbook of Agricultural Economics,* edited by B. Gardner and G. Rausser, 1690–1749. New York: Elsevier North-Holland.

Applebaum, E. 1982. "The Estimation of the Degree of Oligopoly Power." *Journal of Econometrics* 19 (2): 287–99.

Ashenfelter, O., and D. Sullivan. 1987. "Nonparametric Tests of Market Structure: An Application to the Cigarette Industry." *Journal of Industrial Economics* 35:483–99.

Bresnahan, T. F. 1989. "Empirical Studies of Industries with Market Power." In *Handbook of Industrial Organization,* edited by R. Schmalensee and D. Willig, 1011–57. Amsterdam: North-Holland.

Brester, G. W. 1999. "Vertical Integration of Production Agriculture into Value-Added Niche Markets: The Case of Wheat Montana Farms and Bakery." *Review of Agricultural Economics* 21 (1): 276–85.

Brester, G. W., and B. K. Goodwin. 1993. "Vertical and Horizontal Price Linkages and Market Concentration in the U.S. Wheat Milling Industry." *Review of Agricultural Economics* 15 (3): 507–19.

Brorsen, W., J.-P. Chavas, W. R. Grant, and L. D. Schnake. 1985. "Marketing Margins and Price Uncertainty: The Case of the U.S. Wheat Market." *American Journal of Agricultural Economics* 67 (3): 521–28.

Chen, C. H., and K. C. Li. 1998. "Can SIR Be as Popular as Multiple Linear Regression?" *Statistica Sinica* 8:289–316.

Chen, P., and A. Smith. 2010. "Dimension Reduction Using Inverse Regression and Nonparametric Factors with an Application to Stock Returns." Working Paper, University of California, Davis.

Corts, K. 1999. "Conduct Parameters and the Measurement of Market Power." *Journal of Econometrics* 88:227–50.

Gardner, B. 1983. "Efficient Redistribution through Commodity Markets." *American Journal of Agricultural Economics* 65 (2): 225–54.

Gasmi, F., J. Laffont, and Q. Vuong. 1992. "Econometric Analysis of Collusive Behavior in a Soft-Drink Market." *Journal of Economics and Management Strategy* 1 (2): 277–311.

Genesove, D., and W. P. Mullin. 1998. "Testing Static Oligopoly Models: Conduct and Cost in the Sugar Industry, 1890–1914." *RAND Journal of Economics* 29 (2): 355–77.

Green, E., and R. Porter. 1984. "Noncooperative Collusion under Imperfect Price Information." *Econometrica* 52 (1): 87–100.

IBISWorld. 2007. *Flour Milling and Malt Manufacturing in the U.S.* IBISWorld Industry Report no. 31121, September 19. Santa Monica, CA: IBISWorld.

Johnson, D. G. 1979. *Forward Prices in Agriculture.* Reprint, New York: Arno Press. First published 1947, Chicago: University of Chicago Press.

Karp, L. S., and J. S. Perloff. 1993. "Open-Loop and Feedback Models of Dynamic Oligopoly." *International Journal of Industrial Organization* 11:369–89.

Kim, C. S., C. Hallahan, G. Schaible, and G. Schluter. 2001. "Economic Analysis of the Changing Structure of the U.S. Flour Milling Industry." *Agribusiness* 17:161–71.

Li, K.-C. 1991. "Sliced Inverse Regression for Dimension Reduction." *Journal of the American Statistical Association* 86 (414): 316–27.

Love, H. A., and G. C. Rausser. 1997. "Flexible Policy: The Case of the United States Wheat Sector." *Journal of Policy Modeling* 19 (2): 207–36.

Nadaraya, E. 1964. "On Estimating Regression." *Theory of Probability and Its Applications* 9 (1): 141–42.

Naik, P. A., and C. L. Tsai. 2005. "Constrained Inverse Regression for Incorporating Prior Information." *Journal of the American Statistical Association* 100:204–11.

National Agricultural Statistics Service. 2008. *Agricultural Statistics 2008.* Washington, DC: United States Department of Agriculture, Government Printing Office. http://www.nass.usda.gov/Publications/Ag_Statistics/2008/2008.pdf.

———. 2010. *Quick Stats.* Washington, DC: United States Department of Agriculture. http://quickstats.nass.usda.gov/.

Perloff, J. M., L. Karp, and A. Golan. 2007. *Estimating Market Power and Strategies.* Cambridge, UK: Cambridge University Press.

Putnam, J. J., and J. E. Allshouse. 1999. "Food Consumption, Prices, and Expenditures 1970–97." Statistical Bulletin no. 965. Washington, DC: Food and Rural Economics Division, Economic Research Service, U.S. Department of Agriculture. http://www.ers.usda.gov/publications/SB965/.

Rude, J., and K. Meilke. 2004. "Developing Policy Relevant Agrifood Models." *Journal of Agricultural and Applied Economics* 36 (2): 369–82.

Russo, C. 2008. "Modeling and Measuring the Structure of the Agrifood Chain: Market Power, Policy Incidence and Cooperative Efficiency." PhD diss., University of California, Davis.

Russo, C., R. E. Goodhue, and R. J. Sexton. 2009. "Agricultural Support Policies in Imperfectly Competitive Markets: Does Decoupling Increase Social Welfare?" Working Paper, University of Cassino.

———. Forthcoming. "Agricultural Support Policies in Imperfectly Competitive Markets: Why Market Power Matters in Policy Design." *American Journal of Agricultural Economics.*

Stewart, J. 2008. "The Economics of American Farm Unrest, 1865–1900." In *EH.Net Encyclopedia,* edited by R. Whaples. Santa Clara, CA: Economic History Association. http://eh.net/encyclopedia/article/stewart.farmers.

Stiegert, K. 2002. "The Producer, the Baker, and a Test of the Mill Price-Taker." *Applied Economics Letters* 9 (6): 365–68.

U.S. Department of Agriculture (USDA). 2007. *Agricultural Census.* Washington, DC: United States Department of Agriculture. http://www.agcensus.usda.gov/Publications/2007/Online_Highlights/Custom_Summaries/Data_Comparison_Major_Crops.pdf.

———. 2009. *Wheat: Policy.* Washington, DC: United States Department of Agriculture. http://www.ers.usda.gov/Briefing/Wheat/Policy.htm#marketingalaldp.

———. 2010. *Agricultural Outlook.* Washington, DC: United States Department of Agriculture. http://www.ers.usda.gov/publications/agoutlook/aotables/.

Vocke, G., E. W. Allen, and M. Ali. 2005. *Wheat Backgrounder.* Report no. WHS-05k-01. Washington, DC: Economic Research Service, United States Department of Agriculture, December.

Wallace, T. D. 1962. "Measures of Social Costs of Agricultural Programs." *Journal of Farm Economics* 44 (2): 580–94.

Warman, M. 1994. *Cooperative Grain Marketing: Changes, Issues, and Alternatives.* ACS Research Report no. 123. Washington, DC: Agricultural Cooperatives Service, United States Department of Agriculture, April.

Watson, G. S. 1964. "Smooth Regression Analysis." *Sankhyā: The Indian Journal of Statistics, Series A* 26:359–72.

Wilson, W. W. 1995. "Structural Changes and Strategies in the North American Flour Milling Industry." *Agribusiness* 11 (5): 431–39.

3

The Politics and Economics of the U.S. Crop Insurance Program

Bruce A. Babcock

3.1 Introduction

Agricultural subsidies have been part of U.S. agriculture since the 1930s, when support for agriculture was a major part of the national economic stabilization effort of the New Deal. With the exception of the World War II period, the structure of subsidies from the 1930s until 1996 consisted of price supports that were defended by commodity storage and supply controls. Both were largely eliminated in 1996. The main form of support for agriculture today is a combination of $5 billion per year of fixed direct payments that are decoupled from production, and crop insurance, which has turned into a $7 billion program.

It could be argued that deadweight losses from the current system of subsidies are lower than the past combination of supply control and price supports if government revenue is raised efficiently and if the crop insurance program provides farmers with an efficient risk transfer mechanism that the private sector cannot provide. Glauber and Collins (2002), in their review of the history of the crop insurance program, note that private crop insurance markets have routinely failed in the past because of the systemic nature of crop losses. However, they also note that another explanation for the lack of a private crop insurance market may be a lack of demand. If the private sector cannot provide adequate crop insurance, then it is possible that government provision of this missing risk transfer market could improve welfare. But if private insurance markets do not exist because of a lack of demand, then government creation of a crop insurance industry likely leads to large welfare losses.

Bruce A. Babcock is professor of economics and director of the Center for Agricultural and Rural Development at Iowa State University.

In this chapter, I first explore in some detail whether there is empirical evidence of unmet demand for crop insurance that could justify government creation of the market. I use two years of cross-section data to make this determination. The two years span a large change in the actuarial fairness crop insurance premiums that allows revelation of the farmer demand for actuarially fair insurance. Determination of whether crop insurance contracts that are offered to farmers are actuarially fair is not trivial because the rate-making methods used by the responsible government agency were not consistent with actuarial fairness. One contribution of this chapter is the construction of a data set that estimates the degree to which corn, soybean, and wheat farmers were offered insurance contracts that were overrated, underrated, and were actuarially fair.

The empirical section of the chapter finds support for the notion that a substantial number of farmers will increase the amount of actuarially fair insurance that they buy. Although this result accords well with standard expected utility theory, it runs counter to current crop insurance policy and much of the current literature that maintains that farmers need to be heavily subsidized before they will increase their purchase of insurance. One explanation for the large premium subsidies can be found by looking at interest groups who gain from large premium subsidies. A detailed examination of the interest groups who capture rents from the program provides insight into the political forces that have worked to defend and expand the program subsidies. The program serves as a case study that follows closely the predictions of Becker's (1983) theory of legislation as a reflection of economic payoffs from the application of pressure by affected interest groups.

The chapter begins with a general overview of the program. This overview is needed to understand how interest groups extract rents from the program. The construction of a unique data set is then detailed to show how expected producer returns for increased crop insurance coverage can be estimated. The demand for increased insurance coverage is estimated for the subset of the data that includes actuarially fair insurance contracts and for the entire sample. The chapter then moves to an exploration of competition within the crop insurance program and identifies which interest group, other than farmers, captures the lion's share of rents in the program. A brief examination of legislative history that corroborates who benefits from the program along with an assessment of current crop insurance policy that builds on the chapter's results concludes.

3.2   Overview of the U.S. Crop Insurance Program

Policy objectives of the U.S. crop insurance program can be found from the public statements of policymakers. Objectives include to help farmers manage financial risk and to eliminate the need for Congress to pass supplemental ad hoc disaster assistance programs. Congress dramatically restructured the crop insurance program in 2000 with the Agricultural Risk

Protection Act (ARPA). The reform was justified by President Clinton in his statement upon signing the Agricultural Risk Protection Act (ARPA) of 2000: "I have heard many farmers say that the crop insurance program was simply not good value for them, providing too little coverage for too much money. My FY 2001 budget proposal and this bill directly address that problem by making higher insurance coverage more affordable, which should also mitigate the need for ad hoc crop loss disaster assistance such as we have seen for the last three years." And in 2006 testimony before the House Subcommittee on Agriculture, Rural Development, Food and Drug Administration, and Related Agencies, former U.S. Department of Agriculture (USDA) undersecretary J. B. Penn said, "One of the overarching goals of the crop insurance program has been the reduction or elimination of ad hoc disaster assistance."

By all accounts, Congress has seemingly succeeded in its objective to help farmers manage risk. Coverage is provided to more than 350 commodities in all fifty states and Puerto Rico. And more than 80 percent of eligible acres are now insured under the program. The Congressional Budget Office projects crop insurance outlays in excess of $7 billion for the foreseeable future.

Table 3.1 provides summary program information since 2001. The first column shows that insured acreage has increased substantially since 1998.

Table 3.1   **National crop insurance data**

Year	Insured acreage[a] (million acres)	Total premiums[a] ($ million)	Total indemnities[a]	Premium subsidies[a] ($ million)	Administrative and operating subsidies[b]	Net underwriting gains[c]
1998	182	1,876	1,678	946	445	279
1999	197	2,310	2,435	955	499	272
2000	206	2,540	2,595	951	554	282
2001	211	2,962	2,960	1,772	634	348
2002	215	2,916	4,067	1,741	627	−47
2003	217	3,431	3,261	2,042	734	378
2004	221	4,186	3,210	2,477	887	692
2005	246	3,949	2,367	2,344	829	965
2006	242	4,579	3,503	2,682	930	887
2007	272	6,562	3,547	3,824	1,339	1,682
2008	272	9,857	8,664	5,693	2,011	1,163
2009[d]	264	8,948	5,164	5,425	1,619	2,146

[a]Taken from *Summary of Business* reports of the Risk Management Agency of the U.S. Department of Agriculture.

[b]Calculated from exhibit 5, table 5.1 of Grant Thornton LLP and "Total premiums" in this table.

[c]Underwriting gains through 2008 taken from exhibit 1 in Grant Thornton LLP.

[d]Current as of July 11, 2010. Administrative and operating subsidies estimated by multiplying 2009 estimated premium by 18.1 percent, reflecting the cut in administrative and operating subsidies reimbursement enacted in 2009. Underwriting gains in 2009 estimated from historical relationship between gross and net underwriting gains.

Most of this increase came about because pasture land became eligible for insurance during this period. Total premium is the amount of premium that companies are credited with by the USDA's Risk Management Agency (RMA), which administers the program. The amount of premiums that farmers actually pay equals total premiums less premium subsidies. Both have grown tremendously over this period. Indemnities are the amount of insurance claims paid to farmers. Administrative and Operating (A&O) costs are paid to crop insurance companies as cost compensation. Net underwriting gains are the amount of gross underwriting gains that companies keep after federal reinsurance gains and losses are calculated. Taxpayer costs equal subsidies plus net underwriting gains plus total indemnities paid to farmers minus farmer-paid premiums. Each of these is discussed in turn.

3.2.1   Administrative and Operating Subsidies

In 1980, Congress decided that delivery of the crop insurance program should be given to the private sector so that the program could be expanded as rapidly as possible. Companies had an incentive to expand sales because they were essentially paid a sales commission. For each dollar of premium they brought in, companies were given a percentage. That percentage, called A&O, was reduced by 2.3 percentage points (from approximately 20.7 percent of premium) beginning in 2009.

3.2.2   Net Underwriting Gains

A gross underwriting gain occurs in the crop insurance program when premiums exceed indemnities. In these years, crop insurance companies get to keep a portion of the difference. The portion they keep is called the net underwriting gain. For example, in 2004, premiums exceeded claims by $979 million. Companies were allowed to keep $848 million of this difference. In years in which premiums are less than insurance claims, companies may have to pay a portion of the difference, an underwriting loss. In 2002, claims exceeded premiums by $1.15 billion. Companies had to pay the government $52 million of this amount.

The 2002 and 2004 examples nicely illustrate why, on average, crop insurance companies expect to generate underwriting gains. In years in which underwriting gains are positive, companies get to keep a larger proportion of the gain than they have to pay the government in years in which there are underwriting losses. The mechanism by which net gains and losses are determined is the Standard Reinsurance Agreement (SRA).

Companies generate net gains from the SRA in three ways. The first is by determining which of their customers are most likely to generate claims and then giving the premium from these customers and responsibility for any subsequent losses directly to the government. The average customer retained by a company, therefore, has a better risk profile than the average customer in the overall pool. Thus, average claims from the retained pool will be lower than the overall average, and the company will tend to make money.

However, the overall risk of loss from retained customers is still too large for companies to be willing to take on all losses. Hence, the SRA is designed to have the government take on a portion of company losses when claims exceed premiums in exchange for companies giving the government some of their gains when premiums exceed claims. In exchange for companies taking on some of the risk of the crop insurance program, the government is allowing companies to generate some gains. It is almost as if crop insurance companies are selling taxpayers an insurance policy. In years where crop losses are high, taxpayer losses are reduced because some of the losses are covered by the "policy." The "premium" that taxpayers pay for this policy are the underwriting gains that companies garner in years where crop losses are small. Whether taxpayers are getting a good deal by this bargain depends on the size of the premium paid in good years relative to the payments received in bad years.

Table 3.2 summarizes one set of conservative estimates of the potential gains and losses to private crop insurance companies from operation of the current SRA. These estimates are based on loss experience from 1993 to 2005 and likely understate the actual underwriting gains that companies currently expect to make. The table presents four equally likely scenarios regarding crop insurance claims. With $8 billion in premiums, companies should expect to make $868 million per year in net underwriting gains, which is simply the average net underwriting gain across the four scenarios. In exchange for paying companies this average annual amount, taxpayers reduce their loss exposure by $440 million with a 25 percent probability. This simple example demonstrates that taxpayers would be much better off self-insuring by having the federal government simply and directly take over risk rather than sharing the risk at such a high cost.

The third way that companies make money from the SRA is that gains and losses are calculated for each state separately. Given the asymmetry of net gains and losses, separate ceilings on losses for each state will result

Table 3.2   **Potential gains and losses to crop insurance companies under the Standard Reinsurance Agreement**

Insurance claim loss scenario	Loss ratio (indemnity over premium)	Ratio of gain to total premium	Total gain to companies (US$ million)
Very low	0.53	0.238	1,904
Moderately low	0.72	0.136	1,088
Moderately high	0.76	0.115	920
Very high	1.28	−0.055	−440
Average	0.82	0.108	868

Source: Estimated by author.

in lower overall losses that more than compensate the benefits of separate ceilings on gains.

3.2.3   Producer Premium Subsidies

The last taxpayer cost category is premium subsidy. Farmers must pay for crop insurance, but they pay only a portion of the amount needed to cover insured losses. Throughout the 1980s and 1990s, farmers were reluctant to buy enough crop insurance to satisfy Congress. So to get farmers to buy more insurance, ARPA dramatically decreased the portion that farmers must pay. Currently, farmers pay about 41 percent of the amount needed to cover insured losses. This large subsidy means that most farmers will get substantially more back from the program than they pay into it. As will be shown, the large increases in premium subsidies under ARPA also allows for estimation of producer demand for actuarially fair insurance contracts because before the change, most farmers needed to pay more than actuarially fair amounts for low deductible policies, and after the change, many farmers needed to pay less than actuarially fair amounts for increased coverage.

3.3   Demand for Crop Insurance

One justification for federal provision of the crop insurance program is the inability of the private sector to meet risk averse farmers' demands for insurance. The primary reason given for this inability is lack of poolability in insurance claims (Miranda and Glauber 1997). Indeed, crop risk bears many of the characteristics of an uninsurable risk. As noted by Miranda and Glauber, the coefficient of variation of loss ratios is much higher than most other forms of property and casualty insurance. One solution to this lack of poolability is to reinsure in capital markets with catastrophic bonds. In 2009, in excess of $3.4 billion of cat bonds were issued (Johansmeyer 2010), which implies that the total amount of at-risk capital now exceeds $12 billion (GC Securities 2009). Whether the current crop insurance liability of $80 billion could all be reinsured in capital markets is untested because the U.S. taxpayer currently provides reinsurance for the program. Thus, there is at least some possibility that, without public intervention, there could be underprovision of insurance in agriculture. However, a necessary condition for underprovision is an excess demand for insurance by farmers at the level of insurance that the private sector would offer.

The evidence for unmet demand seems weak. Goodwin (1993) showed that demand for insurance is driven primarily by increases in expected indemnities in excess of premiums. His analysis covered a period of quite low participation (1985 to 1990) when adverse selection was likely widespread. Goodwin (1993) estimated expected indemnities by a ten-year average ratio of indemnities received to premiums paid. Just, Calvin, and Quiggin (1999) modeled farmer participation in crop insurance by comparing farmers' expectations of indemnities with insurers' expectations of indemnities and found that asymmetric information was significant in explaining which farmers purchased crop insurance in 1988. They find that risk aversion played a minor role in explaining the incentive to buy insurance. Rather, the opportunity to obtain positive expected returns provides the primary motivation.

These two studies provide little evidence of unmet demand for actuarially fair insurance. However, both studies were conducted when crop insurance participation was quite low. For example, in 1988, only 24.5 percent of eligible acres were enrolled in the program (Food and Agricultural Policy Research Institute [FAPRI] 1999). Thus, it is not surprising that adverse selection was important in determining who participated in the program. More recently, however, participation has become much higher. In 1998 and 2002 (the two years studied here), participation increased to 67.9 percent and 80.2 percent of eligible acres, respectively. The importance of adverse selection in the participation decision should be much lower with such a large share of producers now participating in the program. The question that motivated the study by Just, Calvin, and Quiggin (1999)—whether the pursuit of higher expected returns or risk aversion is what drives crop insurance demand—is still key to determining whether producer welfare gains from provision of a missing risk transfer market can justify government provision of crop insurance.

Consider a risk-averse, expected-utility-maximizing farmer who grows a single crop. The farmer can choose to insure crop yield at any coverage level α. The insurance contract is of the form $I = \max(\alpha\bar{y} - y, 0)$, where I is the insurance indemnity, $\bar{y}$ equals expected yield, and y is realized crop yield. Normalizing crop price to unity, and denoting the distribution of yield as $g(y)$, $0 \leq y \leq y\text{max}$, and the insurance premium as $r(\alpha)$, expected utility for the producer with this insurance contract is

$$EU = \int_0^{\alpha\bar{y}} U[\alpha\bar{y} - c - r(\alpha)]g(y)dy + \int_{\alpha\bar{y}}^{y\text{max}} U[y - c - r(\alpha)]g(y)dy.$$

If $r(\alpha)$ is actuarially fair, then $r(\alpha) = \int_0^{\alpha\bar{y}}(\alpha\bar{y} - y)g(y)dy$. Risk averters will demand full insurance if insurance is actuarially fair (Arrow 1974) or if premiums are subsidized at lower-than-actuarially fair levels. In either case, expected utility is increasing in $\alpha < y\text{max}/\bar{y}$. Risk averters will also demand full insurance even if insurance premiums are higher than actuarially fair levels if risk premiums are large enough. Risk neutral producers, however, buy insurance only if expected indemnities exceed premiums paid.

The U.S. crop insurance program limits α to a maximum of 0.85, a minimum of 0.5, and increments of 0.05. Thus, if risk-averse farmers were offered actuarially fair yield insurance, then all would tend to buy the maximum amount of coverage with $\alpha = 0.85$. This observation would indicate excess demand for insurance by risk-averse producers and evidence that

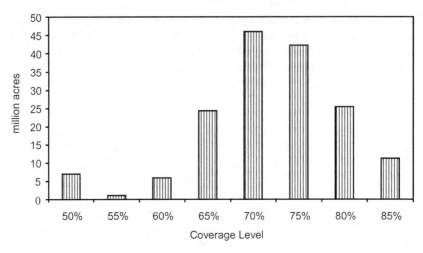

Fig. 3.1   Acres insured at different coverage levels for corn, soybean, and wheat producers in 2009

Source: Summary of Business reports from the Risk Management Agency (RMA) of the U.S. Department of Agriculture.

demand for crop insurance is sufficient to potentially justify government support for the market.

However, it is easily verified that farmers do not, in fact, all buy 85 percent coverage. Figure 3.1 shows that the 85 percent coverage level was actually one of the least popular coverage levels among corn, soybean, and wheat farmers in 2009.

The paucity of 85 percent policies in 2009 by itself is not sufficient evidence to conclude that there is no excess demand for crop insurance. An alternative explanation is that the incremental price that farmers must pay for 85 percent policies is higher than actuarially fair levels. Although USDA's RMA is supposed to set premium rates that are actuarially fair, the RMA places a surcharge on low-deductible policies in an effort to combat adverse selection. Loss experience data that RMA considered in the early 2000s convinced the RMA that farmers who were more likely to suffer crop losses tended to purchase low deductible policies. In addition, the percent premium subsidy declines as farmers move to higher coverage levels. Babcock and Hart (2005) demonstrated that many of the insurance contracts offered to farmers for 85 percent coverage decrease expected returns relative to 80 percent coverage.

The hypothesis to be tested here is whether expected utility is increasing in α when farmers are offered actuarially fair insurance. Evidence in support of this hypothesis would be a large preponderance of farmers moving to high coverage levels when offered increased coverage at an actuarially fair price. To test this hypothesis requires data that measure farmers' coverage-level purchase decisions that are matched up with an estimate of the incremental (to coverage level) actuarial fairness of the crop insurance contracts that are offered to farmers. It is not sufficient to look at participation decisions to answer this question because large premium subsidies make expected returns to participation positive for most farmers. However, wide variations in incremental actuarial fairness across crops, space, and time allows identification of farmer demand for actuarially fair insurance. Insurance data from corn, soybeans, and wheat in 1998 and 2002 are used to test the hypothesis. Because this time period had similar participation rates to current participation, the results from this analysis are more relevant for the current program than results that relied on pre-1995 data.

3.3.1 Construction of a Data Set Measuring Variation in Actuarial Fairness of Low Deductible Policies

Although the RMA has an objective of setting premiums at actuarially fair levels for all coverage, most of its efforts at doing so are concentrated on determining actuarially fair rates when $\alpha = 0.65$. The RMA uses historical loss data as the basis for determining premiums for a crop in a county. To account for the different coverage levels that farmers purchase, the RMA recalculates historical losses as if all farmers purchased 65 percent policies. Babcock, Hart, and Hayes (2004) demonstrated that the RMA's method that was in place until 2003 for determining the incremental premium cost for crop insurance for non-65 percent policies did not lead to actuarially fair incremental premiums even if the 65 percent premium was actuarially fair. For high-risk crops and regions, the incremental cost was too high, leading to higher-than-actuarially-fair premiums. For the lowest-risk crops and regions, the incremental cost was too low.[1]

As shown in figures 3.2 and 3.3, there exists tremendous spatial variation in the riskiness of corn and soybean production. Risk here is measured by the average premium rate charged in a county for a 65 percent policy that protects against yield loss (Actual Production History [APH] insurance). This variation in riskiness combined with the Babcock, Hart, and Hayes (2004) result implies that there were large variations in the degree to which incremental crop insurance premiums were actuarially fair.

The premium rates illustrated in figures 3.2 and 3.3 express premium as a percent of liability. These rates vary across counties by a factor of more than seven with the lowest rates in the central Corn Belt and the highest rates in the Southeast. To measure the incremental farmer cost and expected indemnity of increasing α, the costs and expected indemnities could be calculated

---

1. The RMA changed their rating methods not in response to the publication of the Babcock, Hart, and Hayes (2004) paper but rather to an external review of their rating methods in response to concerns by the crop insurance industry that the incremental cost of Revenue Assurance, which incorporated the Babcock, Hart, and Hayes rating methods, was too low. The external review concluded that the incremental cost of Revenue Assurance premium rates were correct. The RMA responded to this finding by making the incremental costs of Crop Revenue Coverage and Actual Production History correspond to those of Revenue Assurance. A note of disclosure: the author of this article was responsible for developing the rating methods for Revenue Assurance.

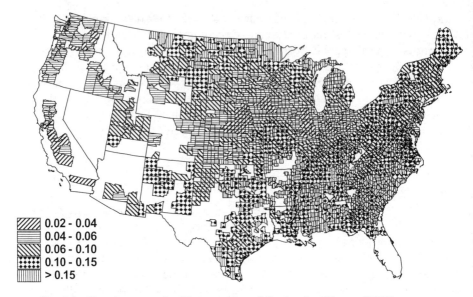

	0.02 - 0.04
	0.04 - 0.06
	0.06 - 0.10
	0.10 - 0.15
	> 0.15

Fig. 3.2   Premium rates for 65 percent Actual Production History (APH) corn insurance for the 2002 crop year

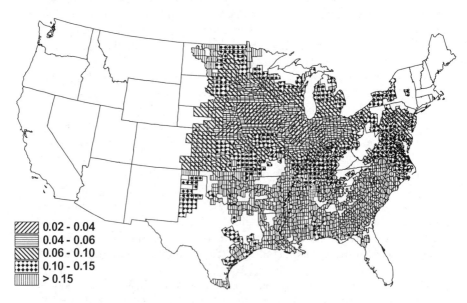

	0.02 - 0.04
	0.04 - 0.06
	0.06 - 0.10
	0.10 - 0.15
	> 0.15

Fig. 3.3   Premium rates for 65 percent deductible Actual Production History (APH) soybean insurance for the 2002 crop year

at all coverage levels. However, in the late 1990s, many farmers could not buy policies with $\alpha > 0.75$. Thus, to allow data from 1998 to be combined with data from 2002, incremental expected returns of moving from $\alpha = 0.65$ to $\alpha = 0.75$ are calculated.

Output prices, expected yields, and production costs are independent of α. Thus, the change in expected returns from increasing crop insurance coverage equals the change in expected indemnities (I) minus the change in producer paid premiums (PP).

(1) $\qquad \Delta\pi = E(I_{75}) - E(I_{65}) - (PP_{75} - PP_{65}) = \Delta I - \Delta PP,$

where the subscript denotes α. If premiums are actuarially fair and unsubsidized, then $\Delta\pi = 0$.

Incremental Costs of Crop Insurance Coverage

To estimate ΔPP in equation (1) requires an accounting of the actual subsidies and premiums charged. The 2002 ARPA changed the subsidy structure but not the premium structure, so we need to estimate ΔPP both before and after ARPA. It is straightforward to calculate ΔPP. Denoting 65 percent and 75 percent premium rates (premium divided by liability) as rate65 and rate75, the premium subsidy rates at 65 percent and 75 percent as psub65 and psub75, and the insurance price as p, the change in the producer premium for the APH plan of insurance is

(2) $\qquad \Delta PP = (1 - psub75) \cdot rate75 \cdot p \cdot 0.75 \cdot \bar{y}$
$\qquad\qquad\qquad - (1 - psub65) \cdot rate65 \cdot p \cdot 0.65 \cdot \bar{y}.$

Both before and after ARPA, 75 percent coverage premium rates (dollars of premium per dollar of liability) for the APH program for corn, soybeans, and wheat equal the 65 percent coverage premiums multiplied by the constant 1.538. Therefore,

(3) $\qquad \Delta PP = p \cdot \bar{y} \cdot rate65$
$\qquad\qquad\qquad \cdot [1.538 \cdot 0.75 \cdot (1 - psub75) - 0.65 \cdot (1 - psub65)],$

which under pre-ARPA conditions equals approximately $0.5 \times p \times \bar{y} \times$ rate65. After ARPA, premium subsidy rates were increased from 41.7 percent to 59 percent for 65 percent coverage and from 23.5 percent to 55 percent for 75 percent coverage. Thus, ΔPP after ARPA is approximately $0.25 \times p \times \bar{y} \times$ rate65, which demonstrates that ARPA cut the incremental cost of moving to $\alpha = 0.75$ in half for all U.S. corn, soybean, and wheat farmers.

Incremental Expected Indemnities from Higher Coverage Levels

If unsubsidized premium rates were set at actuarially fair levels for all α, then the change in expected indemnities from moving to $\alpha = 0.75$ would equal the change in unsubsidized premium:

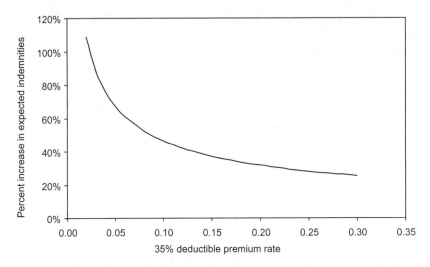

Fig. 3.4   Increase in expected indemnities from moving to 75 percent from 65 percent coverage

(4) $\Delta I = p \cdot \overline{y} \cdot (0.75 \cdot \text{rate75} - 0.65 \cdot \text{rate65})$.

Babcock, Hart, and Hayes (2004) demonstrate that a negative relationship must exist between rate65 and the ratio of rate75 to rate65 if both rates are to be actuarially fair. That is, in high-risk regions, the incremental rate increase as one moves to a 75 percent policy is much lower than the incremental increase in low-risk regions. But crop insurance rates for APH had, until 2003, a constant ratio of rate75 to rate65. Thus, the actual premiums charged cannot be used to estimate ΔI as in equation (4).

Figure 3.4 shows the relationship between ΔI, expressed as a percent change, and alternative premium rates for $\alpha = 0.65$. This relationship was estimated using Monte Carlo integration of yield draws from a beta distribution. The beta parameters were calibrated to generate draws that lead to average (across the yield draws) indemnities that equal each alternative premium rate on the horizontal axis. The beta distribution is regularly used to model yield distributions for crop insurance because of its flexibility. The relationship between the change in expected indemnities and the 65 percent premium rates (35 percent deductible rates) illustrated in figure 3.4 is robust across alternative functional forms for the yield distribution. Thus, if the 65 percent premium rate is actuarially fair, then applying the figure 3.4 relationship will lead to an estimate of ΔI given any 65 percent premium rate.

What we want to determine is whether farmers who are offered incrementally actuarially fair insurance increase their insurance coverage. The data that will be used include only those farmers who have already decided to purchase crop insurance. The purchase decision means that farmers have already decided that crop insurance generates net benefits, either in the form of higher expected returns or reductions in risk.

With $\alpha = 0.65$, farmers received a 41.7 percent premium subsidy in 1998 and a 59 percent subsidy in 2002. These large premium subsidies imply that nearly all farmers are offered a base level of coverage that generates positive expected returns. Thus, it is not surprising that even risk-neutral farmers would find it beneficial to participate in the program. But the figure 3.4 negative relationship combined with the RMA's rule of a constant ratio between premiums rates at 75 percent and 65 percent means that there was a wide variation across crops and regions in expected returns from increasing α.

The change in premium and expected indemnities measures the degree to which insurance is actuarially fair at the margin. I measure demand for incremental insurance as the ratio of acres insured at $\alpha > 0.65$ to acres insured at $\alpha = 0.65$. The unit of observation is the county, as in Goodwin (1993). A unit of observation is a county-crop combination. Demand for more insurance is calculated in 1998 and 2002 for each county where corn, soybeans, and wheat were insured.

The number of insured acres at each coverage level for all insurance products is available from the RMA's *Summary of Business* reports. Data for 1998 and 2002 were used for a number of reasons. The ARPA was passed in June of 2000. Its subsidy provisions went into effect immediately, but farmers had already made their decisions about which coverage level to purchase, so there would be little or no impact from ARPA in 2000. Crop year 2001 data could have been used but it takes time for farmers and the industry to learn about significant changes in policy. Insurance agents must be notified and trained, quoting software must be adjusted, and then farmers must be made aware of the impacts of change. Hence, the 2002 data should more fully reflect awareness of the ARPA policy changes and subsequent changes in coverage levels.

We could also extend the analysis to 2003 and 2004 crop year data, but beginning with the 2003 crop year, the RMA began to implement a new set of premium rates and surcharges at higher coverage levels. The RMA phased these changes in so it would be difficult to accurately calculate the actual insurance offers being made to farmers in these years.

A special 25 percent premium reduction program was implemented late in 1999, reducing producer paid premiums by an additional 25 percent. Undoubtedly, some proportion of agents and their farmer clients were aware of this program, but many were not. Thus, assuming that all farmers made their 1999 coverage-level decisions with full information would be incorrect. A premium reduction program was also in place in 1998, but it was announced after farmers had made their crop insurance decisions. Thus, we can assume that 1998 decisions reflect their prior knowledge about premium and subsidy rates.

In 1998, 71 percent of corn, soybean, and wheat acres were insured with

APH. In 2002, only 34 percent of acres were insured with APH. To account for the rapid movement of producers away from APH insurance to revenue insurance products over this time period, the analysis is extended to estimating the actuarial fairness of the offer price for increased Crop Revenue Coverage (CRC) from 65 percent to 75 percent. First, an explanation of how expected returns to incremental APH coverage is explained.

I estimate the actuarially fair cost of moving to 75 percent coverage from 65 percent coverage as being equal to the product of the average 65 percent premium rate offered to producers in a county who chose to buy at least 65 percent coverage and one plus the percentage increase in expected indemnities shown in figure 3.4. The procedure used to measure the average 65 percent premium rate for those producers who purchased at least 65 percent coverage is best explained with an example.

Table 3.3 presents 2002 corn data for Cass County, Illinois. At each coverage level, the average premium rate is calculated by dividing total premium by total liability. The average rate at each coverage level is then converted to the corresponding average 65 percent rate by dividing it by the appropriate rate relativity factor used by RMA. These rate relativity factors are 1.0 for 65 percent coverage, 1.215 for 70 percent coverage, 1.538 for 75 percent coverage, 1.954 for 80 percent coverage, and 2.462 for 85 percent coverage. The result of this multiplication is reported in the last column of table 3.3. The average 65 percent rate is then calculated by taking the acreage-weighted average of the results in the last column. In this example, the average rate is 0.0591.

Given this estimate of the average rate, we can estimate the average expected gain from moving to 75 percent coverage. Using the beta distribution that generated the relationship in figure 3.4, the actuarially fair 75 percent premium rate is 0.0825. Then using the preceding expressions for ΔI and ΔPP after ARPA, we have $\Delta I = 0.02346 \times p \times \bar{y}$ and $\Delta PP = 0.014927 \times p \times \bar{y}$. Thus, the change in expected profits is $0.008533 \times p \times \bar{y}$. This change in expected profits is normalized by dividing through by the estimate of ΔI. The result then represents the change in expected profits as a percent change.

Table 3.3   Data for Cass County used to calculate average 65 percent premium rates

Coverage level (percent)	Insurance plan	Insured acres (acres)	Total liability ($)	Total premium ($)	Average rate	Average 65% rate
65	APH	2,446	456,555	28,563	0.0626	0.0626
70	APH	113	21,138	1,269	0.0600	0.0494
75	APH	341	75,912	4,590	0.0605	0.0393
80	APH	36	8,525	651	0.0764	0.0391
85	APH	0	0	0	NA	NA

Source: Summary of Business reports from USDA RMA, http://www3.rma.usda.gov/apps/sob/.
Note: APH = actual production history; NA = not available.

In this example, the result is 0.36, which means that the change in expected profit amounts to 36 percent of the change in expected indemnity. In this case, the percent subsidy is positive. If the change in expected profits is negative, then farmers would be forced to pay more than actuarially fair amounts for lower deductible policies. The preceding procedure was implemented for every county in 1998 and 2002 in which corn, soybeans, or wheat was insured under the APH plan of insurance.

It is instructive to calculate the percent subsidy for Cass County before ARPA. Assuming that the average 65 percent premium rate in 1998 was 0.0591, the change in expected profit is $-0.006296 \times p \times \bar{y}$, which translates to a -27 percent subsidy. That is, Cass County farmers were being asked to pay 27 percent more than the actuarially fair incremental cost for 75 percent coverage in 1998. This switch from a 27 percent tax to a 36 percent subsidy is illustrative of the magnitude of the change in actuarial fairness of the choices facing U.S. farmers after passage of ARPA.

Calculating the change in expected profits from higher coverage levels with CRC would seem more difficult than with APH because the CRC rating structure contains three separate components (yield risk, revenue risk, and price risk), and a portion of the change in expected indemnities is due to price variability. However, examination of the relationship between 65 percent APH premium rates and CRC premium rates at the 65 percent, 75 percent, and 85 percent coverage levels reveals an exact linear relationship. Thus, the 65 percent APH premium rate can be used to estimate the change in producer premium for CRC as coverage increases from 65 percent to 75 percent. What remains is how to calculate the change in expected indemnities for CRC.

Because CRC premiums used the same constant rate relativities that were used to determine APH premium rates, they cannot be used to calculate expected indemnities. What is needed is an independent measure of the change in expected indemnities that is based on a revenue distribution, much like the beta yield distribution was to generate figure 3.4. The rating equations from a competing revenue insurance product, Revenue Assurance (RA), that had lower market penetration than CRC during this period can be used to estimate the change in expected indemnities. The coverage provided by RA with the harvest price option is nearly identical to CRC. The RA rating equations are based on Monte Carlo integration of revenue draws as discussed in Babcock and Hennessy (1996).[2]

Before moving to a discussion of results, it is useful to pause and consider the accuracy of my measure of actuarial fairness. First, as stated in the preceding, my measure is accurate insofar as the RMA's premium rate for 65 percent coverage is actuarially fair. Coble et al. (2010) recommend that the RMA continue to use its loss-cost method of rate making, which

2. The RA rating equations are available upon request from the author.

suggests that they conclude there is no better alternative. Loss experience in the period 2000 to 2008 has led some to conclude that premium rates are higher than actuarially fair in Corn Belt states and lower than actuarially fair in Great Plains states (Woodard et al. 2008). If true, then my estimate of what is an actuarially fair offer for marginal coverage in the Corn Belt would actually be a subsidized offer. And for Great Plains farmers, my estimate of an actuarially fair offer would actually be an offer that is priced too high. Evidence of this bias would be a greater take-up of actuarially fair offers in Corn Belt counties relative to Great Plains counties.

The next source of possible error in measurement arises from my using the county as my observational unit. Farmers in a county face different 65 percent coverage rates because they are judged to present different amounts of risk to the insurance pool. I implicitly assume that there is a single representative farmer in a county that faces an incremental price of higher coverage. If my estimate of actuarial fairness is accurate for the mean farmer in a county and farmers in a county are actually distributed around that mean farmer, then about half of farmers were offered more subsidized margin premiums than I measure, and half were offered less-subsidized marginal premiums. There are two factors that mitigate against this being a source of inaccuracy. First, the variability in risk across counties is much larger than the variability in risk within a county. Second, I account for intra-county distributions by defining actuarially fair premiums as being within 10 percentage points of what my measure takes to be as being absolutely fair.

As with all empirical studies, sample selection bias warrants discussion. My sample consists of all farmers who chose to insure their corn, soybeans, and wheat in 1998 and 2002. Between 20 percent and 30 percent of acreage was not insured in this time period. Nonparticipating farmers felt that they had more cost-effective means of risk management than crop insurance, which could imply that they were being offered insurance on too unfavorable terms. Thus, my analysis is only relevant for those farmers who chose to participate. That the increase in premium subsidies in 2000 brought increased participation suggests that participating farmers in 2002 were, on average, of lower risk than farmers who participated in 1998. To the extent that the average 65 percent premium offered to these farmers was actuarially fair, then this difference in level of risk is accounted for.

Finally, some explanation for why nobody else has conducted this type of analysis is needed. I am not the first to estimate the actuarial fairness of insurance offers to farmers. For example, Just, Calvin, and Quiggin (1999) estimated differences between farmers' perceptions of actuarial fairness and the actual premiums they were being charged. They used these differences to predict which farmers chose to buy insurance. One explanation for the differences in actuarial fairness was a lack of rating data caused by low participation. Nobody, to my knowledge, has focused on putting together a data set to calculate the actuarial fairness of the offer that farmers faced for higher coverage levels. Perhaps this is not too surprising because most crop insurance demand studies have focused on the participation decision, rather than on the demand for higher coverage. My focus on marginal coverage arose because in 2002, I was in the unique position of having to defend the large discrepancies between how fast RA premium rates increased as farmers chose to buy more coverage relative to how fast premium rates for competing products increased. Farmers in high-risk counties were moving rapidly to RA from CRC because RA marginal premiums were sometimes less than half what they would have had to pay for CRC. Because the crop insurance industry obtained far fewer premium dollars because of this movement, they forced the RMA to review whether RA premium rates were too low. The work that went into demonstrating to the RMA and the industry that RA marginal rates were accurate is what provided me with the background to put together this data set.

3.4 Results

Table 3.4 provides summary statistics of the entire data set. As shown, the number of counties where farmers purchased APH and CRC insurance is higher in 2002 than 1998 for each crop. The average percent subsidy in 1998 ranged from a low of –83 percent for CRC wheat to a high of –60 percent for CRC corn. This meant that, on average, farmers were forced to pay more

Table 3.4   Data summary by crop, year, and insurance type

	1998		2002	
	APH	CRC	APH	CRC
Corn				
No. of observations	1,878	1,026	2,022	1,730
Average subsidy[a]	–65%	–60%	17%	–1%
Acre ratio[b]	0.09	0.18	0.39	0.62
Soybeans				
No. of observations	1,520	947	1,707	1,384
Average subsidy	–62%	–83%	18%	–11%
Acre ratio	0.10	0.21	0.50	0.67
Wheat				
No. of observations	1,621	806	1,664	1,403
Average subsidy	–72%	–82%	16%	–7%
Acre ratio	0.16	0.16	0.37	0.60

Note: APH = actual production history; CRC = crop revenue coverage.

[a]Ratio of change in expected returns from insuring at 75 percent coverage level to change in expected indemnities from moving to 75 percent coverage from 65 percent coverage.

[b]Ratio of acres insured at greater than 65 percent coverage to acres insured at 65 percent coverage or above.

than actuarially fair amounts for higher levels of yield insurance in 1998. In contrast, in 2002, farmers, on average, were offered higher levels of insurance at close-to-actuarially fair premiums.

Histograms that show the distribution of the degree to which buying additional coverage was subsidized in 1998 and 2002 across counties are presented in figures 3.5 and 3.6 for APH and CRC, respectively. For counties with more than one crop, the average percent subsidy across crops is used. As shown, the change in subsidy structure that took place in 2000 dramatically moved these distributions to the right. This move coincides with the large increase in the percent of acres insured above the 65 percent coverage level as shown in table 3.4. These histograms also show that there are many more counties that were offered actuarially fair increases in insurance in 2002, thus allowing for a better estimation of how many farmers will increase coverage when offered an actuarially fair premium.

A straightforward method for estimating the demand for actuarially fair insurance is to use only those observations where the offer is actuarially fair. Table 3.5 presents results taking counties that fall in the range of −10 percent to +10 percent percent subsidy as being actuarially fair. Overall, almost 50 percent of acreage is insured at higher coverage levels by farmers when offered actuarially fair premiums for the higher coverage. This

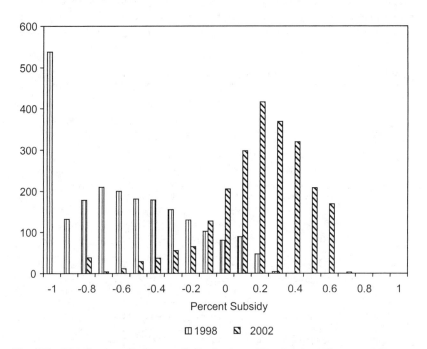

Fig. 3.5   **Distribution of percent subsidy for Actual Production History (APH) across counties in 1998 and 2002**

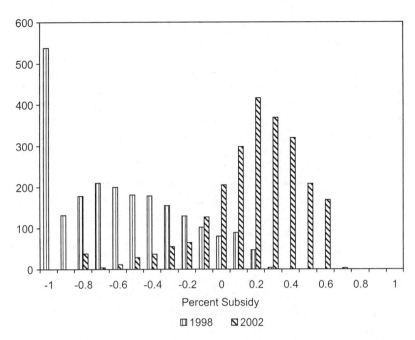

Fig. 3.6   Distribution of percent subsidy for Crop Revenue Coverage (CRC) across counties in 1998 and 2002

Table 3.5   Share of acres insured at higher coverage under actuarially fair insurance

	Acre ratio[a]	No. of observations
Entire sample	0.47	3,425
1998	0.27	562
2002	0.51	2,863
APH	0.30	1,565
CRC	0.62	1,860
APH-1998	0.26	528
APH-2002	0.32	1,039
CRC-1998	0.40	36
CRC-2002	0.62	1,824

Note: APH = actual production history; CRC = crop revenue coverage.
[a]Defined in table 3.4.

proportion seems to have increased in 2002 relative to 1998, but a large part of this apparent increase is caused by the increase in acreage insured under CRC. Farmers are about twice as likely to take the offer of actuarially fair insurance when they buy CRC than when they buy APH. Given the low number of observations of available in 1998, and the small increase from 1998 to 2002 in the proportion of acreage insured at higher coverage levels for APH, it is not apparent that the demand for actuarially fair insurance increased in 2002.

To determine whether the point estimates of differences in demand for actuarially fair insurance are statistically different can be accomplished by regressing the proportion of observed acres insured at higher-than-65 percent coverage level on indicator variables that control for plan of insurance, year, and crop. Because this proportion cannot exceed 1.0 or fall below 0.0, a two-limit Tobit regression procedure is appropriate to use. Two sets of regression results are reported in table 3.6. The first column of results use only 2002 data to indicate if the difference between the demand for actuarially fair insurance by producers who buy CRC is statistically different than for producers who buy APH. As shown, the results support this difference. In addition, the demand for actuarially fair insurance for soybeans is statistically higher than the demand from farmers who buy insurance for their corn and wheat crops. This sample was defined by the percent subsidy. Within the range of plus or minus 10 percent in percent subsidy, the dollar per acre value of the subsidy could still affect the demand for insurance. Thus, the dollar value of the subsidy was included in the regressions. As one would expect, increases in the dollar value of subsidy increases the demand for insurance.

The second set of results is used to test if the demand for actuarially fair insurance increased in 2002. This test is best run using only APH observations because the first regression results indicate that the demand is higher for those who buy CRC, and this demand is primarily exhibited in 2002.

Table 3.6	Regression results for observations where percent subsidy is between +/–10 percent		
Sample		2002 only	APH only
Constant		0.713***	0.241***
		(0.0157)	(0.0217)
APH dummy		–0.377***	
		(0.0164)	
Corn dummy		–0.107***	–7.84E-03
		(0.0190)	(0.0232)
Wheat dummy		–0.099***	8.79E-03
		(0.0200)	(0.0255)
Dollar subsidy		0.112***	0.123***
		(0.0264)	(0.0381)
2002 dummy			0.026
			(0.0208)
Sigma		0.404***	0.375***
		(0.0067)	(0.0084)

Notes: Standard errors are shown in parentheses. APH = actual production history.
***Denotes an asymptotic *t*-statistic that is significant at the 1 percent level.

The hypothesis that demand did not change between 1998 and 2002 cannot be rejected.

These results indicate that the demand for actuarially fair insurance is most important to those producers who buy CRC. Furthermore, more than half of U.S. corn, wheat, and soybean acreage that was able to be insured at higher insurance levels at actuarially fair rates was insured at the higher rates for producers who chose CRC. This insurance product is designed not only to reduce the risk of revenue falling short of a targeted amount, but it is also designed to offset the added financial risk of forward contracts. Many farmers hesitate to forward contract because if they do not produce enough to deliver against the contract, and if market prices have risen above the forward contract price, then they must enter the market to buy expensive grain to fulfill their contract. Thus, the risk management benefits of CRC should be expected to be higher. The results indicate that this is so.

The final regression is to estimate the response of farmers' decisions about buying additional coverage as the percent subsidy increases. Table 3.7 provides the regression results for CRC only. Following Greene (1990, 738), predicted values from a two-limit Tobit model were calculated and reported in figure 3.7. For all three crops, the expected proportion of acres that would be insured at greater than the 65 percent coverage is above 50 percent. That the predicted value from the regression crosses the zero-subsidy point above 50 percent implies decreasing amounts of acreage being insured at higher coverage levels as subsidies increase.

As Babcock and Hart (2005) demonstrate, most farmers are now offered higher revenue insurance coverage (up to 80 percent) at producer premiums that generate positive expected returns. The results shown in figure 3.7 indicate that many farmers will buy higher levels of revenue insurance even if it was offered to them at actuarially fair levels. Given that farmers have shown that they will buy significant amounts of actuarially fair insurance, it is somewhat of a puzzle why premium subsidies for incremental crop insur-

Table 3.7   **Regression results for all crop revenue coverage observations**

Sample	Crop revenue coverage only
Constant	0.734***
	(0.0174)
Percent subsidy	0.854***
	(0.168)
Corn dummy	−0.152***
	(0.023)
Wheat dummy	−0.094***
	(0.024)
Sigma	0.385***
	(0.0077)

***Denotes an asymptotic t-statistic that is significant at the 1 percent level.

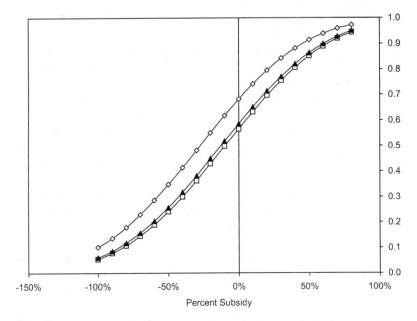

Fig. 3.7   Expected response of increased coverage to increased incremental subsidies

ance are so large. The results of this chapter suggest that a fixed amount of subsidy per acre to entice an adequate number of farmers to enter the program, and then a contract schedule of actuarially fair premiums would result in a large number of farmers buying a significant amount of insurance. One answer to this puzzle is that, when it passed ARPA in 2000, Congress was not aware that farmers were being forced to pay higher-than-actuarially fair incremental rates: all that was observed was that most farmers purchased the 65 percent coverage level. In response, Congress decided to increase premium subsidies to induce farmers to buy higher coverage levels. In this effort to increase premium subsidies, Congress was urged along by those interest groups who benefit from farmers buying high levels of insurance.

Stigler (1971) explains why government regulations often seem to chiefly benefit those who are regulated by the regulators being captured by the industry that they are regulating. And there is ample evidence that the crop insurance industry has a large amount of influence on the RMA. But Becker's (1983) theory of interest group influence over the political process is perhaps a better starting point for an explanation because the House and Senate Agricultural Committees have taken an unusually direct role in dictating how the crop insurance program is run.

Becker (1983) argues that the outcome of the legislative process reflects the pressures brought to bear by interest groups. The influence of powerful interest groups is only limited by increasing deadweight losses from their

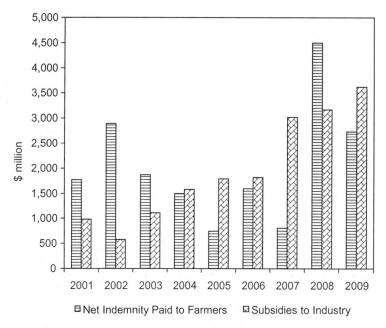

Fig. 3.8   **Revenue from the crop insurance program**

favored policies and by increased pressure from less-favored groups whose interests are increasingly adversely affected by policies that are adopted. Figure 3.8 shows the magnitude of the annual payments (calculated from table 3.1) made to the two beneficiaries of taxpayer support for crop insurance: farmers and the crop insurance industry. The next section explains competition in the crop insurance program to explain who in the industry actually benefits from the industry subsidies shown in figure 3.8.

3.5   Crop Insurance Competition

In this section, a descriptive model of competition in the public/private partnership that characterizes the crop insurance industry is developed. Data are used to validate the main prediction of the model that crop insurance agents are the residual claimants of rents that accrue to the industry.

There are six main interest groups in the crop insurance industry: taxpayers, Congress, government regulators, farmers, crop insurance agents, and crop insurance companies. Taxpayer subsidies created and provide continual support to the industry. Congress reacts to political pressure by passing laws that regulate and subsidize the industry. Regulators implement those laws. Farmers buy crop insurance from a crop insurance agent. Crop insurance agents decide which crop insurance company will receive each farmer's business. Agents make money by earning commission on each policy that they sell. The variable cost of selling policies is much less than the commission on most policies, so the more policies they sell, the more money agents make. Thus, agents have an incentive to compete with other agents for a farmer's business. Crop insurance companies make money from underwriting gains and from A&O reimbursements. The noncommission variable costs are much lower than A&O and expected underwriting gains in almost all states. These variable expenses include loss adjustment costs and labor. Thus, crop insurance companies make more money the more policies that they can obtain. This creates an incentive for companies to compete for agents' books of business. Agents, companies, and farmers have an incentive to lobby Congress to pass laws that work to their favor. As always, because the aggregate cost of the crop insurance program is spread among all taxpayers, opposition to the program by taxpayer groups is unfocused and relatively ineffective.

There exist three sources of competition in this descriptive model. The first is the competition between groups that lobby the House and Senate Agricultural Committees for part of the baseline agricultural budget. This competition pits the crop insurance industry against advocates for more farmer subsidies and advocates for greater spending on nutrition spending. For example, in 2007, the crop insurance industry argued against the reforms of commodity subsidies advocated by the National Corn Growers Association because the reforms would have transferred a large portion of the risk from the crop insurance program directly to taxpayers. More recently, a *Des Moines Register* (January 17, 2010) article about proposed cuts to crop insurance reported that "One of the industry's leading allies in Congress, Sen. Charles Grassley, R-Ia., said he believes the administration wants to use the $4 billion in savings it would get from the cuts to increase spending for child nutrition programs, including school lunches." The outcome of this first competition is the aggregate amount of support that the industry receives from taxpayers.

The second source of competition is the competition between agents for farmers' insurance business. This competition cannot include price competition because of laws passed by Congress at the behest of agents. For example, the chief lobbying arm of crop insurance agents, the Independent Insurance Agents & Brokers of America, sought a ban on a program that allowed crop insurance companies to reduce farmer-paid premiums by passing on a portion of agent commissions to farmers. Congress delivered a final ban on this type of price competition in the 2008 Farm Bill.

Because there is no possibility of price competition, agents must compete in terms of service. The types of service that can be offered include educating farmers about the types of insurance coverage offered, lowering the farmer cost of filling out required forms, and keeping farmers informed of any new information that may prove useful. All of these services are of second-order importance to farmers because either they are one-time benefits or because they do not directly increase farmers' profits. By default, a farmer's business remains with an agent year after year. So unless an agent convinces a farmer to switch agents, no switch will take place.

To a farmer, the benefit of switching must be greater than the cost of switching. The cost of switching includes a bit of paperwork, any social costs involved in moving business from one local neighbor or business person to another who may not be local, and the cost of searching for an agent that can provide superior service. Because there are positive switching costs and only indirect benefits, the incentive for most farmers to switch is not very large, although exact measurement of the incentive would be difficult. Consequently, the payoff from agent investments designed to induce farmers to switch will not be high. In equilibrium, each agent invests an optimal amount to keep their business and to perhaps attract new business, and each farmer has found the agent where the benefits of further switching are outweighed by the costs. Entry costs for new agents, involving the successful passing of a test on crop insurance are nominally low. But the real entry costs are actually quite high because new entrants will find it difficult to build up their book of business by inducing an adequate number of farmers to switch. Thus, each agent has essentially a captive book of business.[3]

Crop insurance companies use price to compete for agents' books of business. The price of an agent's business is the agent commission. Because most agents act independently of companies, they are free to sell their book to the highest bidder. There are approximately fifteen companies bidding for business. The maximum bid that a company will likely make is the difference between the expected revenue that an agent's book of business will bring in minus all noncommission variable costs. With sufficient competition, commission rates will exactly equal this difference. Thus, price competition between companies along with regulatory barriers that limit competition in premium rates make agents the residual claimant in the crop insurance industry. Evidence for agents being the residual claimant is that agent commissions vary dramatically across states. High commissions are paid where companies expect to make large underwriting gains; low commissions are paid in states with low or negative underwriting gains. The other possible residual claimant would be the owners of crop insurance companies, the management and staff of crop insurance companies, or loss adjusters. There may be a limited number of specialized crop insurance actuaries who are efficient at managing the risk profile of crop insurance companies and who, therefore, may capture increased rents. But there is no reason to think that

3. An indicator of the degree to which an agent has a captive book of business is the market value of the book. If the cost of switching were low and the benefits were high, then the market value of an agent's book of business would be low. But anecdotal evidence suggests that the market value of an agents' crop insurance book of business is multiples of the annual profit that can be generated from the book.

the salaries paid to nonspecialized staff or executives should be above market rates of compensation. The Grant Thornton (2009) study on returns to the owners of crop insurance companies shows that the companies are actually making less than the return they have obtained from other lines of property and casualty insurance. The same Grant Thornton study provides convincing evidence that it is the crop insurance agents that are the residual claimant.

Figure 3.9 breaks shows how per-unit crop insurance costs have changed since 2001. Agent commissions are calculated by dividing total commission paid by the number of crop insurance policies. Other costs include crop insurance company salaries, computers, travel, and overhead. These costs are also expressed on a per-policy basis. Loss-adjustment costs are calculated by dividing total loss-adjustment costs by the number of units on which a loss was paid. The average annual growth in other costs is 5.61 percent. The average annual cost in loss adjustment costs is 6.1 percent per year. The Bureau of Labor Statistics reports that the average annual increase in civilian compensation over this time period was 3.4 percent. This suggests that some rents have accrued to claims adjusters and to staff of the crop insurance companies. But the growth and magnitude of compensation paid

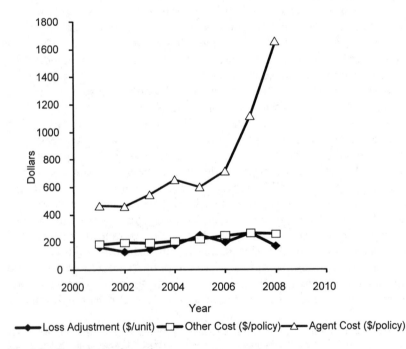

Fig. 3.9 Breakdown of crop insurance program costs
Source: Calculated from Grant Thornton reports prepared for the National Crop Insurance Services.

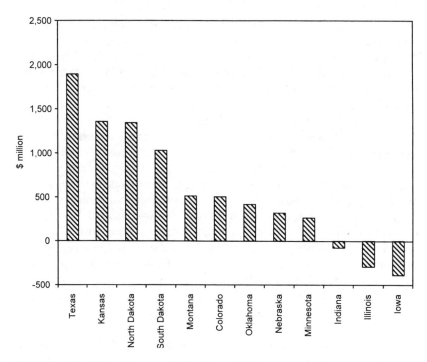

Fig. 3.10   Net indemnities received by farmers: 2000 to 2007
Source: Calculated by author from *Summary of Business* reports from the Risk Management Agency (RMA) of the U.S. Department of Agriculture.

to agents is convincing evidence that agents are the primary beneficiary of increased government support to the crop insurance program, as would be predicted if companies bid for agents' captive books of business.

Agents are not the only group that receives rents from the crop insurance program. As shown in table 3.1, farmers have garnered about $18 billion more in payments than they have paid in premiums since 2001. However, not all regions of the United States have benefited equally from the program. Figure 3.10 shows that the Great Plains states have benefited by a much greater amount from the crop insurance program than have the Corn Belt states. This difference in benefits means that there is a natural alliance between crop insurance companies and members of Congress who represent areas outside of the Corn Belt.[4]

4. Evidence for of this alliance is contained in a press released from the National Corn Growers Association, quoted in a farm policy blog (http://www.farmpolicy.com/?p=520): "The National Corn Growers Association (NCGA) is pleased the Senate Agriculture Committee included a revenue option in the 2007 farm bill, but is disappointed by the committee's action to strip a key component of the optional revenue-based countercyclical program, the integration with federal crop insurance. It is a missed opportunity to provide a better risk management

The outcome of this type of pressure group politics is a crop insurance program that operates for the benefit of crop insurance agents and that has generated positive expected returns for farmers in the Great Plains. The results of this study, combined with the figure 3.10 results, indicate that Corn Belt farmers have received more in risk management benefits that they have received in expected returns.

Becker's (1983) theory of competition between pressure groups shows that a limiting factor on the benefits bestowed on more powerful groups is that opposition to these benefits increases by losing groups the more that is given to winners. An implication of this part of his theory is that politicians will work to neutralize opposition by giving them benefits that do not come at the expense of the favored powerful group. In this case, in the new Farm Bill that was passed in 2008, Corn Belt farmers were given a new subsidy program called ACRE (Average Crop Revenue Election), which is a farm program that is not integrated with crop insurance. Thus, Corn Belt farmers obtained a new subsidy program, and Great Plains farmers and crop insurance agents were able to keep the crop insurance program.

3.6   Conclusions

The crop insurance program has grown so dramatically over the last ten years because private interests supported expansion of the program and Congress accommodated these private interests under the guise of expanding farmers' ability to manage production and price risk. But taxpayer costs of the program—along with the associated transfers to private industry—grew so large that some small steps were taken in 2010 to rein in costs.[5] The policy question going forward is whether reform of the program could cut costs further while still meeting farmers' demands for efficient risk transfer tools, thereby increasing program efficiency. The results from this study provide insight into these questions.

The demand for crop insurance is motivated by both expected returns and risk reduction. This study presents estimates of demand for crop insurance that isolate the demand for risk reduction from the demand for expected returns by construction of a data set from which the farmer response to

tool in the new farm bill," said NCGA President Ron Litterer. An amendment accepted by the committee on a voice vote stripped the crop insurance integration from the revenue package. Corn growers support an optional revenue program starting in 2010. The ACR [Average Crop Revenue] plan initially put forward by Agriculture Committee Chairman Tom Harkin (D-Iowa) would have allowed farmers to insure part of their farm revenues directly through the government, costing private crop insurance companies an estimated $2.2 billion over five years, according to the Congressional Budget Office. Along with an outcry from the industry, Harkin's plan ran into objections from Western senators who feared that insurance costs would rise as corn growers in the rainy Midwest shifted to the government plan.

5. Caps were placed on agent commission rates and overall A&O reimbursements with an objective of reducing government costs by about $600 million per year.

actuarially fair contract offers can be estimated. Results indicate that a large number of producers find that the risk reductions offered by revenue insurance generate significant value, suggesting that some public support for crop insurance may be useful if private provision of crop insurance is infeasible because of a lack of reinsurance markets.

These results also indicate that large premium subsidies are not needed to induce farmers to buy higher amounts of insurance because a significant proportion of farmers would buy actuarially fair insurance. However, significant adverse selection problems would arise if premium subsidies were withdrawn because it is difficult to offer insurance contracts that are individually actuarially fair. One method that would limit adverse selection problems would be to give farmers a lump-sum payment if they participate at some minimum coverage level in the program and then offer them the opportunity to buy more insurance at unsubsidized premium rates. Adverse selection would then be limited to the purchase of incremental coverage rather than to the entire insurance pool. Such a reform could reduce program costs substantially.

The larger question of whether the private sector could meet the crop insurance demands of farmers is difficult to answer because the private sector has been crowded out of the market for crop insurance. Total federal crop insurance liability in 2010 exceeded $78 billion, with a significant portion of this risk located in the Corn Belt states of Iowa, Illinois, Indiana, Minnesota, Nebraska, and Missouri. This liability is small relative to hurricane liability, but recent difficulties in the market for insurance coverage against hurricane losses suggests that there is no assurance that private reinsurers would step in to facilitate private provision of the type of crop insurance that is currently being provided.[6]

References

Arrow, Kenneth. 1974. *Essays in the Theory of Risk Bearing.* New York: North-Holland.

Babcock, B. A., and C. E. Hart. 2005. "ARPA Subsidies, Unit Choice, and Reform of the U.S. Crop Insurance Program." Briefing Paper no. 05-BP 45. Ames, IA: Center for Agricultural and Rural Development, Iowa State University, February.

Babcock, B. A., C. E. Hart, and D. J. Hayes. 2004. "Actuarial Fairness of Crop Insurance Rates with Constant Rate Relativities." *American Journal of Agricultural Economics* 86:563–75.

Babcock, B. A., and D. Hennessy. 1996. "Input Demand under Yield and Revenue Insurance." *American Journal of Agricultural Economics* 78:416–27.

6. According to a recent (February 23, 2011) report in the *Wall Street Journal,* 25 percent of the Florida market for property insurance amounted to $451 billion, with a 1 percent chance that losses from a hurricane could exceed $20 billion.

Becker, Gary S. 1983. "A Theory of Competition among Pressure Groups for Political Influence." *Quarterly Journal of Economics* 98 (3): 371–400.

Coble, K. H., T. O. Knight, B. K. Goodwin, M. F. Miller, and R. Rejesus. 2010. "A Comprehensive Review of the RMA APH and COMBO Rating Methodology Draft Final Report." Washington, DC: Risk Management Agency, U.S. Department of Agriculture. http://www.rma.usda.gov/pubs/index.html#actuarial.

Food and Agricultural Policy Research Institute (FAPRI). 1999. *Rainbow Book. A Summary of the November 1999 FAPRI Baseline.* Columbia, MO: University of Missouri.

GC Securities. 2009. "Cat Bonds Persevere in Tumultuous Market." *GC Capital Ideas,* February 4. http://www.gccapitalideas.com/2009/02/04/cat-bonds-persevere-in-tumultuous-market/.

Glauber, J. W., and K. J. Collins. 2002. "Crop Insurance, Disaster Assistance, and the Role of the Federal Government in Providing Catastrophic Risk Protection." *Agricultural Finance Review* 62:81–101.

Goodwin, B. K. 1993. "An Empirical Analysis of the Demand for Multiple Peril Crop Insurance." *American Journal of Agricultural Economics* 75:423–34.

Grant Thornton. 2009. "Federal Crop Insurance Program: Profitability and Effectiveness Analysis, 2009 Update." Prepared on behalf of the National Crop Insurance Services, Inc. Chicago: Grant Thornton LLP. www.ag-risk.org/NCIS PUBS/SpecRPTS/GrantThornton/Grant_Thornton_Report-2009_FINAL.pdf.

Greene, W. H. 1990. *Econometric Analysis.* New York: Macmillian.

Johansmeyer, T. 2010. "Catastrophe Bond Market Hits Target, Records Possible in 2010." *Blogging Stocks,* January 4. http://www.bloggingstocks.com/2010/01/04/catastrophe-bond-market-hits-target-records-possible-in-2010/.

Just, R. E., L. Calvin, and J. Quiggin. 1999. "Adverse Selection in Crop Insurance: Actuarial and Asymmetric Information Incentives." *American Journal of Agricultural Economics* 81:834–49.

Miranda, M. J., and J. Glauber. 1997. "Systemic Risk, Reinsurance, and the Failure of Crop Insurance Markets." *American Journal of Agricultural Economics* 79:205–15.

Risk Management Agency (RMA). Various years. "Summary of Business Report" www.rma.usda.gov/FTP/Reports/Summary_of_Business/sumbtxt.zip.

Stigler, G. 1971. "The Theory of Economic Regulation." *Bell Journal of Economics and Management Science* 2 (1): 3–18.

Woodard, J. D., G. D. Schnitkey, B. J. Sherrick, N. Lozano-Gracia, and L. Anselin. 2008. "A Spatial Econometric Analysis of Loss Experience in the U.S. Crop Insurance Program." Working Paper, University of Illinois.

Supply and Effects of
Specialty Crop Insurance

Ethan Ligon

4.1 Introduction

The federal government has played a role in providing crop insurance to producers of particular sorts of crops across the United States since 1938, soon after Franklin Delano Roosevelt announced the creation of an institution to provide such insurance. Roosevelt's rationale for the program had explicitly to do with smoothing supply, as "neither producers nor consumers are benefited by wide fluctuations in either prices or supplies of farm products."[1]

The original system Roosevelt proposed was for wheat and allowed payment of both premiums and indemnities in either cash or in kind, at least in part because in-kind payments by farmers could be used to establish buffer stocks of wheat. What became the Federal Crop Insurance Corporation (FCIC) no longer accepts or makes in-kind payments, and the federal government no longer makes any effort to reduce variation in prices by managing buffer stocks of wheat or other commodities. It seems that the original motivation for the program—to smooth food supply and prices—

Ethan Ligon is associate professor of agricultural and resource economics and a member of the Giannini Foundation of Agricultural Economics at the University of California, Berkeley.

This research was supported in part by the National Research Initiative of the Cooperative State Research, Education, and Extension Service, USDA, Grant CA-B*-AEC-7424-CG, and in part by funding from the Giannini Foundation of Agricultural Economics. Alana LeMarchand provided excellent research assistance, particularly in the preparation of the data set. My thanks also to Ron Lundine of U.S. Department of Agriculture/Risk Management Agency (USDA/RMA) for his help in understanding the process by which new crop insurance programs are created.

1. See *New York Times,* February 19, 1937.

has changed. The motivation now has to do with providing an orderly way to improve producer welfare by providing payments to producers in states of nature when either yields or prices are low.

It's been possible to purchase policies to cover low yields of wheat in many states since federal crop insurance began in 1938. However, both the areas and the crops for which policies are available have expanded over time. Insurance to cover low yields of "program" crops other than wheat emerged in many states in the years subsequent to 1938 and expanded beyond the program crops with the passage of the 1980 Federal Crop Insurance Act. Only since the late nineties, however, have policies become available for insuring against losses associated with the production of most fruits and vegetables. The number and variety of such products have expanded dramatically over the last decade, following legislative changes made in 1994, 1996, and 2000 designed to encourage the use of crop insurance by farmers.

To grasp the scale of the change, consider just the case of California, where a predominance of fruit and vegetable crops are grown. A given insurance product is specific to a particular crop and county of production. Figure 4.1 shows both the number of county-crops in a given year according to the National Agricultural Statistics Service (NASS) and the number of county-crop insurance products offered. From the figure, one can see that in 1981 there were just a handful of contracts offered (twenty-eight; for almonds, citrus, grapes, raisins, and processing tomatoes). There was a sharp increase in 1989, to nearly 500 products, and then an explosion in 1990, followed by an even larger explosion in 1995. The number of products has

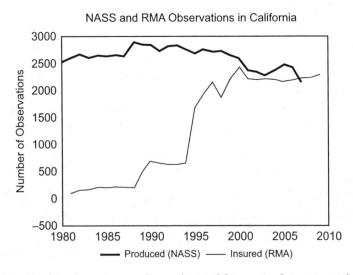

Fig. 4.1   **Number of county-crop observations and frequency of county-crop insurance contracts, by year**

grown since and now amounts to about 2,300 products across California's fifty-eight counties.

There are two types of justifications typically offered for the provision of crop insurance. The first has to do with concern for producers' welfare. This is not a trivial concern, especially for fruits and vegetables, because these commodities may involve much more risk than do cereal crops. The second has to do with consumer welfare—the idea is that by providing insurance to a risk-averse producer, one can induce those individual producers to act as though they were more nearly risk-neutral and more willing to make production and management decisions consonant with the interests of consumers. Further, such programs could be expected to encourage entry by new producers as insurance should lower the costs of production by risk-averse producers and, thus, lower prices.

Specialty crops, particularly fruits and vegetables, differ in several important respects from traditional commodity crops in ways that may affect both demand for insurance and the difficulty of supplying insurance. Let us first consider some demand-side issues. First, prices for many perishable fruits and vegetables have much greater variation than do prices for storable commodities. One might expect this to create increased demand for crop insurance that could deal with this price risk. However, second, a predominance of fruits and vegetable crops in California are marketed via vertical contracts with intermediaries, and, in many cases, these contracts already play an important role in the producer's risk management (Wolf, Hueth, and Ligon 2001). The existence of these alternative arrangements for managing risk ought to tend to reduce demand for federal crop insurance. Third, because production of many specialty crops is concentrated within a relatively small geographical area, spatial (e.g., weather) shocks that affect production in this area will have a much larger effect on aggregate supply than would a similar shock for a commodity with more geographically dispersed production. As a consequence, negative shocks to yield will cause positive shocks to price—it's not even clear that the average producer will be harmed by such production shocks because the increase in price may easily exceed the decrease in aggregate production. Thus, demand for yield insurance for any commodity with a combination of geographic concentration of production and inelastic short-run demand should be expected to be very low.

Turning to the supply side, the sheer diversity of specialty crops both across commodities and across space for a particular commodity makes the design of appropriate insurance products more demanding than it may be for commodity crops. Further, the well-developed organizations that serve, for example, wheat farmers in other states, and that may serve as an important channel for identifying and marketing to relevant producers will be absent for many (though not all) specialty crops. Related, to the extent that designing an insurance product for a particular crop involves some level of fixed costs (e.g., the costs of the five-year feasibility and pilot programs the

Risk Management Agency [RMA] conducts), then the return to the invest-ment made in these fixed costs may be lower in a state where there are many diverse crops with geographically concentrated production.

If the extension of federal crop insurance programs to cover fruit and vegetable production has affected either producer or consumer welfare, then we would expect to see this reflected in output and prices. We have high frequency (weekly) data available for wholesale prices of a wide range of fruits and vegetables in California and elsewhere in the country. We have monthly production data for many crops by California county. And then, finally, we have data on the expansion of crop insurance programs across counties, years, and crops.

This chapter uses data on crop insurance policies to explore the variation in the timing of their introduction in different locations for different crops. Aside from simply seeking to describe the data, we're interested in using these data to try and understand something about the supply of insurance (the topic of section 4.3). In section 4.4, we tackle the central question of the chapter: what effect does the introduction of crop insurance programs have on output of the insured crops and on prices of those crops? Section 4.5 concludes.

4.2   Data on Insurance for Specialty Crops in California

4.2.1   Data Sources

For the results and discussion of specialty crop insurance in California found in this chapter, we rely principally on two different sources of data. First, data on agricultural production and prices collected by the National Agricultural Statistics Service (NASS), which maintains a database of agri-cultural production and prices since 1980.[2] These data include informa-tion for produce as well as for livestock and other crops. Second, the Risk Management Agency (RMA) that administers the FCIC insurance policies maintains a database of insurance policies sold for qualifying agricultural products.[3]

Using data from these two sources, we construct a database that matches data on insurance supply and demand with data on production and prices. The unit of observation in the resulting data set is a county-crop-year: because the number of California counties hasn't changed over the period 1981 to 2007 (the period our analysis covers) and the crops NASS has collected data on haven't much changed, we have a balanced data set of 190 crops over twenty-six years and fifty-seven counties (only urban San

2. See http://www.nass.usda.gov/Statistics_by_State/California/Publications/AgComm/indexcac.asp.
3. See http://www.rma.usda.gov/data/sob/scc/.

Francisco County is missing). However, as not all crops are grown in every county, the total number of crop-county pairs is 1,053, and the total number of crop-county-year observations is 29,485. Because NASS and RMA use slightly different methods of identifying crops, we had to construct a concordance to match up data from these respective sources: details may be found in appendix A.

4.2.2   Brief Descriptive History

Crop Insurance in California

Though a program of federal crop insurance began in the United States in 1938, until 1981, the operations of the FCIC were extremely limited in two ways. First, prior to 1981, the FCIC only insured program commodities such as grains, dairy, and oilseeds, and, second, crop insurance consisted mainly of free disaster coverage. However, 1980 saw the passage of the Agricultural and Food Act, which was meant to replace free coverage with an experimental buy-up insurance, which required participants to pay an insurance premium for coverage and which was to be made available for a much broader variety of crops (beyond commodity crops).

Demand in California for the insurance products offered in the eighties was weak. Demand everywhere was weak—despite subsidies that made the expected return to insurance policies large and positive for the average enrolled producer, only 25 percent of eligible acreage was enrolled by 1988 (Glauber 2004). But because of inadequate data with which to rate policies for specialty crops, insurance products simply didn't exist to cover more than a very small share of agricultural production in California. Figure 4.2 shows a time series of the number of crops for which policies were offered in California by year: in 1981, there were only thirteen such crops (basically, the program crops plus policies for almonds, citrus, grapes, raisins, and tomatoes).

Further, prior to 1985, insurable yields for a particular farm depended on average yields in the county, and adequate data to estimate the distribution of county-level yields even for the small number of insurable crops were limited to a handful of California counties.

After the passage of two ad hoc disaster bills (in 1988 and 1993: Risk Management Agency 2009), Congress passed the Federal Crop Insurance Reform Act of 1994 (FCIRA). The principal goals of the Act were to expand coverage to cover more (especially specialty) crops, and to increase participation by creating a new category of mandatory.[4] Prior to 1994, the insurance policies available offered varied levels of coverage as a function

---

4. A list of specialty and nonspecialty crops can be found in appendix B. More precisely, having at least cat insurance became a criterion for producer eligibility for a range of other federal programs.

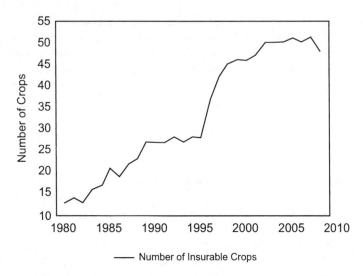

Fig. 4.2   Number of insurable crops in California, by year

of the premium amount paid. The catastrophic (cat) coverage offered in 1994 established a low baseline level of coverage with no premium (though producers were charged a flat nominal administrative fee).[5] The results of this legislative change for use of crop insurance in California can be seen in figure 4.3. In 1995, there was no very large change in demand for the buy-up policies, but a huge increase in demand for the new quasi-mandatory cat policies. This huge increase went a considerable way toward achieving the goal of increasing overall producer participation. However, the increase in participation evident in figure 4.3 for California was almost entirely due to the new mandatory cat insurance—no policy for new California crops was developed by the RMA between 1991 and 1997, at which time programs for apricots and nectarines were developed (see table 4.1).[6]

A second act of Congress, the Federal Agriculture Improvement and Reform Act of 1996 (FAIR), gave the option of forgoing cat insurance in exchange for forfeiture only of eligibility for federal disaster benefits. The Act also created the RMA, whose function was to administer FCIC crop insurance, including researching crops to make insurance available on more crops.

5. Compensation was for "losses exceeding 50 percent of an average yield paid at 60 percent of the price established for the crop for that year."

6. Of the many specialty crops that aren't covered (in at least some counties), some disaster insurance is available based on countywide production, rather than on a given producer's production history. These specialty crops are instead covered by the "Noninsured Crop Disaster Assistance Program," which was also created by the 1994 act.

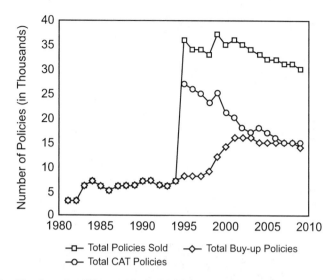

Fig. 4.3   Number of policies sold in California, by category and year

Notable Features of California Agriculture

Among the important agricultural states, California is notable for the very large share of specialty crops in the total value of its agricultural production. As an examination of figure 4.4 makes clear, fruits and vegetables collectively accounted for over half the total value of California agricultural production in 2007, with a collective value of roughly twenty billion dollars. It's not only that the nominal value of fruits and vegetables have been increasing sharply since the 1980s; their share in the total value of California agricultural production has also increased over time and has exceeded half of total value since about 2000. The only other class of agricultural commodities to increase its share over this period of time is dairy, so between figure 4.4 and figure 4.5, we see a picture of increasing specialization, with the three highest value categories of agricultural commodities accounting for an increasing share of total production over time.

What accounts for this increased specialization? The increased specialization evident in these figures occurs over the same period in which insurance for specialty crops is introduced. In a study of program crops, O'Donoghue, Roberts, and Key (2009) find that the expansion of crop insurance associated with the 1994 FCIRA led to modest increases in on-farm specialization, either because producers substituted toward crops whose expected returns increased with the introduction of subsidized insurance or because insurance reduced demand for crop diversification for risk-management reasons. One possibility is that similar mechanisms are at work here and that with

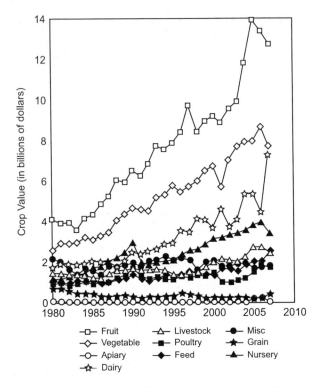

Fig. 4.4   **Total market values for California agricultural production**

the introduction of insurance, the improvement in the (insured) distribution of returns to growing fruits and vegetables led farmers to substitute toward these commodities.

This hypothesis is consistent with figure 4.6, which shows not only a steady increase in the total value of Californian agricultural production over time, but also that this increase in value is essentially entirely attributable to the increase in the value of insurable crops (i.e., crops produced in a county where insurance is available for that crop). So one might be tempted to infer that the expansion of crop insurance to cover specialty crops over this period led to an increase in the value of these crops.

However, this inference is not so straightforward. The problem is that an increasing number of crops became insurable at an increasing number of locations over this period. Furthermore, as discussed in section 4.3, insurance was wasn't randomly assigned to new crop-counties over time; rather, the total value of the crop in a particular location was the key variable that led the RMA to create or expand new programs. So the increase in the value of insurable crops evident in figure 4.6 could easily be entirely a consequence of the way the *supply* of insurance changed over time and not have anything

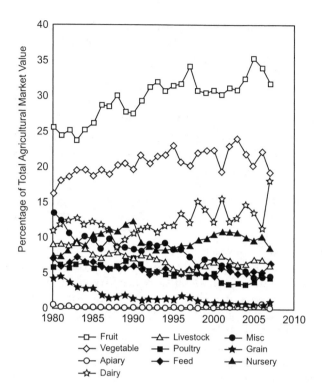

Fig. 4.5 Market shares of California agricultural production

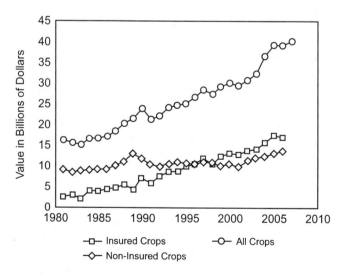

Fig. 4.6 Total market value of California agricultural production

to do with either demand for that insurance or with the effects of insurance on crop specialization or production. Sorting out these different possible reasons for the increase in the value of insurable crops is the central goal of this chapter.

4.3   Supply of Insurance for Specialty Crops in California

We have data on a total of 190 different agricultural commodities. These are all produced in California and result from merging of NASS and RMA data sets. Of these 190, the RMA classifies all but seventeen as "specialty" crops.

There are 173 fruit and vegetable specialty crops grown in California. Of these, twenty-seven are covered by a crop-specific insurance program in one or more California counties.

Table 4.1 shows how new insurance policies are offered for different crops at different times. The numbers that appear in each cell indicate the number of California counties (of which there are fifty-eight in total) for which insurance policies are offered for a given crop. So, for example, we see that insurance for walnuts was first offered in 1985, debuted in ten counties, and by 2007 was offered in twenty-five counties.

If the decision to offer insurance for a particular crop in a particular region was left to competitive firms, each seeking to earn a profit through the development of new policies, then we'd expect the supply of insurance to depend on the equilibrium price. However, for crop insurance in the United States, the decision to offer insurance for a particular crop in a particular county is a bureaucratic one, made not by the insurance firms that sell the product, but rather by the RMA. It's not entirely clear what the objectives of the RMA are, but it does seem clear that maximizing profits from the provision of insurance is not among its principal objectives: the net cost of crop insurance to the U.S. Treasury is well in excess of 3.5 billion dollars per year (General Accounting Office 2007).[7]

Regardless of the RMA's objectives in creating new insurance products, we know quite a lot about their decision rule as they have developed a rather clear procedure for determining whether to offer insurance for a particular specialty crop in a particular region (General Accounting Office 1999, appendix III).

7. It's possible that the RMA weighs the costs of this subsidy against what the costs of disaster relief would be in absence of crop insurance. In 2002, when insured acreage nationwide was roughly 80 percent of the total, with average coverage of roughly 60 percent, Congress allocated $2.1 billion in supplemental disaster assistance. Had the crop insurance program not existed, the costs of providing the same levels of compensation to growers as those that they actually received would have been roughly $4.4 billion; thus, in 2002, the crop insurance program saved the U.S. Treasury about $2.3 billion dollars. But because disasters of this scale seem to occur less often than every other year, it's not at all clear that ad hoc disaster relief is less cost-effective than are existing crop insurance programs.

Table 4.1　　　　　**Number of counties with insurance products for different crops**

First insured	Crop name	1981	1985	1990	1994	1995	1996	1997	2002	2007
1981	Almonds	4	15	16	16	16	16	16	16	16
	Citrus fruit	3	4	8	8	11	12	12	15	15
	Grapes	8	15	16	17	24	26	26	31	31
	Raisins	7	7	6	6	6	6	6	6	6
	Tomatoes	6	14	11	11	15	15	15	14	12
1982	Dry beans		15	15	15	18	16	18	16	3
1984	Walnuts		10	15	16	25	25	25	24	25
1985	Potatoes		2	1	3	6	5	5	5	13
1986	Peaches			11	13	13	13	11	12	11
	Prunes			10	11	12	14	14	14	13
1988	Apples			2	1	9	10	12	17	12
	Figs			3	3	3	3	3	2	2
1989	Pears			3	4	8	7	7	7	5
1990	Fresh market tomatoes			2	3	6	5	6	5	4
	Fresh plums			6	4	7	8	8	7	6
1997	Apricots							11	11	10
	Nectarines							7	7	6
1998	Avocados								6	5
	Sweet potatoes								1	5
1999	Cherries								2	2
	Wild rice								3	3
2000	Strawberries								2	3
2002	Raspberry and blackberry								1	
2003	Nursery									3
	Onions									1
2007	Alfalfa seed									1
2008	Pecans									2

Notes: On the left is the first year the crop was introduced; generally, policies were sold every year following. Selected years afterward (on the right) are simply a snapshot of subsequent years, including more detail around the Federal Crop Insurance Reform Act of 1994 and Federal Agriculture Improvement and Reform Act of 1996. Each entry for a year and crop represents only the number of counties in which policies were sold for that year and crop.

There are three basic criteria that must all be satisfied for a product to be developed. First, the crop must be "economically significant"; second, there must be "producer interest"; and third, offering the product must be "feasible" (General Accounting Office 1999, appendix III).

The FCIC regards a particular crop economically significant in a particular area only if the total market value of the crop is at least one of the following:

1. $3 million in the agricultural statistics district (of which there are nine in California) where it will be covered.
2. $9 million in the state where it will be covered.

3. \$15 million in the RMA administrative region (of which there are ten nationwide).

4. \$30 million nationally.

Producer interest in insurance is considered to be indicated by high levels of noninsured disaster payments as well as recommendations by RMA regional offices. For a pilot program to be initiated, projected producer participation in the program must be at least 10 percent.

Offering an insurance product may be infeasible if, for example, there are inadequate data to evaluate the actuarial soundness of the product; if mechanisms to market the product are lacking; or if the proposed product itself is too complicated (General Accounting Office 1999).

Once the RMA has decided to try to develop a new insurance product, the process of development takes about five years to complete, including two years of feasibility studies and three years to carry out a pilot program.

Operationally, the criteria for economic significance described in the preceding don't offer sufficient guidance about what crops to develop programs for, as very many crops in many locations satisfy those criteria, and the RMA presumably lacks the resources to develop programs for all of these at once.[8] To deal with these constraints, the RMA has developed a list of crops ranked according to market value. We understand from conversations with analysts within the RMA that this list provides primary guidance about what crop to focus on next and that the RMA seldom initiates new programs for more than a single crop per year.

We wish to test the hypothesis that the RMA's decisions regarding what crops to insure in what counties depend on the value of the crop in different counties. Our approach is to model the probability of a policy being offered for a particular crop-county-year. Let d_{ijt} be equal to one if a policy for crop j is offered in county i in year t, and equal to zero otherwise.

We imagine that there are characteristics of counties or crops that are essentially fixed in the short run but that may affect the probability of a crop policy being introduced in that county. Obvious features of counties that could matter include the overall importance of agriculture in that county or the effectiveness of insurance salespeople operating in that particular area. Features of crops that are fixed and may affect the probability of policy introduction may include features of the commodity itself, which may make it infeasible to introduce insurance, or involve commodity-specific grower associations, which are more or less enthusiastic about the introduction of insurance policies for their particular crop (a correspondent at the RMA tells us that lettuce growers in California have resisted the introduction of crop insurance).

---

8. Over the period 1982 to 2008, there has been, on average, less than one new California crop program developed per year, and in no year have there been more than two new crop programs introduced.

Table 4.2   **Factors affecting the probability of new crop insurance programs across different counties**

Variable	Specification			
	(1)	(2)	(3)	(4)
County fixed effects	Yes	Yes	Yes	Yes
Crop fixed effects	No	Yes	Yes	Yes
Year fixed effects	No	No	Yes	Yes
Value rank-year interactions	No	No	No	Yes
Log-likelihood	−14,061.02*	−6,627.50*	−4,382.37*	−4,341.32*
Degrees of freedom	57	162	24	27

*Significant at the 95 percent level.

Let R_{jt} denote the RMA's ranking of the crop value in year t (with the lowest-value crop receiving a ranking of 1). We estimate

(1) $$\text{Prob}(d_{ijt}=1)=\alpha_i+\gamma_j+\eta_t+\left(\sum_{s=1980}^{t}\delta_s\right)R_{jt}+v_{ijt},$$

where the $\{\alpha_i\}$ are a collection of county fixed effects, the $\{\gamma_j\}$ are a collection of crop fixed effects, and the $\{\eta_t\}$ are a collection of year effects. The term $(\sum_{s=1}^{t}\delta_s)R_{jt}$ allows there to be a time varying but cumulative effect of crop ranking on probability of a policy being offered.

We use a logistic model to estimate equation (1), with results reported in table 4.2. Each successive column adds an additional collection of variables and reports the resulting log-likelihood so that column (1), for example, presents a measure of fit for a regression of policy offerings on just a set of county fixed effects, column (2) adds crop fixed effects, and so on. The reported log-likelihood ratios allow us to construct likelihood ratio tests of the null hypothesis that the coefficients associated with the newly added variables are all equal to zero.

Each of the collections of county, crop, and year effects jointly are significant and explain a great deal of variation in whether a policy is offered. Though no individual term in the rank-year interactions is statistically significant, these are collectively extremely important in terms of explaining variation. We interpret this as evidence that even *after* throwing out variation at the level of the county, the crop, and the year that our characterization of the RMA's supply decision is useful in predicting what crop-counties will have insurance products developed for them.

4.4   Effects of Insurance on Output and Prices

The consequences of crop insurance programs for consumer welfare can be presumed to depend on two different channels: first, the cost of the programs to taxpaying consumers; and second, via the effect the programs have on prices and quantities of agricultural commodities purchased by consumers.

It's reasonably straightforward to document the direct costs of FCIC programs for U.S. taxpayers. From the General Accounting Office report cited in the preceding (General Accounting Office 2007), we have a figure of roughly $3.5 billion per year, or roughly $30 dollars per year for each U.S. household. There are numerous elaborations on these costs available in the literature and on estimates of the welfare losses involved in having the government involved in effecting these transfers from taxpayers to producers (e.g., Gardner and Kramer 1986; Wright and Hewitt 1994; Glauber 2004).

In comparison, the literature on the ultimate effects of crop insurance on prices and quantities is surprisingly small, and small relative to the literature on demand for crop insurance or its effects on farmer behavior. Young, Vandeveer, and Schnepf (2001) is an exception: using a computable general equilibrium model, they estimate the effects of crop insurance subsidies on prices and supply of eight program crops. They find a small shift (a 0.4 percent increase in planted acres) toward production of those crops, but because demand for those same program crops is inelastic, prices tend to fall by a much larger proportion. Overall, they compute that the roughly $1.5 billion dollars spent in crop insurance premium subsidies led to an increase in farm income of roughly one billion dollars.

Here, by exploiting variation in the timing of the introduction of crop insurance policies across crops and counties and then combining this with county-level data on prices and output, we're in a position to try to deliver some tentative estimates of the effects of crop insurance on the observable variables most germane to consume welfare. The findings of O'Donoghue, Roberts, and Key (2009) lead us to expect that the introduction of crop insurance programs will, other things equal, lead to some substitution toward the insured crop and, hence, produce an increase in output. Some crops we examine may be produced only in very limited areas and have no close substitutes, and for these crops, we might expect increased supply to have a large effect on price. However, most of the different crops we examine are highly disaggregated and most have close substitutes or can be grown in other counties, states, or countries. Accordingly, we'd expect demands to generally be highly elastic. If this is correct, then increased supply will have at most a modest effect on prices.

We begin by considering a simple reduced-form supply relationship, which takes the form

$$(2) \qquad \log q_{ijt} = \alpha_i + \gamma_j + \eta_t + \beta d_{ijt} + \varepsilon_{ijt},$$

where (as before) the $\{\alpha_i\}$ are county dummies; the $\{\gamma_j\}$ are crop dummies; and the $\{\eta_t\}$ are year dummies. A couple of features of this equation are worthy of note. First, in a supply equation, we'd ordinarily expect prices to feature prominently on the right-hand-side of the equation, but prices do not appear explicitly in equation (2). The reason is that we implicitly assume that prices will vary only across crops, counties, and time, and so any variation in price will be captured by some combination of the dummy variables that appear prominently in equation (2).

Second, the crop dummies are particularly important here as they allow us to avoid the problem that the output of different crops are measured in different units. So long as these incommensurate units (e.g., cartons of mature green tomatoes, pounds of almonds) are unchanging over time, the combination of taking logs and adding crop-specific dummies allows us to compare the dimensionless percentage changes output across crops.

However, the key coefficient of interest for us is β, which captures the effects of introducing crop insurance for a given county-crop on supply. This coefficient can be interpreted as an elasticity—the introduction of insurance for a particular crop in a particular county can be expected to increase production by a factor β.

The problem with estimating equation (2) as it stands, of course, is that the introduction of crop insurance is endogenous. Indeed, making the point that crop insurance depends importantly on observables such as value rank was the main point of section 4.3. However, we can use the results of section 4.3 to address the problem of endogeneity here. In particular, if one were to take the estimates the conditional probabilities of a program being introduced for a given crop-county from the estimation reported in table 4.2, we could treat this as a sort of first stage in a two-stage-least squares estimator of the effects of crop insurance on supply. For this strategy, we would let $\hat{d}_{ijt}$ denote the estimated probability of introduction and then use these estimates in a second stage:

$$(3) \qquad \log q_{ijt} = \alpha_i + \gamma_j + \eta_t + \beta\hat{d}_{ijt} + \varepsilon_{ijt}$$

In effect, the interactions between rank and years that appear in equation (1) would act as instruments for the endogenous introduction of crop insurance.

In practice, using a logit-first stage with a least-squares second stage would make inefficient use of the information contained in the first-stage right-hand-side variables and complicate estimation of standard errors. Accordingly, we adopt a generalized method of movements (GMM) or three-stage-least-squares (3SLS) approach to estimation. A nice consequence of this approach is that because we have more excluded instruments than coefficients to estimate, we can also test the specification and validity of our instruments as well as remain quite agnostic as regards the covariance structure of residuals.[9]

---

9. This flexibility does come at a price: with a full set of rank-year interactions, the GMM optimal weighting matrix can't be reliably estimated using our finite sample. Accordingly, we use a smaller set of decade-rank interactions as excluded instruments in the estimates presented here.

Table 4.3   Estimated average supply response to crop insurance and the average of the reciprocal of price elasticity of demand for crops grown in California

Variable	(1)	(2)
Policy available	1.38*	
	(0.23)	
Policy × nontree		0.50
		(0.49)
Policy × tree		1.64*
		(0.27)
Excluded instruments	d	d
χ^2	11.12	6.80
p-value	(0.05)	(0.13)
Estimator	GMM	GMM

Note: d = decade-rank interactions. The top panel presents generalized method of moments (GMM) estimates of coefficients in column (2). County, crop, and year fixed effects are included but not reported. Figures in parentheses are standard errors.
*Significant at the 95 percent level of confidence.

Our estimate of the value of β in equation (2) appears in the first column of table 4.3. We find that the introduction of insurance for a given crop has a highly significant effect on the quantity supplied—there's no doubt great variation across commodities in terms of this supply response, but our estimate is that, on average, there's a 138 percent increase in output for crops with crop insurance, compared with uninsured crops. However, our ability to test the underlying specification is useful here: we're able to *reject* the hypothesis that our instruments are valid in this specification at a 5 percent level of confidence.

In a search for the reasons for the rejection, it was suggested to us that the effects of crop insurance on supply might differ dramatically between annual and perennial crops, on the logic that sunk costs for perennial crops implied that producers of such crops would have to bear considerably greater risk in the absence of insurance. We explore this idea by introducing an interaction between whether a crop was a "tree crop" or not.[10]

The second column of table 4.3 shows the result. In place of an indicator for "Policy available," we interact a pair of indicator variables for tree and nontree crops with the policy availability indicator; the pair of coefficients associated with these variables then become estimates of the elasticities we seek. And, indeed, introducing this heterogeneity of response makes a remarkable difference. For tree crops, the estimated supply elastic-

10. Not all California crops can be unambiguously identified as annual or perennial because, for some crops, this depends on location. However, all tree crops are perennial, and so we've used this as a method of distinguishing effects.

ity increases somewhat, to 164 percent. But for nontree crops, the estimate (while still positive) is not statistically distinguishable from zero.

The estimated elasticity of 1.64 is not very precisely estimated—a 95 percent confidence interval about this estimate is [1.11, 2.17]. But, even if imprecise, the elasticity tells us that in counties where crop insurance for tree crops was introduced it stimulated a doubling or tripling of production over the twenty-seven years for which we have production data. But note that this should *not* be interpreted as evidence of an overall increase in output across all crops—equation (2) doesn't allow us to distinguish between increases in total output across crops and substitution between crops. It's entirely possible that the introduction of subsidized insurance actually leads producers to substitute away from higher-value crops (or perhaps lower-value crops better suited to a particular farm), reducing the total value of production.[11]

We won't pursue the issue of the effects of crop insurance on the total value of agricultural output here for want of data (our analysis here relies heavily on variation across crops, and so aggregating across these has a high cost in terms of both the statistical power and size of any tests we might conduct). Instead, we'll return to a consideration of the demand side, on the grounds that any positive effect of crop insurance on consumer surplus must come via a reduction in the prices of insured commodities.

Accordingly, we specify an inverse demand function for produce of type j from county i in year t according to

$$(4) \qquad \log p_{ijt} = \alpha_i + \gamma_j + \eta_t + \beta \log q_{ijt} + \varepsilon_{ijt}.$$

There's some abuse of notation here, as we're reusing variables that entered the supply equation (2). Hopefully, context makes it clear that these are all in fact different quantities. Only the quantity supplied q_{ijt} is common across equations (2) and (4). As in our specification of the supply equation, we have a set of county fixed effects, a set of crop fixed effects, and a set of time effects.

As in equation (2), the crop dummies γ_j are critical allowing us to make comparisons of price across crops measured using different units. The time effects play an even more important role here than previously because they capture changes over time in the value of the dollar—we've left the values of prices p_{ijt} in nominal terms, so that the $\{\eta_t\}$ terms capture the effects of inflation on prices.

In this case, the key variable of interest is quantities—what we'd like to

---

11. A casual investigation of this hypothesis involves substituting total crop value for total crop production in equation (2). This yields estimates suggesting that the introduction of insurance results in an increase in tree-crop value smaller than the increase in tree-crop production, while the value of nontree crops actually falls significantly in response to the introduction of insurance. However, tests of the overidentifying restrictions result in a rejection of this specification; further investigation is left for future research.

know is how changes in the quantity supplied affect prices. But, of course, these quantities are endogenous—if we didn't already know this from examination of equation (2), we could see that we're contending with the classic problem of separately identifying supply and demand relationships. But our estimation of equation (2) suggests a strategy to address this endogeneity: by using predicted values of (log) quantities from equation (2) in place of actual quantities, we obtain

(5) $\log p_{ijt} = \alpha_i + \gamma_j + \eta_t + \beta \widehat{\log q}_{ijt} + \varepsilon_{ijt}$

Then estimates of the coefficient β can be interpreted as the average of the reciprocal of price elasticities (thus, values close to zero imply high elasticities). Because this is the only parameter of interest, we don't need a table to report it: we estimate an inverse demand elasticity of –0.056, with a standard error (computed using the heteroskedasticity-consistent method of White [1980] of 0.003). Thus, we find a negative elasticity, consistent with the law of demand, and significantly different from zero. Indeed, our estimate is reasonably precisely estimated—a 95 percent confidence interval about the estimate is [–0.050, –0.062], suggesting that demand is quite elastic. This is consistent with the hypothesis that such highly disaggregated commodities are likely to permit a great deal of substitution. As before, recall that this is an *average* reciprocal elasticity—for commodities that are only grown in a few counties in California or which possess no close substitutes, price elasticity may be much smaller.

4.5 Conclusion

In this chapter, we've gathered evidence on the process by which crop insurance programs are created and used this evidence to estimate the supply of crop insurance programs across counties, crops, and years. We've found that an administrative rule that gives priority to crops with the highest ranking value has considerable predictive power, though crop- and county-specific variables also play an important role.

We've used our predictions regarding the introduction of crop insurance to deal with issues of the endogeneity of the supply of crop insurance programs and estimate the effects of the introduction of crop insurance programs on both the supply of and demand for different crops.

Our estimates regarding the effects of crop insurance on the supply of and demand for insured crops indicate that effects differ for tree and nontree crops, perhaps as a consequence of the much larger investments at risk in crops of the former type. We find what we think is a rather large effect on supply for tree crops but no significant effect for nontree crops. However, we can't say whether the large effect of insurance on tree-crop production is principally due to more efficient production or substitution away from other crops.

We find a significant negative effect of crop insurance on prices for insured crops, though the magnitude of the effect is small compared to the effect on supply (around −5 to −6 percent). This last finding is consistent with the view that demand for such highly disaggregated commodities is likely to be highly elastic. A consequence is that crop insurance for these specialty crops has little benefit for consumers, *even* if it generates a large supply response.

Appendix A
Detailed Data Description

The production and insurance data obtained from the NASS and RMA websites are organized differently.[12] First, the production data (which date from 1980) use a unique commodity, county and year as the unit of observation, while the insurance data group data by crop, year, county, and insurance plan. Second, the RMA definitions of crops are less specific (and broader reaching) than the NASS definitions; thus, there are many more production commodities than insurance crops. (See figure 4A.1.)

Production

Output information is reported as acres harvested, tons produced, and total market value, as appropriate for the commodity type (animal commodities, for example, only include information for total market value). The number of counties with production data stayed primarily constant year over year, ranging from fifty-seven counties (1980 to 1988) to fifty-nine counties (2004 to 2007).

Production Types

The raw data have been further organized by an external classification by broad production type (see figure 4A.2):

1. Fruit
2. Vegetable
3. Apiary
4. Dairy
5. Livestock
6. Poultry
7. Feed
8. Miscellaneous
9. Grain
10. Nursery

---

12. This appendix was written with Alana LeMarchand. Additional details and discussion may be found in her Berkeley undergraduate honors thesis of 2009.

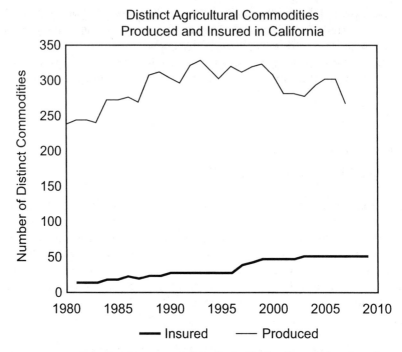

Fig. 4A.1   Number of distinct commodities in the California agricultural market, as defined by the National Agricultural Statistical Service (NASS)

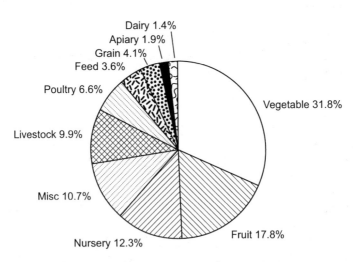

Fig. 4A.2   Distinct agricultural commodity names in California, as defined by the National Agricultural Statistics Service (NASS), grouped by type

There are many more unique commodities in the fruit and vegetable categories than in the other categories although this is not necessarily related to the actual aggregate market value of goods of different types. Analysis of the share of actual market value of each production category indicated that the number of commodities in each category is not correlated with market share.

Insurance

While the insurance data include such supplementary information as premium and coverage level, the most pertinent information is which commodities are insured and the type of insurance plans offered. The total number of commodities insured since 1989 is sixty-three, but there have never been more than fifty-one commodities insured in a single year. The number of insured crops began at twenty-three and increased with time, including an abrupt jump in the year 1997 (twenty-eight crops in 1996, thirty-seven crops in 1997).

Insurance Plan Types

There are seven insurance plan types offered. The following description is adapted from material available on the RMA website, including information for less-traditional pilot programs.[13]

AGR: Adjusted Gross Revenue: insures revenue of the entire farm rather than an individual crop by guaranteeing a percentage of average gross farm revenue, including a small amount of livestock revenue. The plan uses information from a producer's Schedule F tax forms and current year expected farm revenue to calculate policy revenue guarantee.

APH: Actual Production History: insure producers against yield losses due to natural causes such as drought, excessive moisture, hail, wind, frost, insects, and disease. The farmer selects the amount of average yield he or she wishes to insure; from 50 to 75 percent (in some areas to 85 percent). The farmer also selects the percent of the predicted price he or she wants to insure; between 55 and 100 percent of the crop price established annually by the RMA. If the harvest is less than the yield insured, the farmer is paid an indemnity based on the difference. Indemnities are calculated by multiplying this difference by the insured percentage of the established price selected when crop insurance was purchased.

ARC: Avocado Revenue Coverage: pilot since 1998.

ARH: Actual Revenue History: pilot for dry beans in 2009.

CRC: Crop Revenue Coverage: provides revenue protection based on price and yield expectations by paying for losses below the guarantee at the higher of an early-season price or the harvest price.

13. See http://www.rma.usda.gov/policies/ and http://www.rma.usda.gov/pilot/2010pilot .html.

DOL: Dollar Plan: provides protection against declining value due to damage that causes a yield shortfall. Amount of insurance is based on the cost of growing a crop in a specific area. A loss occurs when the annual crop value is less than the amount of insurance. The maximum dollar amount of insurance is stated on the actuarial document. Amount of insurance is based on the cost of growing a crop in a specific area. A loss occurs when the annual crop value is less than the amount of insurance. The maximum dollar amount of insurance is stated on the actuarial document.

GRP: Group Risk Plan: policies use a county index as the basis for determining a loss. When the county yield for the insured crop, as determined by the NASS, falls below the trigger level chosen by the farmer, an indemnity is paid. Payments are not based on the individual farmer's loss records. Yield levels are available for up to 90 percent of the expected county yield. GRP protection involves less paperwork and costs less than the farm-level coverage described in the preceding.

PRV: Pecan Revenue: since 2005, began as a pilot.

Qualitative Distribution of Plans in the Data

AGR: Adjusted Gross Revenue: This plan is not crop specific and applies only to the entire production of a farm.

ARH: Actual Revenue History: This plan is sold only beginning in 2009 as a pilot for dry beans.

GRP: Group Risk Plan: This plan is indexed on county production and comprises an insignificant percentage of policies sold.

APH: Actual Production History: This plan is by far the most common plan type and is linked most directly with production volume.

CRC: Crop Revenue Coverage: This plan protects a farmer's crop based on yield and price. It is also more significant in terms of numbers than the AGR, ARH, or GRP plans.

DOL: Dollar Plan: This plan protects against yield shortfall below a certain dollar amount. It is the second most common plan, after the APH.

PRV: Pecan Revenue: This plan applies only to pecans and could only be useful in regressions where policies are linked specifically to crops.

ARC: Avocado Revenue Coverage: This plan applies only to avocados and could only be useful in regressions where policies are linked specifically to crops.

Graphical Presentation of Insurance Plan Distribution

Figure 4A.3 presents data on premiums, liabilities, indemnities, and subsidies for each RMA insurance plan category. Raw data are included below each bar chart in figure 4A.3. "Premium" and "Net Reported Acres" are scaled so as to be more readable.

Figure 4A.4 simply indicates the number of policies offered by plan. It is

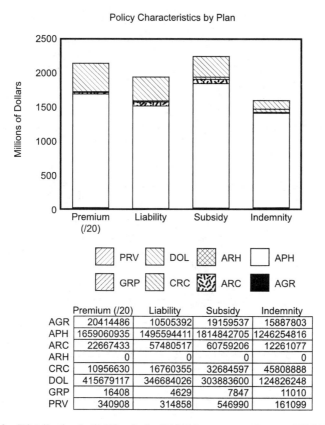

Fig. 4A.3   Distribution in California, by Risk Management Agency (RMA) insurance plan category, of cumulative monetary value of total premiums, liabilities, subsidies, and indemnities

clear from this data that traditional APH policies comprise the great majority of RMA insurance plan activity, with the DOL plan a very distant second. After that, the most significant share of policies comes from the AGR, ARC, and CRC plans. As mentioned, AGR insurance is not crop specific and, thus, is inappropriate for a crop-specific analysis; ARC insurance is only for avocados; CRC is crop specific and applicable to many different crops. ARH, GRP, and PRV are insignificantly small. However, ARC, ARH, and PRV plans may be included in regressions where policies are linked specifically to crops. They might also be studied later on for their influence on the avocado, dry bean, and pecan markets, respectively.

Production-Insurance Correspondence

As shown in the preceding, there are many more production commodities than there are insurance crops. This is due in part to the nature of the insur-

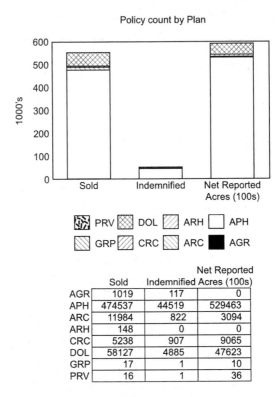

Fig. 4A.4   Cumulative number of policies of each category sold in California since 1980

ance crop designation (more general, spanning several production commodities) and also in part to the fact that many crops are not insured. Correspondences between production and information have been established using the crop and commodity names of each respective data set. There are 100 production commodities found to correspond to fifty-six insurance crops.

All insurance crop designations encompass one or more production commodity designation, except in a few fruit crops. Tangelos, plums, and apricots have two insurance crop identities corresponding to a single production commodity (usually due to a distinction between fresh and processing grade fruit).

The other notable aspect of the link created between production and insurance information is that there are seven commodities that could not as of yet be linked with production commodities. This is due to ambiguous categories definitions (i.e., four types of insurance categories and six types of production categories for oranges). These unlinked insurance commodities

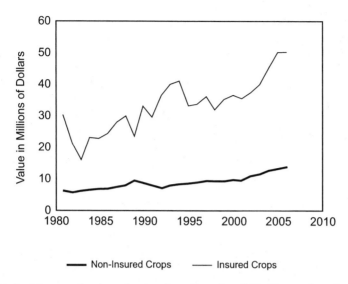

Fig. 4A.5   Mean production value (total market value divided by number of crops) for California agricultural production

include special citrus, processing beans, nursery (container), AGR, stone-fruit, and oranges.

This correspondence permits the comparison of production of insured crops to production of uninsured crops. Figure 4A.5 shows that mean production value of insured crops is above that of uninsured crops over the entire thirty-year period analyzed; overall growth of market value of insured crops is also greater than for uninsured crops (although this may not necessarily be true for percent growth).

Appendix B

List of Specialty and Nonspecialty Crops

Specialty Crops

Almonds
Apples
Avocado/mango trees (Florida)
Avocados
Blueberries
Canning beans
Citrus trees

Citrus
Cranberries
Dry beans
Dry peas
Figs
Florida fruit trees
Grapes (table)
Grapes (wine)
Green peas
Macadamia nuts
Macadamia trees
Nursery
Onions
Peaches
Pears
Pecans
Peppers (fresh)
Plums
Popcorn
Potatoes
Prunes
Raisins
Stonefruit
Sweet corn (fresh)
Sweet corn (processing)
Tomatoes (fresh)
Sweet potatoes
Tomatoes (processing)
Walnuts

Nonspecialty Crops

Barley
Canola
Corn
Cotton
Extra long staple cotton
Flaxseed
Forage production
Forage seeding
Grain sorghum
Hybrid corn seed
Millet
Peanuts
Rice

Rye
Oats
Safflower
Soybeans
Sugarbeets
Sugarcane
Sunflowers
Hybrid sorghum seed
Tobacco
Wheat

Appendix C
Integration of RMA Crop Value List

One such list assembled by the RMA using crop value data from 2005 and 2006 was made available by the RMA correspondent for the previously stated research purposes.[14] The list included information on crops at all stages in the insurance policy process, from those at full regulatory status (those already insured) to those not yet being considered for new policies, and everything in between. However, because the informal list contained data for uninsured crops as well as for insured crops, many crops could not be identified with the unique FCIC crop codes that have been previously used to organize crop information in the research database and to define a correspondence between RMA policy information and NASS production and price information. Indeed, no numeric identifiers were used at all in the list provided. In addition, there were several critical discrepancies and complications that must be resolved before integrating the list information into the database:

- The national crop values reported do not correspond to NASS nationwide reported crop values (only a few were checked, and some were off by 50 percent but not by orders of magnitude). Because the list is significant for this research as an indicator of relative crop value as considered by the RMA, this may not be considered significant.
- One third of the listed crops are missing crop value information for 2006. According to the RMA correspondent, the incompletion of some columns can be considered insignificant. In this case, it may be preferable to use only the 2005 crop value data in order to generate a relative ranking of crops by value.

---

14. This section drawn from the thesis of Alana LeMarchand.

- Three of the crops contained no crop value data whatsoever (for either year). These crops were chicory, collard greens, and kale. The latter two crops are grown in California, so it remains to be determined whether these crops should be thrown out of the list. For now, they will be dropped from the list as insignificant in determining rank by crop value because they comprise less than 2 percent of the 163 observations.
- A few high-value crops were aggregated in the list. Notably, citrus fruit (all oranges, grapefruit, etc.), citrus trees (a pilot in Florida), dry beans (limas, red, navy, etc.), and floriculture (all nonbulb flowers). To appropriately integrate this data into a new table in the existing database, all crops corresponding to each of these categories would need to have the same ranking (or to be aggregated as a single crop to reflect the RMA's consideration of them as a single crop. This is generally typical of RMA reporting compared to NASS reporting: an RMA policy of a certain general crop name will generally correspond to apply to several NASS commodities. It is important to note, however, that there were crops that were subject to aggregation even among varying RMA crop policies, namely citrus fruit and peaches.
- Several crops in the nationwide list are not grown in California and, thus, are not present in the current database. Because these crops will not be significant in the research beyond determining a nationwide crop value rank, they will not be tied or added to the current FCIC and NASS crop lists in the database. Their crop code will be marked null in the database, indicating that they are not California crops.

An initial version of the list has been generated using the preceding modifications. For simplicity's sake, we create a third unique identifier in addition to the NASS and FCIC codes in order to capture the aggregation described in the preceding.

Correspondences were simple to make in most cases, but the following is a list of crops with problematic correspondences, primarily due to lack of specificity of which NASS crops are represented by these RMA crop names because the RMA uses different crop nomenclature than the NASS does (see table 4C.1).

The resulting data have been inserted as three tables with the following fields into the database (see table 4C.2).

The ID field in the first two columns represents the aggregation solution discussed in the preceding. The rank list table may be used to generate crop value rankings (as a temporary auto incremented and indexed table with a MySQL query) based on RMA status, in the event that it would be useful to include or to exclude certain status categories (such as "regulatory," which signifies crops already fully insured).

Table 4C.1

Risk Management Agency crop name	National Agricultural Statistics Service	
	names	codes
Corn	Corn grain	111991
Forage	Pasture forage miscellaneous	194799
Forage seeding	Hay alfalfa, hay green chop	181999, 195299
Hybrid seed corn	Corn seed	171119
Silage: corn/sorghum	Corn silage, sorghum silage, silage	111992, 114992, 195199
Sweet corn (processing, instead of fresh)	Not distinct from "fresh" in NASS data	Null
Trees	No evident correspondence	Null

Note: Null indicates crop not grown in California.

Table 4C.2

	Table Name	
rank_ID	rank_list	rank_status
ID	ID	codep
ins_code	status	name
prod_code	value	
	name	

References

Gardner, B., and R. Kramer. 1986. "Experience with Crop Insurance Programs in the United States." In *Crop Insurance for Agricultural Development,* edited by P. Hazell, C. Pomareda, and A. Valdes, 211–12. Baltimore: Johns Hopkins University Press.

General Accounting Office. 1999. *Crop Insurance: USDA's Progress in Expanding Insurance for Specialty Crops.* GAO/RCED 99-67, April. Report to the Ranking Minority Member, Subcommittee on Risk Management, Research, and Specialty Crops, Committee on Agriculture, House of Representatives. Washington, DC: General Accounting Office.

———. 2007. *Crop Insurance: Continuing Efforts Are Needed to Improve Program Integrity and Ensure Program Costs Are Reasonable.* GAO 07-819T, May. Washington, DC: General Accounting Office.

Glauber, J. 2004. "Crop Insurance Reconsidered." *American Journal of Agricultural Economics* 86 (5): 1179–95.

O'Donoghue, E., M. Roberts, and N. Key. 2009. "Did the Federal Crop Insurance Reform Act Alter Farm Enterprise Diversification?" *Journal of Agricultural Economics* 60 (1): 80–104.

Risk Management Agency. 2009. *A History of the Crop Insurance Program.* Washington, DC: Risk Management Agency.

White, H. 1980. "A Heteroskedastic-Consistent Covariance Matrix Estimator and a Direct Test for Heteroskedasticity." *Econometrica* 48:817–38.

Wolf, S., B. Hueth, and E. Ligon. 2001. "Policing Mechanisms in Agricultural Contracts." *Rural Sociology* 66 (3): 359–82.

Wright, B. D., and J. A. Hewitt. 1994. "All-Risk Crop Insurance: Lessons from Theory and Experience." In *Economics of Agricultural Crop Insurance: Theory and Evidence,* edited by D. L. Hueth and W. H. Furtan, 73–112. Norwell, MA: Kluwer Academic.

Young, C. E., M. L. Vandeveer, and R. D. Schnepf. 2001. "Production and Price Impacts of U.S. Crop Insurance Programs." *American Journal of Agricultural Economics* 83 (5): 1196–1203.

5

Risk Response in Agriculture

Jeffrey LaFrance, Rulon Pope, and Jesse Tack

5.1 Introduction

Farm and food policies affect crop acres, asset management, intensive and extensive margin decisions, and risk management choices in agricultural production. For example, in 1991, less than 25 percent of cropland (82 million acres) was covered by a federally subsidized crop insurance contract, with $11.2 billion in total liability, $740 million in insurance premiums, premium subsidies of 25 percent ($190 million) of gross farm premiums, and total indemnity payments of $955 million. Relative to premiums paid by farmers ($550 million), for each $1.00 in premiums paid by the typical insured farmer, $1.75 in indemnity payments were received.

Even with this relatively profitable insurance program, farmer participation rates remained quite low. This outcome is likely due to the race to the bottom problem in a pooling equilibrium (LaFrance, Shimshack, and Wu 2000, 2001, 2002, 2004). However, Congress responded to the *appearance* of an incomplete insurance market with increased subsidies and many new forms of insurance.

The 1996 Federal Agricultural Improvement and Reform Act and the amendments to the 1938 Federal Crop Insurance Act that are commonly known as the Agricultural Risk Protection Act of 2000 mandated higher subsidy rates, the development and marketing of new insurance products for virtually every crop and livestock product produced in the United States, and

Jeffrey LaFrance is the Distinguished Professor of Agricultural and Resource Economics at Wahington State University, professor at the Paul G. Allen School for Global Animal Health at Washington State University, and professor of agricultural and resource economics at the University of California, Berkeley. Rulon Pope is professor of economics at Brigham Young University. Jesse Tack is assistant professor of agricultural economics at Mississippi State University.

substantial subsidies for crop insurance marketing firms and large private reinsurance companies.

This change in farm policy greatly expanded the federal crop insurance program. In 2003, the Federal Crop Insurance Corporation (FCIC) provided insurance products for more than 100 crops on 217 million acres (2/3 of all cropland). The total insurance liability was $40.6 billion, with $3.4 billion in insurance premiums, subsidies of almost 60 percent of gross premiums ($2.0 billion), and total indemnity payments of $3.2 billion. The current program includes subsidy payments to private companies marketing federal crop insurance equal to 24.5 percent of gross premiums for administration and oversight (A&O), and to private reinsurance companies equal to 13.6 percent of gross premiums. Reinsurance companies also have the right to sell up to 50 percent of their contracts back to the FCIC (that is, to the taxpayer) at cost. The FCIC's Risk Management Agency's (RMA) book of business shows that 20 percent of the insured farms account for nearly 80 percent of indemnity payments. This suggests substantial adverse selection, as well as moral hazard, because the majority of the federally subsidized crop insurance products calculate premiums based on deviations from county-level yield trends. That is to say, FCIC insurance products are based on a pooling equilibrium established at the county level and, in some cases, larger areas known as risk regions.

The net effect is that for each $1.00 in premiums actually paid by farmers, they receive an average of $2.40 in indemnity payments, insurance marketing firms receive $0.40 in A&O subsidies, and reinsurance companies make in the neighborhood of $0.45 in profit due to the combined direct subsidies on premiums and their reinsurance rights with the FCIC, which allow them to cream, or high grade, the insurance pool.

In 2004, the RMA issued a request for proposals to develop subsidized pasture and range insurance for 440 million acres of private, public, and Native American pasture and rangeland in the country. Many agricultural economists at land grant universities across the country actively consult with the RMA and private insurance companies to develop new and expand existing federally subsidized crop insurance products.

Although this is only one example of the ubiquitous nature of federal intervention in U.S. agriculture, there is a large literature on the impacts of subsidized crop insurance on variable input use and the intensive margin (Nelson and Loehman 1987; Chambers 1989; Quiggin 1992; Horowitz and Lichtenberg 1994; Smith and Goodwin 1996; Babcock and Hennessy 1996). The effects of subsidized crop insurance programs on the extensive margin also has been the subject of considerable analysis (Gardner and Kramer 1986; Goodwin and Smith 2004; Keeton, Skees, and Long 1999; and Young et al. 2000; LaFrance, Shimshack, and Wu 2000, 2001, 2002, 2004), all of which conclude that subsidized crop insurance results in additional planting of marginal crop acres. Williams (1988), Turvey (1992), Wu (1999), and

Soule, Nimon, and Mullarkey (2000) examine the impacts of subsidized crop insurance on choices of crop mixes and acreage decisions. Empirical results in this component of the literature suggest that economically marginal land also is environmentally marginal. These results all suggest that subsidized crop insurance tends to increase environmental degradation. Even so, very little of the previous work in this area uses structural models or takes into account the dynamic nature of agricultural decision making under risk.

To better understand these and many other longstanding issues in U.S. agricultural policy, this chapter develops a comprehensive structural econometric model of variable input use; crop mix and acreage choices; investment and asset management decisions; and consumption, savings and wealth accumulation in a stochastic dynamic programming model of farm-level decision making over time. This model develops and establishes clear and intuitively appealing relationships between dynamic life-cycle consumption theory, the theory of the competitive firm subject to risk, and modern finance theory.

We present, discuss, and apply a new class of variable input demand systems in a multiproduct production setting. All of the models in this class can be estimated with observable data; are exactly aggregable; are consistent with economic theory for any von Neumann-Morgenstern expected utility function; and can be used to nest and test exact aggregation, economic regularity, functional form, and flexibility. Implications of monotonicity, concavity in prices, and convexity in outputs and quasi-fixed inputs are developed for a specific subset of this class of models. We then apply this to thirteen variable inputs in U.S. agriculture over the sample period 1960 to 1999.

The results obtained from this empirical variable cost model are used to help develop a structural model of the dynamic decision problems faced by a generic agricultural producer. In this life-cycle model of agricultural decisions under risk, farmers create income and wealth through savings, investment in risky financial assets, own-labor choices both on- and off-farm, and agricultural production and investment activities. This disciplines the economic theory of agricultural production over time and under risk and helps to better identify risk preferences and other model parameters.

While it is beyond the scope of this chapter, one could solve the system of arbitrage equations derived in section 5.4 for optimal farmland, capital, and share allocations and use the estimated parameters to simulate the effects of a variety of different agricultural policy instruments. For example, one could use the parameter estimates to investigate the effect of policies targeting lower food prices, taking into account the supply lag driven by partial adjustment of land and capital over time. In addition, one could use these estimates to investigate how far the extensive margin will expand or contract in response to a variety of policy scenarios including subsidization of corn for ethanol, an increase in the variety of subsidized crop insurance products, and the introduction of new revenue support programs such as the Average Crop Revenue Election (ACRE) program.

Our empirical findings have important implications for the agricultural economy and associated policy instruments. The majority of theoretical and empirical agricultural policy analysis assumes curvature of the farm-agents utility function and the inability of these agents to invest in off-farm revenue opportunities when making on-farm production decisions. While our empirical findings do find evidence of utility function curvature, we also find that this function is much flatter than is typically assumed or estimated. Importantly, this finding suggests that previous studies of risk-reducing policy instruments have likely overstated on-farm impacts. We also find evidence that farmers respond to off-farm revenue opportunities, which has important implications for the current debate on the subsidization of corn for ethanol. Given the role of the general economy in providing farm-agents with off-farm revenue opportunities, this finding suggests that farm-level impacts of ethanol subsidization have been affected by the recent downturn in the economy. This is a subtle point that is not addressed in the literature surrounding this debate but implies that analysis conducted prior to the downturn is no longer relevant.

5.2   The Production Model and Two Results

Five longstanding questions in economics, econometrics, and agricultural economics are the choice of functional form; the degree of flexibility; the conditions required for and regions of economic regularity; consistency with aggregation from micro- to macro-level data; and how best to handle simultaneous equations bias, errors in variables, and latent variables in a structural econometric model. In this chapter, we attempt to deal with all of these issues in a coherent framework for the analysis of a life-cycle model of agricultural production, investment, consumption, and savings decisions.

Analysis of multiproduct behavior of firms is common in economics (Färe and Primont 1995; Just, Zilberman, and Hochman 1988; Shumway 1983; Lopez 1983; Akridge and Hertel 1986). A large literature on functional structure and duality guides empirical formulations and testing based on concepts of nonjointness and separability (Lau 1972, 1978; Blackorby, Primont, and Russell 1977, 1978; Chambers 1984). Nonjoint production processes reduce to additivity in costs (Hall 1973; Kohli 1983). Separability in a partition of inputs or outputs often results in separability in a similar partition of prices (Blackorby, Primont, and Russell 1977; Lau 1978).

The neoclassical model of conditional demands for variable inputs with joint production, quasi-fixed inputs, and production and output price risk is

(1) $\qquad x(w, \bar{y}, z) = \arg \min \{w^\top x : F(x, \bar{y}, z) \leq 0\}$

where $x \in \mathcal{X} \subseteq \mathbb{R}_+^{n_x}$ is an n_x-vector of variable inputs, $w \in \mathcal{W} \subseteq \mathbb{R}_+^{n_x}$ is an n_x-vector of variable input prices, $\bar{y} \in \mathcal{Y} \subseteq \mathbb{R}_+^{n_x}$ is an n_y-vector of planned outputs, $z \in \mathcal{Z} \subseteq \mathbb{R}_+^{n_x}$ is an n_z-vector of quasi-fixed inputs.[1] $F : \mathcal{X} \times \mathcal{Y} \times \mathcal{Z} \to \mathbb{R}$ is the joint production transformation function, which is the boundary of a closed and convex production possibilities set that is characterized by free disposal in inputs and outputs. Let the variable cost function be denoted by $c(w, \bar{y}, z) \equiv w^\top x(w, \bar{y}, z)$. We assume throughout that the production process is subject to supply shocks of the general form

$$(2) \qquad y = \bar{y} + h(\bar{y}, z, \varepsilon), \; E[h(\bar{y}, z, \varepsilon) \mid x, \bar{y}, z] = \mathbf{0}.$$

In either a static or a dynamic setting, it is a simple matter to show that equation (1) is implied by equation (2) and the expected utility hypothesis for *all* von Newman-Morgenstern preferences (Pope and Chavas 1994; Ball et al. 2010).

Planned output is a vector of latent, unobservable variables in production with supply risk. Hence, to estimate the demand system in equation (1) directly, one must either identify and estimate the expectations formation process or address the errors in variables problem associated with using y in place of $\bar{y}$ in the demand equations (Pope and Chavas 1994). One branch of the literature advocates specifying an ex ante cost function where planned output is replaced by cost, which is observable when the variable inputs are committed to the production process (Pope and Chavas 1994; Pope and Just 1996; Chambers and Quiggin 2000; Chavas 2008; Ball et al. 2010; LaFrance and Pope 2010). In a joint production process, this requires making assumptions such that the input demands are functions of input prices, the levels of quasi-fixed inputs, and the variable cost of production,

$$(3) \qquad x(w, \bar{y}, z) = \tilde{x}[w, z, c(w, \bar{y}, z)].$$

This approach makes particular sense in agriculture where outputs and output prices are observed ex post. The main result of LaFrance and Pope (2010) on this question is as follows (a proof of this result is presented in appendix A of this chapter).

PROPOSITION 1. *The following functional structures are equivalent:*

$$(4) \qquad x(w, \bar{y}, z) \equiv \tilde{x}[w, c(w, \bar{y}, z), z];$$

$$(5) \qquad c(w, \bar{y}, z) \equiv \tilde{c}[w, z, \theta(\bar{y}, z)];$$

$$(6) \qquad F(x, \bar{y}, z) \equiv \tilde{F}[x, z, \theta(\bar{y}, z)].$$

---

1. In this section, we use $\bar{y} \in \mathbb{R}_+^{n_y}$ to denote the n_y-vector of planned or expected outputs to simplify notation. In later sections, we modify this notation to $\bar{Y} = a \cdot \bar{y}$, where a is the n_y-vector of acres planted to crops, $\bar{y}$ now is the n_y-vector of expected yields, and $\cdot$ is the Hadamard product. We also define z explicitly in the following.

In other words, outputs must be weakly separable from the variable input prices in the variable cost function. This, in turn, is equivalent to outputs being weakly separable from the variable inputs in the joint production transformation function.

This is a tight result—separability is both necessary and sufficient for the variable inputs to be estimable in ex ante form. Hereafter, we will call any such demand model an *ex ante joint production system.*

A second common issue in the empirical analysis of agricultural supply decisions is that some level of aggregation is virtually unavoidable. Micro-level data needed to study input use, acreage allocations, and asset management choices at the farm level do not exist. Aggregation from micro-level decision makers to macro-level data has been studied extensively in consumer theory.[2] This has received less attention in production economics (Chambers and Pope 1991, 1994; Ball et al. 2010; LaFrance and Pope 2008, 2010).

Recently, LaFrance and Pope (2009) obtained the indirect preferences for all exactly aggregable, full rank systems of consumer demand equations. Their result extends directly to production in the following way. Let $K \in \{1,2,3,4\}$ and define the smooth real-valued function, $\omega: \mathbb{R} \times \mathbb{R} \to \mathbb{R}$, by

$$(7) \quad \omega[\eta(w),\theta] = \begin{cases} \theta, & \text{if } K = 1,2 \text{ or } K = 3 \text{ and } \lambda'(s) = 0, \\ \theta + \int_0^{\eta(w)} [\lambda(s) + \omega(s,\theta)^2] ds, & \text{if } K = 3,4, \text{ and } \lambda'(s) \neq 0, \end{cases}$$

subject to $\omega(0,\theta) = \theta$ and $\partial\omega(0,\theta)/\partial s = \lambda(0) + \theta^2$, where $\eta : \mathcal{W} \to \mathbb{R}$ and $\lambda : \mathbb{R} \to \mathbb{R}$ are smooth, real-valued functions, and η is $0°$ homogeneous. A class of full rank and exactly aggregable ex ante production systems can be characterized as follows.[3]

PROPOSITION 2. *Let* $\pi : \mathcal{W} \to \mathbb{R}_{++}$, $\pi \in \mathcal{C}^\infty$, *be strictly positive valued, increasing, concave, and* $1°$ *homogeneous; let* $\eta : \mathcal{W} \to \mathbb{R}_+$, $\eta \in \mathcal{C}^\infty$, *be positive valued and* $0°$ *homogeneous; let* $\alpha, \beta, \gamma, \delta : \mathcal{W} \to \mathbb{C} = \{a + \imath b, a, b \in \mathbb{R}\}$, $\alpha, \beta, \gamma, \delta \in \mathcal{C}^\infty$, *be* $0°$ *homogeneous and satisfy* $\alpha\delta - \beta\gamma = 1, \imath = \sqrt{-1}$; *and let* $f: \mathbb{R}_{++} \to \mathbb{C}, f \in \mathcal{C}^\infty$, *and* $f' \neq 0$. *Then the variable cost function for any full rank, exactly aggregable, ex ante joint production system is a special case of*

---

2. An important subset of the literature on this topic includes Gorman (1953, 1961, 1981); Muellbauer (1975, 1976); Howe, Pollak, and Wales (1979); Deaton and Muellbauer (1980); Jorgenson, Lau, and Stoker (1980, 1982); Russell (1983, 1996); Jorgenson and Slesnick (1984, 1987); Lewbel (1987, 1988, 1989, 1990, 1991, 2003); Jorgenson (1990); Diewert and Wales (1987, 1988); Blundell (1988); van Daal and Merkies (1989); Jerison (1993); Russell and Farris (1993, 1998); Banks, Blundell, and Lewbel (1997), LaFrance et al. (2002); LaFrance (2004); LaFrance, Beatty, and Pope (2006); and LaFrance and Pope (2009). The focus in the literature has been interior solutions and smooth demand equations. We remain faithful to this approach throughout the present chapter.

3. This result is consistent with exact aggregation as defined by Gorman (1981). One part of our ongoing work is to extend this class to Lau's (1982) definition of exact aggregation, generalizing the left-hand side of equation (8) to $f[c(w,\bar{y},z)/\pi,z]$, wherein cost and quasi-fixed inputs vary across individual economic agents.

$$(8) \quad f\left(\frac{c(w,\bar{y},z)}{\pi(w)}\right) = \frac{\alpha(w)\omega[\eta(w),\theta(\bar{y},z)]+\beta(w)}{\gamma(w)\omega[\eta(w),\theta(\bar{y},x)]+\delta(w)}$$

LaFrance and Pope (2009) present a complete proof of necessity in the case of consumer choice theory. Their proof applies to the current problem with only minor changes in notation. Sufficiency is shown here by considering the structure of the input demands generated by equation (8). This is accomplished simply enough by differentiating with respect to w and applying Shephard's lemma. To make the notation as compact as possible, let a bold subscript w denote a vector of partial derivatives with respect to the variable input prices and suppress the arguments of the functions $\{\alpha, \beta, \gamma, \delta, \eta, \pi\}$ to yield (after a large amount of straightforward but tedious algebra, which is presented in appendix B):

$$(9) \quad x = \frac{\pi_w}{\pi}c + \pi\Big\{[\alpha\beta_w - \beta\alpha_w + (\alpha^2\lambda + \beta^2)\eta_w]\frac{1}{f'}$$
$$-[\alpha\delta_w - \delta\alpha_w + \gamma\beta_w - \beta\gamma_w + 2(\alpha\gamma\lambda + \beta\delta)\eta_w]\frac{f}{f'}$$
$$+[\gamma\delta_w - \delta\gamma_w + (\gamma^2\lambda + \delta^2)\eta_w]\frac{f^2}{f'}\Big\}.$$

Thus, equation (8) generates input demands that have the finitely additive and multiplicatively separable structure of any full rank, exactly aggregable system (Gorman 1981; Lau 1982; Lewbel 1989). Note that there are potentially up to four linearly independent variable cost terms on the right with four associated linearly independent vectors of input price functions. Hence, any system generated by equation (8) will have rank up to, but no greater than four, the highest possible rank (Lewbel 1987, 1990, 1991; LaFrance and Pope 2009).

A third issue when estimating a system of variable input demand equations such as equation (9) is the fact that quasi-fixed inputs, planned outputs, variable input prices, and total variable cost all are jointly determined with the input demands. Consistent estimation under these conditions is addressed in the following empirical application.

5.3   The Econometric Cost Model, Data, and Estimates

Previous work at both state and national levels of aggregation with our data set strongly suggests that full rank three seriously overparameterizes the structural model for this data. As a result, we restrict attention here to a rank two model. In this part of the chapter, we analyze the conditional demands for thirteen variable inputs in U.S. agriculture: pesticides and herbicides; fertilizer; fuel and natural gas; electricity; purchased feed; purchased seed; purchased livestock; machinery repairs; building repairs; custom machinery services; veterinary services; other materials; and labor. The specification of the variable cost function normalized by the farm wage rate is,

$$c_t(\tilde{w}_t, A_t, K_t, a_t, \bar{Y}_t) = [\alpha_{10} + \alpha_1^\top \tilde{w}_t] A_t + [\alpha_{20} + \alpha_2^\top \tilde{w}_t] K_t$$

(10)

$$+ \sqrt{\tilde{w}_t^\top B \tilde{w}_t + 2\gamma^\top \tilde{w}_t + 1} \times \theta(A_t, K_t, a_t, \bar{Y}_t),$$

where $z_t = [A_t K_t a_t]^\top$; A_t is farmland, K_t is the value of farm capital; $a_t = [a_{1t} a_{2t} \ldots a_{n_y t}]^\top$ is the n_y-vector of acres planted to crops; $A_t = a_{0t} + \underline{\iota}^\top a_t$, with a_{0t} denoting farmland that is not devoted to crop production; $\bar{Y}_t = [a_{1t} \bar{y}_{1t} \ldots a_{n_y t} \bar{y}_{n_y t}]^\top$ is the n_y-vector of planned crop production, with each element defined as the product of acres planted to the crop times the expected yield per acre; and $\tilde{w}_t = [w_{1t}/w_{n_x t}, \ldots, w_{n_x-1t}/w_{n_x t}]^\top$ is the $(n_x - 1)$-vector of variable input prices except the farm wage normalized by $w_{n_x t}$.

We treat the n_x^{th} input, labor, asymmetrically with respect to the other inputs both in the structural and stochastic parts of the econometric model. To conserve and simplify notation from this point forward, we drop the tilde (~) over the first $n_x - 1$ input prices, absorb the normalization by w_{n_x} into the notation for variable cost and the $n_x - 1$ first input prices, and define $N = n_x - 1$.

We assume constant returns to scale so that $\theta(A_t, K_t, a_t, Y_t)$ is 1° homogeneous. Define $\alpha_1(w_t) = \alpha_{10} + \alpha_1^\top w_t$, $\alpha_2(w_t) = \alpha_{20} + \alpha_2^\top w_t$, and $\beta(w_t) = (w_t^\top B w_t + 2\gamma^\top w_t + 1)^{1/2}$. The necessary and sufficient conditions for the variable cost function to be increasing and concave in the variable input prices throughout an open set containing the data points are as follows (see appendix C for a complete derivation of the cost function and θ):

Monotonicity in w:

$$\frac{\partial c(w_t, A_t, K_t, a_t, \bar{Y}_t)}{\partial w} = \alpha_1 A_t + \alpha_2 K_t + \frac{\theta}{\beta(w_t)}(Bw_t + \gamma)$$

(11)

$$= \alpha_1 A_t + \alpha_2 K_t + \left[\frac{c_t - \alpha_1(w_t)A_t - \alpha_2(w_t)K_t}{w_t^\top B w_t + 2\gamma^\top w_t + 1}\right](Bw_t + \gamma) \geq 0;$$

Concavity in w:

$$\frac{\partial^2 c(w_t, A_t, K_t, a_t, \bar{Y}_t)}{\partial w \partial w^\top} = \frac{\theta}{\beta(w_t)} B - \frac{\theta}{\beta(w_t)^2}(Bw + \gamma)(Bw + \gamma)^\top$$

(12)

$$= \left[\frac{c(w_t, A_t, K_t, a_t, \bar{Y}_t) - \alpha_1(w_t)A_t - \alpha_2(w_t)K_t}{w_t^\top B w_t + 2\gamma^\top w_t + 1}\right]$$

$$\times \left[B - \frac{(Bw_t + \gamma)(Bw_t + \gamma)^\top}{(w_t^\top B w_t + 2\gamma^\top w_t + 1)}\right],$$

symmetric, negative semidefinite. Setting $B = LL^\top + \gamma\gamma^\top$, where L is a (lower or upper) triangular matrix with nonzero main diagonal elements implies

$$(13) \qquad \begin{bmatrix} B & \gamma \\ \gamma^T & 1 \end{bmatrix} = \begin{bmatrix} L & \gamma \\ 0^T & 1 \end{bmatrix} \begin{bmatrix} L^T & 0^T \\ \gamma^T & 1 \end{bmatrix} = \begin{bmatrix} LL^T + \gamma\gamma^T & \gamma \\ \gamma^T & 1 \end{bmatrix}$$

is positive definite. It follows that $[B - (Bw + \gamma)(Bw + \gamma)^T/(w_t^T Bw_t + 2\gamma^T w_t + 1)]$ is positive semidefinite and that

$$(14) \qquad \begin{bmatrix} w_t^T & 1 \end{bmatrix} \begin{bmatrix} B & \gamma \\ \gamma^T & 1 \end{bmatrix} \begin{bmatrix} w_t \\ 1 \end{bmatrix} = w_t^T Bw_t + 2\gamma^T w_t + 1 > 0 \,\forall\, w_t \in \mathbb{R}_+^{n_x - 1}.$$

Given this, the variable cost function is concave in w if and only if

$$(15) \qquad c_t(\tilde{w}_t, A_t, K_t, a_t, \overline{Y}_t) < [\alpha_0 + \alpha_1^T \tilde{w}_t] A_t + [\alpha_2 + \alpha_2^T \tilde{w}_t] K_t$$

(LaFrance, Beatty, and Pope 2006). Hence, we impose $B = LL^T + \gamma\gamma^T$ during estimation and check the monotonicity conditions in equation (11) at all data points once the model is estimated and find that they are satisfied. We develop the specification for $\theta(A_t, K_t, a_t, \overline{Y}_t)$ in the section on life-cycle consumption and investment decisions and appendix C.

Applying Shephard's lemma to equation (10) and rearranging terms then gives the empirical variable input demand equations in normalized expenditures per dollar of capital as

$$(16) \qquad e_t = W_t \left[\alpha_1 \frac{A_t}{K_t} + \alpha_2 + \left(\frac{(c_t/K_t) - \alpha_1(w_t)(A_t/K_t) - \alpha_2(w_t)}{w_t^T Bw_t + 2\gamma^T w_t + 1} \right)(Bw_t + \gamma) \right] + u_t,$$

where $W_t = \mathbf{diag}[w_{it}]$ is the diagonal matrix with $w_{i,t}$ as the ith main diagonal element, and $e_t = [w_{1,t}x_{1,t} \ldots w_{n_x-1,t}x_{n_x-1,t}]^T$ is the $(n_x - 1)$-vector of normalized expenditures per dollar of capital on all inputs except labor, and we follow standard practice in the empirical analysis of demand systems and add a vector of random errors to the right-hand-side to obtain the empirical model. We assume that the errors terms for the twelve equations estimated follow to an unrestricted first-order autoregressive (AR[1]) process,

$$(17) \qquad u_t = Ru_{t-1} + \varepsilon_t, \, \varepsilon_t \text{ i.i.d. } (0, \Sigma), \, t = 1, \ldots, T.$$

As noted in the preceding section, we apply this model to annual aggregate data on thirteen variable inputs in U.S. agriculture (pesticides and herbicides, fertilizer, fuel and natural gas, electricity, purchased feed, purchased seed, purchased livestock, machinery repairs, building repairs, custom machinery services, veterinary services, other materials, and farm labor). The sample period is 1960 to 1999. These data were compiled by the United States Department of Agriculture's (USDA) Economic Research Service (ERS) and is described in detail in Ball, Hallahan, and Nehring (2004). Farmland, equipment, buildings, and structures are treated as quasi-fixed inputs. Hereafter, this data set is called the Ball data.

Due to the way that several variables are constructed in the Ball data, it is necessary to modify and augment this data for empirical implementation. First, we define the replacement cost of owner-operator labor by the farm wage rate. This implies that the return to owner-operator labor in the Ball data due to management skill is treated as a part of the residual claimant's quasi-rent. Second, we use a direct measure of the value of capital obtained from the ERS rather than the measures constructed in the Ball data. Third, estimates of the price of farmland are taken from state-level surveys conducted by the National Agricultural Statistics Service (NASS), rather than the constructed measures in the Ball data. Finally, we adjust the measure of agricultural land. The Census of Agriculture has reported land in farms in four- to five-year intervals for 1954, 1959, 1964, 1969, 1974, 1978, 1982, 1987, 1992, 1997, 2002, and 2007. These are the total farmland numbers used in the sample years that match the census years. The ERS reports the harvested acres for all major crops by state and year since 1947. This data is used to adjust the farmland measures in the Ball data as follows. First, the difference between total farmland in the Ball data and harvested acres is calculated for each noncensus year by state. Second, in each period between adjacent censuses, the average of this difference is calculated. This mean difference is treated as fixed in each of the three- or four-year intervals between census years and added to harvested acres to obtain the measure of farmland used in this study in those years of our sample period. We normalize costs, expenditures, and acres by capital rather than total land because we are more confident in the capital measure, and Pope, LaFrance, and Just (2007) have shown that deflating by a variable that is subject to measurement error leads to difficult econometric issues.

Estimation is by nonlinear generalized method of moments (GMM), which assumes a parametric 12×12 AR(1) process for the time series component and White/Huber robust covariance matrix estimator that is consistent under heteroskedasticity of an unknown form. The instruments are variable cost per unit of capital, land per unit of capital, and variable input prices all lagged two periods, plus the following general economy variables lagged one period: real per capita disposable personal income; unemployment rate; the real rate of return on AAA corporate thirty-year bonds; real manufacturing wage rate; real index of prices paid by manufacturers for materials and components; and real index of prices paid by manufacturers for fuel, energy and power. Per capita disposable personal income is deflated by the Consumer Price Index (CPI) for all items. The aggregate wholesale price variables are deflated by the implicit price deflator for gross domestic product (GDP). The real rate of return on corporate bonds is calculated as the nominal rate of return minus the midyear annual inflation rate.

Table 5.1 presents the estimated 12×12 AR(1) matrix. The Eigen values of the implied autocovariance structure are well within the stability region, with two real roots and five complex conjugate pairs:

Table 5.1   First-order autocorrelation parameters for the variable cost model

	Pesticide	Fertilizer	Fuel and natural gas	Electricity	Purchased feed	Purchased seed	Purchased livestock	Machinery repairs	Building repairs	Hired machinery	Veterinary services	Other materials
Pesticide	-.235 (.323) [.230]	.280 (.184) [.139**]	-.556 (.303*) [.242**]	-.199 (.235) [.141]	-.188 (.126) [.070***]	-.291 (.394) [.281]	-.029 (.096) [.057]	-.029 (.283) [.223]	-.341 (.482) [.290]	-.283 (.320) [.206]	.690 (.464) [.254***]	.040 (.084) [.068]
Fertilizer	.333 (.524) [.259]	.699 (.298*) [.161***]	-.265 (.491) [.256]	-.664 (.382*) [.191***]	.062 (.204) [.121]	-2.01 (.639***) [.459***]	.154 (.155) [.101]	1.22 (.459***) [.324***]	-1.50 (.782*) [.399***]	.704 (.519) [.273***]	-.339 (.753) [.412]	.046 (.136) [.075]
Fuel and natural gas	.256 (.229) [.072***]	.193 (.130) [.047***]	.503 (.215**) [.085***]	-.263 (.167) [.071***]	-.040 (.089) [.039]	-1.11 (.279***) [.132***]	.034 (.068) [.029]	.577 (.200***) [.107***]	-.159 (.342) [.117]	.321 (.227) [.093***]	.131 (.329) [.135]	-.048 (.060) [.025**]
Electricity	.066 (.256) [.178]	-.147 (.146) [.087*]	-.168 (.240) [.155]	.455 (.187**) [.107***]	-.052 (.100) [.055]	.049 (.312) [.227]	-.061 (.076) [.046]	.254 (.224) [.088***]	.120 (.382) [.209]	-.268 (.254) [.141*]	.067 (.368) [.259]	-.064 (.067) [.035*]
Purchased feed	-.463 (.626) [.376]	.252 (.356) [.257]	.199 (.587) [.433]	-.404 (.456) [.253]	.066 (.244) [.181]	-.018 (.763) [.483]	.308 (.186*) [.157**]	-.770 (.548) [.351**]	-.016 (.934) [.625]	.345 (.620) [.282]	.025 (.900) [.625]	.220 (.163) [.098**]
Purchased seed	-.120 (.276) [.217]	.185 (.157) [.121]	-.282 (.258) [.234]	-.303 (.201) [.155*]	-.043 (.107) [.070]	-.447 (.336) [.236*]	.012 (.082) [.068]	.607 (.241**) [.245**]	-.296 (.411) [.328]	-.185 (.273) [.190]	-.139 (.396) [.273]	-.019 (.072) [.055]
Purchased livestock	.0450 (.649) [.399]	-.244 (.369) [.226]	-.577 (.608) [.391]	-.332 (.472) [.312]	.073 (.252) [.176]	-.222 (.790) [.600]	.362 (.192*) [.158**]	.615 (.568) [.402]	-.769 (.967) [.603]	-.951 (.643) [.328***]	-.975 (.932) [.562*]	-.296 (.169*) [.118**]
Machinery repairs	.030 (.245) [.162]	.135 (.140) [.090]	-.359 (.230) [.187*]	-.008 (.178) [.139]	-.045 (.095) [.059]	-.616 (.299**) [.246**]	.064 (.073) [.059]	.657 (.215***) [.127***]	.0763 (.366) [.206]	-.003 (.243) [.136]	-.024 (.352) [.195]	-.048 (.064) [.049]

(continued)

Table 5.1 (continued)

	Pesticide	Fertilizer	Fuel and natural gas	Electricity	Purchased feed	Purchased seed	Purchased livestock	Machinery repairs	Building repairs	Hired machinery	Veterinary services	Other materials
Building repairs	.084	.064	-.079	-.007	.011	-.105	.009	.376	-.071	-.041	-.337	-.021
	(.160)	(.091)	(.150)	(.116)	(.062)	(.195)	(.047)	(.140***)	(.238)	(.158)	(.230)	(.042)
	[.112]	[.065]	[.144]	[.083]	[.047]	[.132]	[.041]	[.119***]	[.162]	[.094]	[.175*]	[.031]
Hired machinery	.113	-.097	.011	.032	-.043	-.119	-.123	.239	-.060	-.060	-.209	-.043
	(.253)	(.144)	(.237)	(.184)	(.098)	(.308)	(.075)	(.221)	(.377)	(.250)	(.363)	(.066)
	[.252]	[.124]	[.200]	[.150]	[.083]	[.304]	[.079]	[.197]	[.324]	[.219]	[.416]	[.047]
Veterinary services	.041	-.089	-.006	.135	.004	.096	.021	-.129	.329	-.356	.602	.023
	(.149)	(.085)	(.140)	(.108)	(.058)	(.181)	(.044)	(.130)	(.222)	(.148**)	(.214***)	(.039)
	[.076]	[.059]	[.110]	[.049***]	[.035]	[.122]	[.020]	[.079]	[.132**]	[.084***]	[.152***]	[.019]
Other materials	-1.45	-.062	-.085	.272	-.188	2.46	.178	-2.37	-.409	.477	.986	.211
	(.997)	(.567)	(.934)	(.726)	(.388)	(1.21**)	(.295)	(.873***)	(1.49)	(.987)	(1.43)	(.259)
	[.622**]	[.355]	[.568]	[.473]	[.272]	[1.09**]	[.232]	[.881***]	[1.01]	[.633]	[.841]	[.140]

Note: Numbers in parentheses and in square brackets are standard Gaussian and White heteroskedasticity consistent asymptotic standard errors, respectively.

***Significant at the 1 percent level.

**Significant at the 5 percent level.

*Significant at the 10 percent level.

(18) $\lambda_1 = 0.6772;$

$\lambda_2 = 0.1294;$

$\lambda_{3,4} = 0.7594 \pm 0.2998\iota, \text{ modulus} = 0.8165;$

$\lambda_{5,6} = 0.4104 \pm 0.5273\iota, \text{ modulus} = 0.6682;$

$\lambda_{7,8} = 0.3056 \pm 0.3371\iota, \text{ modulus} = 0.4550;$

$\lambda_{9,10} = -0.4222 \pm 0.0832\iota, \text{ modulus} = 0.4304;$

$\lambda_{11,12} = -0.0863 \pm 0.2638\iota, \text{ modulus} = 0.2776.$

A system of twelve linear first-order difference equations has the same dynamic structure as a single twelfth-order linear difference equation. This implies that the time series properties of this model are quite complex. No evidence is found for any additional serial correlation in the data.

The single equation and systemwide first- and second-order Brownian bridge tests for specification error and parameter instability developed in LaFrance (2008) provide no evidence of misspecification or parameter instability. (Appendix D presents and discusses this set of within-sample residual test statistics.)

Table 5.2 presents the parameter estimates for the structural part of the model. To obtain a positive definite B matrix, the lower four main diagonal elements of the Choleski factor L were restricted to 0.01, and the off-diagonal elements in the last four columns were restricted at 0.0. In other words, the estimated symmetric but not curvature restricted B matrix has four negative Eigen values. As a consequence, the standard errors in table 5.2 are conditional on these inequality restrictions. The estimated structural parameters reported in table 5.2 generate a variable cost function that is increasing and weakly concave in all variable input prices throughout the data set. We conclude that this is a coherent and reasonable model of the short-run cost of production in U.S. agriculture.

5.4 Crop Acres, Capital, Savings and Investment, and Consumption in Agriculture

Although the organizational form of farms can vary widely, a recent report by Hoppe and Banker (2006) finds that 98 percent of U.S. farms remained family farms as of 2003. In a family farm, the entrepreneur controls the means of production and makes investment, consumption, and production decisions. In this section, we develop and analyze a model of the intertemporal nature of these decisions. The starting point is a model similar in spirit to Hansen and Singleton's (1983), but generalized to include consumption decisions and farm investments as well as financial investments and production decisions. The additional variable definitions required for this are as follows:

Table 5.2   Estimates of α and B matrix parameters for the variable cost function

Variable	Pesticide	Fertilizer	Fuel and natural gas	Electric	Purchased feed	Purchased seed	Purchased livestock	Machinery repairs	Building repairs	Hired machinery	Veterinary services	Other materials	Labor
Land	−55.21	−64.95	−21.99	−15.54	−67.00	−44.30	−51.87	−20.32	−9.286	−19.31	−8.374	−114.7	36.84
	(74.23)	(81.50)	(29.73)	(23.17)	(101.7)	(58.38)	(77.94)	(31.02)	(11.61)	(28.14)	(12.76)	(171.2)	(30.43)
Capital	3.554	4.140	1.515	1.001	5.044	2.837	3.808	1.615	.6197	1.362	.5819	8.634	−1.564
	(3.854)	(4.233)	(1.572)	(1.223)	(5.314)	(3.040)	(4.102)	(1.637)	(.6152)	(1.479)	(.6678)	(8.920)	(1.596)
Pesticide	3.695***	2.995*	.9374	.7107	5.042***	2.558**	2.832	1.020	.3742	1.201*	.4549	7.442**	−1.390***
	(1.288)	(1.824)	(.8018)	(.6566)	(1.800)	(1.179)	(1.851)	(.7793)	(.3850)	(.6345)	(.3043)	(3.324)	(.5430)
Fertilizer	2.995*	5.119***	1.469**	1.388***	4.709***	2.480*	3.806**	.5221	1.332***	1.425**	.5213	6.914*	−1.501***
	(1.824)	(1.390)	(.6455)	(.4419)	(2.284)	(1.364)	(1.671)	(1.116)	(.2998)	(.5894)	(.3355)	(4.174)	(.5742)
Fuel and natural gas	.9374	1.469**	1.027***	.3949	1.303	.8504	1.764***	.7740***	.5493***	.7398***	.0858	2.189	−.6180***
	(.8018)	(.6455)	(.2538)	(.2106)	(1.068)	(.6167)	(.5496)	(.2860)	(.2480)	(.2268)	(.1824)	(1.736)	(.1851)
Electric	.7107	1.388***	.3949	1.372***	1.559**	−.4480	1.129**	−.2855	.7367*	.0407	−.0248	1.322	−.4507**
	(.6566)	(.4419)	(.2106)	(.4332)	(.6354)	(1.038)	(.5554)	(.7581)	(.3984)	(.4945)	(.2198)	(1.547)	(.2276)
Purchased feed	5.042***	4.709***	1.303	1.559**	7.302***	3.229*	4.087**	1.382	.4051	1.857**	.6293	9.850**	−2.067***
	(1.800)	(2.284)	(1.068)	(.6354)	(2.414)	(1.736)	(2.426)	(1.111)	(.4920)	(.7602)	(.4126)	(4.878)	(.6721)
Purchased seed	2.558**	2.480*	.8504	−.4480	3.229*	5.547***	2.694**	1.608	−.2849	2.494***	.9234***	4.455	−1.408**
	(1.179)	(1.364)	(.6167)	(1.038)	(1.736)	(1.135)	(1.305)	(1.032)	(1.027)	(.8282)	(.2894)	(3.182)	(.2930)
Purchased livestock	2.832	3.806**	1.764***	1.129**	4.087**	2.694**	4.303***	1.161	.3629	1.424**	.6567**	7.340**	−1.585***
	(1.851)	(1.671)	(.5496)	(.5554)	(2.426)	(1.305)	(1.638)	(.9572)	(.4089)	(.5827)	(.3026)	(3.950)	(.5396)
Machinery repairs	1.020	.5221	.7740***	−.2855	1.382	1.608	1.161	6.508*	−2.860	2.635***	.2365	3.098**	−.8805***
	(.7793)	(1.116)	(.2860)	(.7581)	(1.111)	(1.032)	(.9572)	(3.475)	(1.825)	(.8530)	(.3453)	(1.528)	(.2250)
Building repairs	.3742	1.332***	.5493***	.7367*	.4051	−.2849	.3629	−2.860	7.456**	−1.735*	−.6479	.4392	−.4085**
	(.3850)	(.2998)	(.2480)	(.3984)	(.4920)	(1.027)	(.4089)	(1.825)	(3.378)	(1.032)	(.4479)	(.9521)	(.1467)
Hired machinery	1.201*	1.425**	.7398***	.0407	1.857**	2.494***	1.424**	2.635***	−1.735*	2.493**	.3542*	.8858	−.8651***
	(.6345)	(.5894)	(.2268)	(.4945)	(.7602)	(.8282)	(.5827)	(.8530)	(1.032)	(1.055)	(.2006)	(2.280)	(.1362)
Veterinary services	.4549	.5213	.0858	−.0248	.6293	.9234***	.6567**	.2365	−.6479	.3542*	.2980***	1.410**	−.2534**
	(.3043)	(.3355)	(.1824)	(.2198)	(.4126)	(.2894)	(.3026)	(.3453)	(.4479)	(.2006)	(.1159)	(.6818)	(.1150)
Other materials	7.442**	6.914*	2.189	1.322	9.850**	4.455	7.340**	3.098**	.4392	.8858	1.410**	21.99***	−2.845**
	(3.324)	(4.174)	(1.736)	(1.547)	(4.878)	(3.182)	(3.950)	(1.528)	(.9521)	(2.280)	(.6818)	(6.588)	(1.374)
Labor	−1.390***	−1.501***	−.6180***	−.4507**	−2.067***	−1.408**	−1.585***	−.8805***	−.4085**	−.8651***	−.2534**	−2.845**	1.000
	(.5430)	(.5742)	(.1851)	(.2276)	(.6721)	(.2930)	(.5396)	(.2250)	(.1467)	(.1362)	(.1150)	(1.374)	(—)

Note: Numbers in parentheses are White heteroskedasticity consistent asymptotic standard errors.

***Significant at the 1 percent level.

**Significant at the 5 percent level.

*Significant at the 10 percent level.

W_t = beginning-of-period total wealth,
b_t = current holding of bonds with a risk free rate of return r_t,
f_t = current holding of a risky financial asset,
$p_{F,t}$ = beginning-of-period market price of the financial asset,
$\rho_{F,t+1}$ = dividend plus capital gains rate on the financial asset,
$a_{i,t}$ = current allocation of land to the ith crop, $i = 1, \ldots, n_Y$,
A_t = total quantity of farm land,
$p_{L,t}$ = beginning-of-period market price of land,
$\rho_{L,t+1} = (p_{L,t+1} - p_{L,t})/p_{L,t}$ = capital gain rate on land,
$\overline{y}_{i,t}$ = expected yield per acre for the ith crop, $i = 1, \ldots, n_Y$,
$y_{i,t+1}$ = realized yield of the ith crop,
$p_{Y,t+1}$ = end-of-period realized market price for the ith farm product,
$\boldsymbol{q}_t$ = vector of quantities of consumption goods,
$\boldsymbol{p}_{Q,t}$ = vector of market prices for consumer goods,
m_t = total consumption expenditures,
$u(\boldsymbol{q}_t)$ = periodic utility from consumption.

As with all discrete time models, timing can be represented in multiple ways. In the model used here, all financial returns and farm asset gains are assumed to be realized at the end of each time period (where depreciation is represented by a negative asset gain). Variable inputs are assumed to be committed to farm production activities at the beginning of each decision period, and the current period market prices for the variable inputs are known when these use decisions are made. Agricultural production per acre is realized stochastically at the end of the period such that

(19) $$y_{i,t+1} = \overline{y}_{i,t}(1 + \varepsilon_{i,t+1}), i = 1, \ldots, n_Y,$$

where $\varepsilon_{i,t+1}$ is a random output shock with $E(\varepsilon_{i,t+1}) = 0$. Consumption decisions are made at the beginning of the decision period, and the current market prices of consumption goods are known when these purchases are made. Utility is assumed to be strictly increasing and concave in $\boldsymbol{q}_t$. The total beginning-of-period quantity of land is $A_t = \boldsymbol{\iota}^T \boldsymbol{a}_t$, with ι denoting an n_Y-vector of ones. Homogeneous land is assumed with a scalar price, $p_{L,t}$.

To simplify our derivations, we require an uncommon piece of matrix notation. The Hadamard/Schur product of two $n \times m$ matrices $\mathbf{A}$ and $\mathbf{B}$ is the matrix whose elements are element-by-element products of the elements of $\mathbf{A}$ and $\boldsymbol{B}$, $\boldsymbol{A} \bullet \boldsymbol{B} = \boldsymbol{C} \Leftrightarrow c_{ij} = a_{ij} b_{ij} \; \forall i, j$. This definition assists the derivation of the arbitrage conditions present in what follows.

Revenue at $t + 1$ is the random price times production

(20) $$R_{t+1} = \sum_{i=1}^{n_Y} (p_{Y,t+1} \overline{Y}_{i,t} a_{i,t}(1 + \varepsilon_{i,t+1})) \equiv (\boldsymbol{p}_{Y,t+1} \bullet \boldsymbol{a}_t \bullet \overline{\boldsymbol{y}}_t)^T(\boldsymbol{\iota} + \boldsymbol{\varepsilon}_{t+1}).$$

Wealth is allocated at the beginning of period t to investments, the variable cost of production, and consumption,

(21) $$W_t = b_t + f_t + p_{L,t} A_t + K_t + c_t(w_t, a_t, K_t, \overline{Y}_t) + m_t.$$

Although some costs occur at or near harvest (near $t + 1$), we include all costs in (21) at time t because they are incurred before revenues are received. Consumer utility maximization yields the indirect utility function conditioned on consumer good prices and consumption expenditure,

$$(22) \qquad \upsilon(\boldsymbol{p}_{Q,t}, m_t) \equiv \max_{q \in R_+^{n_Q}} \left\{ u(\boldsymbol{q}) : \boldsymbol{p}_{Q,t}^\mathsf{T} \boldsymbol{q} = m_t \right\}.$$

Realized end of period wealth is

$$(23) \qquad W_{t+1} = (1 + r)b_t + (1 + \rho_{F,t+1})f_t + (1 + \rho_{L,t+1})p_{L,t}A_t$$
$$+ (1 + \rho_{K,t+1})K_t + (\boldsymbol{p}_{Y,t+1} \bullet \boldsymbol{a}_t \bullet \bar{\boldsymbol{y}}_t)^\mathsf{T} (\boldsymbol{\iota} + \boldsymbol{\varepsilon}_{t+1}),$$

where $\rho_{K,t+1}$ is the proportional change in the value of capital held at the beginning of the production period. Thus, the decision maker's wealth is increased by net returns on assets and farm revenue. The owner/operator decision maker's intertemporal utility function is assumed to be

$$(24) \qquad U_t(\boldsymbol{q}_1, \ldots, \boldsymbol{q}_T) = \sum_{t=0}^T (1 + r)^{-t} u(\boldsymbol{q}_t).$$

The producer is assumed to maximize von Neumann-Morgenstern expected utility of the discounted present value of the periodic utility flows from goods consumption.

By Euler's theorem, constant returns to scale implies linear homogeneity of the variable cost function in capital, land, and output. For the variable cost function derived and estimated in this chapter, this implies

$$(25) \qquad
\begin{aligned}
c_t(\boldsymbol{w}_t, \boldsymbol{a}_t, A_t, K_t, \bar{\boldsymbol{Y}}_t) \equiv & \frac{\partial c_t(\boldsymbol{w}_t, \boldsymbol{a}_t, A_t, K_t, \bar{\boldsymbol{Y}}_t)}{\partial \boldsymbol{a}_t^\mathsf{T}} \boldsymbol{a}_t + \frac{\partial c_t(\boldsymbol{w}_t, \boldsymbol{a}_t, A_t, K_t, \bar{\boldsymbol{Y}}_t)}{\partial A_t} A_t \\
& + \frac{\partial c_t(\boldsymbol{w}_t, \boldsymbol{a}_t, A_t, K_t, \bar{\boldsymbol{Y}}_t)}{\partial K_t} K_t + \frac{\partial c_t(\boldsymbol{w}_t, \boldsymbol{a}_t, A_t, K_t, \bar{\boldsymbol{Y}}_t)}{\partial \bar{\boldsymbol{Y}}_t^\mathsf{T}} \bar{\boldsymbol{Y}}_t.
\end{aligned}$$

The vector of expected crop outputs satisfies

$$(26) \qquad \bar{\boldsymbol{Y}}_t = \bar{\boldsymbol{y}}_t \bullet \boldsymbol{a}_t,$$

where $\bar{y}_{j,t}$ is the expected yield per acre, and $a_{j,t}$ is the number of acres planted for the jth crop. The variable cost function might depend on time due to technological change or other dynamic forces, and the subscript t indicates this possibility. To distinguish quasi-fixed from variable inputs and to account for the possibility of hysteresis in agricultural investments, we allow for adjustment costs for total farmland and capital,

$$(27) \quad C_{\text{Adj}}(A_t - A_{t-1}, K_t - K_{t-1}) = \frac{1}{2} \gamma_A (A_t - A_{t-1})^2 + \frac{1}{2} \gamma_K (K_t - K_{t-1})^2,$$

with $\gamma_A, \gamma_K \geq 0$.

This problem is solved by stochastic dynamic programming working backward recursively from the last period in the planning horizon to the first. In the last period, the optimal decision is to invest or produce nothing and consume all remaining wealth, that is, $m_T = W_T$. Denote the last period's optimal value function by $v_T(W_T, A_{T-1}, K_{T-1})$. Then $v_T(W_T, A_{T-1}, K_{T-1}) = v(\mathbf{p}_{Q,T}, W_T)$ is the optimal utility for the terminal period. For all other time periods, stochastic dynamic programming yields the Bellman backward recursion (Bellman and Dreyfus 1962). For an arbitrary $t < T$, the Lagrangean for the problem at time t is

$$(28) \quad \ell_t = v(\mathbf{p}_{Q,t}, m_t) + (1 + r)^{-1} E_t \{ V_{t+1}[(1 + r)b_t + (1 + \rho_{F,t+1})f_t$$
$$+ p_{L,t+1}A_t + (1 + \rho_{K,t+1})K_t + (\mathbf{p}_{y,t+1} \cdot \bar{\mathbf{y}}_t \cdot \mathbf{a}_t)^{\mathsf{T}}(\mathbf{\iota} + \mathbf{\varepsilon}_{t+1}), A_t, K_t]\}$$
$$+ \lambda_t \{ W_t - m_t - b_t - f_t - p_{L,t}A_t - K_t - c_t(w_t, \mathbf{a}_t, A_t, K_t, \bar{\mathbf{y}}_t \cdot \mathbf{a}_t)$$
$$- \frac{1}{2}\gamma_A(A_t - A_{t-1})^2 - \frac{1}{2}\gamma_K(K_t - K_{t-1})^2 \} + \mu_t(A_t - \mathbf{\iota}^{\mathsf{T}}\mathbf{a}_t),$$

where $E_t(\bullet)$ is the conditional expectation at the beginning of period t given information available at that point in time, λ_t is the shadow price for the beginning-of-period wealth allocation constraint, and μ_t is the shadow price for the land allocation constraint. The first-order, necessary and sufficient Kuhn-Tucker conditions are the two constraints and the following:

$$(29) \quad \frac{\partial \ell_t}{\partial m_t} = \frac{\partial v_t}{\partial m_t} - \lambda_t \leq 0, m_t \geq 0, m_t \frac{\partial \ell_t}{\partial m_t} = 0;$$

$$(30) \quad \frac{\partial \ell_t}{\partial b_t} = E_t \left(\frac{\partial V_{t+1}}{\partial W_{t+1}} \right) - \lambda_t \leq 0, b_t \geq, b_t \frac{\partial \ell_t}{\partial b_t} = 0;$$

$$(31) \quad \frac{\partial \ell_t}{\partial f_t} = (1 + r)^{-1} E_t \left[\frac{\partial V_{t+1}}{\partial W_{t+1}} (1 + \rho_{F,t+1}) \right] - \lambda_t \leq 0, f_t \geq 0, f_t \frac{\partial \ell_t}{\partial f_t} = 0.$$

$$(32) \quad \frac{\partial \ell_t}{\partial A_t} = (1 + r)^{-1} E_t \left(\frac{\partial V_{t+1}}{\partial W_{t+1}} p_{L,t+1} + \frac{\partial V_{t+1}}{\partial A_t} \right)$$
$$- \lambda_t \left[p_{L,t} + \frac{\partial c_t}{\partial A_t} + \gamma_A(A_t - A_{t-1}) \right] + \mu_t \leq 0, A_t \geq 0, A_t \frac{\partial \ell_t}{\partial A_t} = 0;$$

$$(33) \quad \frac{\partial \ell_t}{\partial K_t} = (1 + r)^{-1} E_t \left[\frac{\partial V_{t+1}}{\partial W_{t+1}} (1 + \rho_{K,t+1}) + \frac{\partial V_{t+1}}{\partial K_{t+1}} \right]$$
$$- \lambda_t \left[1 + \frac{\partial c_t}{\partial K_t} + \gamma_K(K_t - K_{t-1}) \right] \leq 0, K_t \geq 0, K_t \frac{\partial \ell_t}{\partial K_t} = 0;$$

$$(34) \quad \frac{\partial \ell_t}{\partial \mathbf{a}_t} = (1 + r)^{-1} E_t \left[\frac{\partial V_{t+1}}{\partial W_{t+1}} (\mathbf{p}_{Y,t+1} \cdot \bar{\mathbf{y}}_t) \bullet (\mathbf{\iota} + \mathbf{\varepsilon}_{t+1}) \right]$$
$$- \lambda_t \left(\frac{\partial c_t}{\partial \mathbf{a}_t} + \frac{\partial c_t}{\partial \bar{\mathbf{Y}}_t} \cdot \bar{\mathbf{y}}_t \right) - \mu_t \mathbf{\iota} \leq 0, \mathbf{a}_t \geq 0, \mathbf{a}_t^{\mathsf{T}} \frac{\partial \ell_t}{\partial \mathbf{a}_t} = 0;$$

(35) $$\frac{\partial \ell_t}{\partial \bar{y}_t} = (1 + r)^{-1} E_t \left[\frac{\partial V_{t+1}}{\partial W} p_{Y,t+1} \cdot a_t \cdot (\iota + \varepsilon_{t+1}) \right] - \lambda_t \frac{\partial c_t}{\partial \bar{Y}_t} \cdot a_t \leq 0,$$

$$\bar{y}_t \geq 0, \bar{y}_t^\top \frac{\partial \ell_t}{\partial \bar{y}_t} = 0.$$

We also have the following implications of the envelope theorem:

(36)
$$\frac{\partial V_t}{\partial W_t} = \lambda_t;$$

$$\frac{\partial V_t}{\partial A_{t-1}} = \lambda_t \gamma_A (A_t - A_{t-1});$$

$$\frac{\partial V_t}{\partial K_{t-1}} = \lambda_t \gamma_K (K_t - K_{t-1});$$

where the variables $\{\lambda_t, A_t, K_t\}$ are all evaluated at their optimal choices.

Combining the Kuhn-Tucker conditions with the results of the envelope theorem and assuming an interior solution for consumption, bonds, and risky financial assets, we obtain the standard Euler equations for smoothing the marginal utility of consumption and wealth,

(37) $$\frac{\partial \upsilon_t}{\partial m_t} = E_t \left(\frac{\partial \upsilon_{t+1}}{\partial m_{t+1}} \right) = \frac{\partial V_t}{\partial W_t} = E_t \left(\frac{\partial V_{t+1}}{\partial W_{t+1}} \right) = \lambda_t = E_t(\lambda_{t+1}),$$

and the standard arbitrage condition for excess returns to risky financial assets,

(38) $$E_t \left[(\rho_{F,t+1} - r) \frac{\partial V_{t+1}}{\partial W_{t+1}} \right] = 0.$$

The complementary slackness of the Kuhn-Tucker condition in equation (35) implies that for each crop we have the supply condition under risk,

(39) $$E_t \left\{ \frac{\partial V_{t+1}}{\partial W_{t+1}} \left[p_{Y,t+1} - (1 + r) \frac{\partial c_t}{\partial \bar{Y}_{i,t}} \right] \right\} \bar{Y}_{i,t} = 0, i = 1, \ldots, n_y.$$

For each crop produced in positive quantity, this reduces to the well-known result that the conditional covariance between the marginal utility of future wealth and the difference between the ex post realized market price the marginal cost of production must vanish. The multiplicative factor $1 + r$ is multiplied by ex ante marginal cost so that these two economic values are measured at a common point in time—in the present case at the end of the production period.

To obtain the arbitrage condition for the level of investment in agriculture, we combine the linear homogeneity property of the variable cost function in $(a_t, A_t, K_t, \bar{Y}_t)$ from equation (25) with complementary slackness in Kuhn-Tucker conditions in equations (33) to (37),

(40)
$$0 = \frac{\partial \ell_t}{\partial a_t^T} a_t + \frac{\partial \ell_t}{\partial A_t} A_t + \frac{\partial \ell_t}{\partial K_t} K_t,$$

which, after considerable rearranging and combining of terms, gives

(41) $E_t \left(\frac{\partial V_{t+1}}{\partial W_{t+1}} \{ s_{K,t}(\rho_{K,t+1} - r) + s_{L,t}(\rho_{L,t+1} - r) + \pi_{t+1} \right.$

$$+ s_{K,t}\gamma_K(K_{t+1} - (2 + r)K_t + (1 + r)K_{t-1})$$

$$\left. + s_{A,t}\gamma_A[A_{t+1} - (2 + r)A_t + (1 + r)A_{t-1}] \} \right) = 0,$$

where $s_{K,t} = K_t/(p_{L,t}A_t + K_t)$ is capital's share of the value of the investment in agriculture in period t, $s_{L,t} = p_{L,t}A_t/(p_{L,t}A_t + K_t)$ is land's share of the value of the investment in agriculture in period t, $s_{A,t} = A_t/(p_{L,t}A_t + K_t)$ is the ratio of the quantity of land to the value of the investment in agriculture at the beginning of the production period, and

(42)
$$\pi_{t+1} = \frac{R_{t+1} - (1 + r)c_t}{p_{L,t}A_t + K_t}$$

is the ex post net return to crop production over the variable cost of production relative to the ex ante value of agricultural investment so that it is measured as a rate of return to agricultural production. The first three terms inside of the square brackets of equation (41) represent the total sum of the excess returns to agriculture, including the rate of net return to crop production over variable costs. The last two terms in square brackets capture the effects of adjustment costs for farm capital and farmland. This has the standard one period ahead and one period behind second-order difference structure common to quadratic adjustment cost models in dynamic optimization problems.

To implement this system of Euler equations, we assume that the indirect utility function for consumption goods is a member of the certainty equivalent class,

(43)
$$\upsilon(p_{Qt}, m_t) = \frac{m_t}{\pi_C(p_{Qt})} - \frac{1}{2}\beta \left[\frac{m_t}{\pi_C(p_{Qt})} \right]^2,$$

where $0 \le \beta < \pi_C(p_{Qt})/m_t, \forall t$ and $\pi_C(p_{Qt})$ is the CPI for all items. Then the marginal utility of money in each period is

(44)
$$\lambda_t = \frac{1 - \beta[m_t/\pi_C(p_{Qt})]}{\pi_C(p_{Qt})}.$$

This allows us to identify the effects of risk aversion separately from those of adjustment costs and hysteresis in agricultural investment decisions. We assume that the preferences of agricultural producers are of the same class as all other individuals in the economy. This allows use of the observable variable per capita personal consumption expenditure, rather than the latent variable wealth, to model the empirical arbitrage equations.

5.4.1 Empirical Arbitrage Equations and Data

Let $n \leq n_y$ be the number of crops included in the empirical model. The specification that we choose for $\partial c_t / \partial \overline{Y}_{i,t}$ is (see appendix C for a complete derivation),

(45)
$$\frac{\partial c_t}{\partial \overline{Y}_{i,t}} = \widehat{\beta(w_t)} \left(\theta_i + \sum_{j=1}^{n_x} \theta_{ij} \frac{\overline{Y}_{j,t}}{K_t} \right), \text{ with}$$

$$\widehat{\beta(w_t)} = \sqrt{\tilde{w}_t^{\mathsf{T}} \hat{B} \tilde{w}_t + 2\hat{\gamma}^{\mathsf{T}} \tilde{w}_t w_{n_x,t} + w_{n_x,t}^2}.$$

We use the estimated $\widehat{\beta(w_t)}$ obtained from the ex ante variable input demand system, and $\tilde{w}_t = [w_{1,t} \dots w_{n_x-1,t}]^{\mathsf{T}}$ is the vector of variable input prices other than the farm wage.[4] The $n + 3$ empirical arbitrage/Euler equations, therefore, are

(46) Consumption / Bonds:

$$\beta(m_{t+1} - m_t) = u_{1,t+1},$$

Risky Assets:

$$(1 - \beta m_{t+1})(\rho_{F,t+1} - r) = u_{2,t+1},$$

Crops:

$$(1 - \beta m_{t+1}) \left[p_{Y_i,t+1} - (1+r)\widehat{\beta(w_t)} \left(\theta_i + \sum_{j=1}^{n} \frac{\theta_{ij} Y_{j,t+1}}{K_t} \right) \right] = u_{i,t+1},$$

$$i = 3, \dots, n + 2,$$

Agriculture:

$$(1 - \beta m_{t+1})\{ s_{K,t}(\rho_{K,t+1} - r) + s_{L,t}(\rho_{L,t+1} - r) + \pi_{t+1}$$
$$+ s_{A,t}\gamma_A(A_{t+1} - (2 + r)A_t + (1 + r)A_{t-1})$$
$$+ s_{K,t}\gamma_K[K_{t+1} - (2 + r)K_t + (1 + r)K_{t-1}]\} = u_{n+3,t+1}.$$

The estimation method for this part of the modeling exercise again is nonlinear three-stage least squares/generalized method of moments (NL3SLS/GMM) with a parametric AR(1) correction for autocorrelation and White/Huber heteroskedasticity consistent estimated covariance matrix. We restrict the parameter matrix $\Theta = [\theta_{ij}]$ to be positive semidefinite by estimating it in Choleski factored form, $\Theta = QQ^T$, where Q is a lower triangular matrix.

---

4. All prices, costs, and revenues, including the value of farm capital, are deflated by the CPI in this part of the empirical analysis.

5.4.2 Empirical Results

We analyze acreage and supply decisions under risk for ten crops with the greatest value in the United States in 2006: soybeans, corn, cotton, hay, potatoes, rice, sugar beets, sugarcane, tobacco, and wheat. Crop revenues includes the value of government payments that is been imputed in the Ball data, to at least partially capture the effects of farm-level price, income, and other subsidy and stabilization programs on the distribution of realized farm revenues. The ten crops analyzed in this study account for 94 to 95 percent of total farm revenue from crop production and an even larger share of crop acreage. In addition to the ten crop production decisions under risk, we estimate Euler equations for the excess return to investing in agriculture, personal consumption expenditures, and the rate of return to stocks as measured by the Standard & Poor 500 index.

To ensure a consistent definition of real values in this component of the model, we deflate all nominal prices, revenues, costs, and other values by that year's CPI for all items. We scale all aggregate economic data—for example, the total value of agricultural investment in U.S. agriculture—by the U.S. population to measure these variables all in per capita units. As noted in the preceding, real per capita personal consumption expenditures represents the Euler equation for the marginal utility of money over time.

Table 5.3 presents the unrestricted 13×13 AR(1) coefficient matrix. Similar to the variable cost function model, Eigen values of the implied autocovariance structure are well within the stability region, with five real roots and four complex conjugate pairs:

$\lambda_1 = 0.8960;$
$\lambda_2 = -0.5510;$
$\lambda_3 = 0.2829;$
$\lambda_4 = 0.1800;$
$\lambda_5 = -0.0471;$
$\lambda_{6,7} = 0.2966 \pm 0.6058\iota, \text{modulus} = 0.6745;$
$\lambda_{8,9} = 0.5789 \pm 0.0674\iota, \text{modulus} = 0.5828;$
$\lambda_{10,11} = -0.0048 \pm 0.5157\iota, \text{modulus} = 0.5157;$
$\lambda_{12,13} = -0.4497 \pm 0.1827\iota, \text{modulus} = 0.4854.$

Also similar to the properties of the cost function estimates, there is no evidence of any additional serial correlation in the error terms, and all of the systemwide and single equation Brownian bridge tests fail to reject the null hypothesis of no model specification errors or parameter instability at all standard levels of significance.

Table 5.4 presents the parameter estimates for the conditional mean components of the arbitrage model. To obtain a positive definite Θ matrix, the lower four main diagonal elements of the Choleski factor Q were restricted to 0.01, and the off-diagonal elements in the last four columns were restricted

Table 5.3   First-order autocorrelation parameters for the arbitrage/Euler equations

	Soybeans	Corn	Cotton	Hay	Potatoes	Rice	Sugar beets	Sugarcane	Tobacco	Wheat	Agriculture	Financial asset	Consumption
Soybeans	-.418	.696	-.813	.027	.162	.065	-.062	.036	-.670	.396	-1.25	-.037	-.078
	(.334)	(.321**)	(1.23)	(.064)	(.207)	(.193)	(.094)	(.115)	(1.00)	(.382)	(5.83)	(1.54)	(2.46)
	[.202*]	[.203***]	[.587]	[.049]	[.130]	[.121]	[.057]	[.065]	[.536]	[.190**]	[3.30]	[.940]	[1.54]
Corn	.341	.277	1.46	-.025	-.0044	.074	-.085	.045	-.383	.0085	1.90	-1.36	-1.38
	(.264)	(.253)	(.974)	(.050)	(.163)	(.153)	(.074)	(.090)	(.795)	(.301)	(4.60)	(1.21)	(1.94)
	[.140**]	[.122*]	[.470***]	[.034]	[.089]	[.080]	[.033**]	[.054]	[.298]	[.129]	[2.45]	[.629**]	[.986]
Cotton	-.054	.022	-.261	.014	-.045	-.064	-.00012	.013	.070	.011	-.558	.070	.436
	(.066)	(.063)	(.243)	(.012)	(.040)	(.038)	(.018)	(.022)	(.198)	(.075)	(1.15)	(.304)	(.486)
	[.055]	[.051]	[.206]	[.014]	[.027*]	[.038]	[.013]	[.018]	[.131]	[.055]	[1.10]	[.210]	[.361]
Hay	-1.71	2.62	-3.11	.244	.302	-.824	.167	.103	-1.40	1.67	27.6	2.55	-12.2
	(1.48)	(1.42*)	(5.49)	(.285)	(.920)	(.862)	(.418)	(.512)	(4.48)	(1.69)	(25.9)	(6.85)	(10.9)
	[.788*]	[1.08**]	[2.54]	[.177]	[.459]	[.559]	[.359]	[.318]	[2.37]	[.983*]	[21.2]	[3.57]	[7.16*]
Potatoes	-.081	.146	1.06	.159	.238	.033	.036	.139	-.904	-.191	1.61	2.99	.696
	(.358)	(.344)	(1.32)	(.068**)	(.222)	(.207)	(.100)	(.123)	(1.08)	(.409)	(6.26)	(1.65*)	(2.64)
	[.230]	[.182]	[.712]	[.041***]	[.162]	[.125]	[.051]	[.079*]	[.484*]	[.233]	[3.45]	[1.17**]	[.799]
Rice	.092	-.428	1.72	-.104	.230	-.021	-.148	-.067	-.188	.216	-1.19	.396	-2.34
	(.474)	(.455)	(1.75)	(.091)	(.293)	(.274)	(.133)	(.163)	(1.42)	(.541)	(8.28)	(2.18)	(3.49)
	[.295]	[.312]	[1.02*]	[.054*]	[.167]	[.186]	[.092]	[.083]	[.800]	[.326]	[6.26]	[1.42]	[2.07]
Sugar beets	-.044	-.693	6.03	-.065	-.793	.492	.072	.0049	1.15	-.172	13.9	2.15	-1.80
	(1.00)	(.966)	(3.71)	(.193)	(.622)	(.583)	(.282)	(.346)	(3.03)	(1.14)	(17.5)	(4.63)	(7.42)
	[.707]	[.749]	[3.63*]	[.156]	[.462*]	[.529]	[.277]	[.283]	[2.44]	[.976]	[10.33]	[4.02]	[7.27]

	(1)	(2)	(3)	(4)	(5)	(6)	(7)	(8)	(9)	(10)	(11)	(12)	(13)
Sugarcane	-.404	-.658	4.33	-.083	-1.17	.367	.413	-.482	.135	.148	-4.40	-6.93	-6.35
	(.821)	(.788)	(3.03)	(.157)	(.508**)	(.476)	(.230*)	(.282*)	(2.47)	(.938)	(14.3)	(3.78*)	(6.05)
	[.413]	[.518]	[2.21**]	[.067]	[.342***]	[.339]	[.191**]	[.196**]	[1.69]	[.562]	[7.58]	[2.25***]	[4.90]
Tobacco	-.0075	-.096	.866	-.0052	.018	-.028	.00048	-.0051	.103	-.175	.240	.467	-.881
	(.130)	(.124)	(.480*)	(.024)	(.080)	(.075)	(.036)	(.044)	(.392)	(.148)	(2.27)	(.600)	(.960)
	[.050]	[.045]	[.183***]	[.0078]	[.028]	[.024]	[.011]	[.018]	[.135]	[.047***]	[.912]	[.175***]	[.311***]
Wheat	-.377	.095	-1.59	.020	-.099	.138	-.090	.018	.085	.572	-3.57	-.909	3.15
	(.342)	(.328)	(1.26)	(.065)	(.212)	(.198)	(.096)	(.118)	(1.03)	(.391)	(5.98)	(1.57)	(2.52)
	[.200*]	[.173]	[.682**]	[.044]	[.095]	[.095]	[.051*]	[.068]	[.520]	[.181***]	[2.00*]	[.973]	[1.22***]
Agriculture	-.0015	.0092	.047	-.0014	.010	-.00046	-.0035	.0040	.048	.018	.505	.032	.0077
	(.012)	(.012)	(.046)	(.0024)	(.0078)	(.0073)	(.0035)	(.0043)	(.038)	(.014)	(.220**)	(.058)	(.093)
	[.0067]	[.0091]	[.025*]	[.0014]	[.0044**]	[.0045]	[.0023]	[.0030]	[.026*]	[.0089***]	[.151***]	[.040]	[.040]
Financial asset	-.036	-.082	.114	.0011	.0021	.048	.0016	-.0057	-.235	-.010	.761	.383	-.191
	(.047)	(.045)	(.175)	(.0091)	(.029)	(.027)	(.013)	(.016)	(.143)	(.054)	(.831)	(.219)	(.351)
	[.031]	[.036**]	[.099]	[.0057]	[.016]	[.021]	[.012]	[.011]	[.102]	[.039]	[.787]	[.143]	[.222]
Consumption	.016	.027	-.054	.017	.015	-.010	-.0012	.010	.022	.017	.119	-.033	.392
	(.036)	(.034)	(.133)	(.0069)	(.022)	(.020)	(.010)	(.012)	(.108)	(.041)	(.629)	(.166)	(.265)
	[.017]	[.012]	[.048]	[.0035]	[.0067]	[.010]	[.0038]	[.0048]	[.045]	[.013]	[.227]	[.081]	[.134]

Note: Numbers in parentheses and in square brackets are standard Gaussian and White heteroskedasticity consistent asymptotic standard errors, respectively.

***Significant at the 1 percent level.
**Significant at the 5 percent level.
*Significant at the 10 percent level.

Table 5.4   Estimates of the θ matrix parameters for the arbitrage/Euler equations

Variable	Soybeans	Corn	Cotton	Hay	Potatoes	Rice	Sugar beets	Sugarcane	Tobacco	Wheat
Constant	0.448166***	0.216785***	0.082527***	3.612791***	0.4599***	1.048429***	2.33634***	1.584347***	0.145584***	0.409247***
	(0.036632)	(0.024595)	(0.011218)	(0.197076)	(0.038374)	(0.064534)	(0.172455)	(0.161094)	(0.013593)	(0.033842)
Soybeans	78.34646***	-21.4571***	-1.80305	387.3322***	-32.5804*	41.71316**	-248.589***	-207.111***	3.799219	-0.08132
	(10.91386)	(3.992182)	(4.006355)	(88.54985)	(17.18041)	(20.3605)	(61.53115)	(52.65531)	(2.398324)	(7.195695)
Corn	-21.4571***	10.15897***	-2.01571	-83.6499***	0.028554	6.208851	79.67549***	65.14134***	1.87953**	-4.10214
	(3.992182)	(2.021877)	(1.236323)	(31.09819)	(4.755907)	(6.176972)	(17.03617)	(17.90792)	(0.767446)	(3.139472)
Cotton	-1.80305	-2.01571	35.56266***	-15.4651	115.6149***	-77.2543***	-215.395***	-369.447***	-3.23171*	0.82913
	(4.006355)	(1.236323)	(1.236323)	(59.45403)	(19.98758)	(29.06395)	(70.0208)	(65.29259)	(1.670253)	(2.366206)
Hay	387.3322***	-83.6499***	-15.4651	3920.242***	13.57345	-532.877**	523.7243	876.7413	-106.728***	101.2035*
	(88.54985)	(31.09819)	(59.45403)	(839.7853)	(205.9008)	(265.5106)	(593.9006)	(765.733)	(30.21382)	(55.11529)
Potatoes	-32.5804*	0.028554	115.6149***	13.57345	406.3396***	-353.176***	-427.114*	-916.150***	-27.1575***	16.57514*
	(17.18041)	(4.755907)	(19.98758)	(205.9008)	(110.575)	(80.82956)	(239.1877)	(247.4911)	(7.104441)	(8.869283)
Rice	41.71316**	6.208851	-77.2543***	-532.877**	-353.176***	577.8639***	-421.694*	-196.946	77.80513***	-67.5467***
	(20.3605)	(6.176972)	(29.06395)	(265.5106)	(80.82956)	(155.478)	(253.3207)	(271.0725)	(12.01163)	(11.23272)
Sugar beets	-248.589***	79.67549***	-215.395***	523.7243	-427.114*	-421.694*	3761.957***	4775.894***	-124.254***	112.5914***
	(61.53115)	(17.03617)	(70.0208)	(593.9006)	(239.1877)	(253.3207)	(1065.617)	(1221.091)	(21.47398)	(30.83312)
Sugarcane	-207.111***	65.14134***	-369.447***	876.7413	-916.150***	-196.946	4775.894***	6539.396***	-132.38***	135.5798***
	(52.65531)	(17.90792)	(65.29259)	(765.733)	(247.4911)	(271.0725)	(1221.091)	(1500.074)	(16.05683)	(31.36628)
Tobacco	3.799219	1.87953**	-3.23171*	-106.728***	-27.1575***	77.80513***	-124.254***	-132.38***	12.70347***	-11.9499***
	(2.398324)	(0.767446)	(1.670253)	(30.21382)	(7.104441)	(12.01163)	(21.47398)	(16.05683)	(1.405087)	(2.30313)
Wheat	-0.08132	-4.10214	0.82913	101.2035*	16.57514*	-67.5467***	112.5914***	135.5798***	-11.9499***	12.13339**
	(7.195695)	(3.139472)	(2.366206)	(55.11529)	(8.869283)	(11.23272)	(30.83312)	(31.36628)	(2.30313)	(5.291329)

Note: Numbers in parentheses are White heteroskedasticity consistent asymptotic standard errors.

***Significant at the 1 percent level.

**Significant at the 5 percent level.

*Significant at the 10 percent level.

at 0.0. In other words, the estimated symmetric but not curvature restricted Θ matrix has four negative Eigen values. As a consequence, the standard errors in table 5.2 are conditional on these inequality restrictions. The estimated structural parameters reported in table 5.4 generate a system of ten linear marginal cost functions that are increasing in planned output levels throughout the sample period. We conclude that this is a coherent and reasonable model of U.S. agricultural production.

The point estimate for the curvature parameter in the quadratic indirect utility function is $\hat{\beta} = 6.571 \times 10^{-5}$, with an estimated classical Gaussian asymptotic standard error of 2.793×10^{-7} and an estimated White/Huber robust standard error of 5.914×10^{-7}, both implying a highly significant risk aversion parameter. On the other hand, the point estimates for the quadratic adjustment cost parameters are mixed. The point estimate for adjustment costs on farmland is $\hat{\gamma}_A = -4.455 \times 10^{-6}$, with an estimated classical Gaussian asymptotic standard error of 1.985×10^{-6} and an estimated White/Huber robust standard error of 1.477×10^{-6}. In both cases, this is statistically different from zero at the 5 percent significance level, although economically, the sign is not what we would expect a priori. The point estimate for adjustment costs in farm capital is $\hat{\gamma}_K = 4.012 \times 10^{-11}$, with an estimated classical Gaussian asymptotic standard error of 2.738×10^{-11} and an estimated White/Huber robust standard error of 2.383×10^{-11}. While this has the expected sign, the classical standard error implies this is not statistically different from zero at the 10 percent level of significance, while the robust standard error implies that it marginally is significant at the same level. We suspect that either this level of aggregation across agents cannot capture these effects or else there is at most only a small level of adjustment cost in the farm sector. On the other hand, if there is no adjustment cost mechanism in U.S. agriculture, and if the quadratic indirect utility model is correctly specified, then the Euler equations estimated here are theoretically and empirically correct even with national aggregate data.

5.5   Conclusions

This chapter has developed and analyzed a new structural model of variable input use, production, acreage allocations, capital investment, and consumption choices in the U.S. farm sector. The theoretical framework identifies and incorporates the restrictions that are necessary and sufficient to estimate variable input use using only observable data and to aggregate from micro units of behavior to county-, state-, region-, or country-levels of data and analyses. We defined, specified, and estimated a dynamic life-cycle model of decision making under risk. We disciplined the model and associated parameter estimates for risk aversion in agricultural production and investment decisions with the interactions that naturally occur among the available alternative investment and savings opportunities in the economy.

Current work applies this to state-level data, which should mitigate the issues related to aggregating across different production regions, climates, and output choice sets. We incorporate input and output specific technological change in the empirical model, which should help address issues due to specification errors and structural change that cannot be captured in the aggregate setup. We are specifying and estimating the variable input use decisions and the asset management choices simultaneously to exploit cross-equation parameter restrictions and increase the efficiency of our parameter estimates. And last, the data set is in the final stages of being updated to the twenty-first century, which will make the model and empirical analysis more timely and relevant to current farm policies.

One of the central issues guiding agricultural policy is how risk affects choice and welfare. Here, that is manifest in the movement toward general equilibrium found in the cross-moment equations in equation (46) and the cost structure in equation (45). This provides a rich mechanism for policy analysis. The conventional agricultural focus is how policies affect the risk environment and thereby production choice and welfare. Thus, for example, in a partial equilibrium model of the farm sector, one often studies the effects of a particular policy on the risk environment on the portfolio of crop choice (Chavas and Holt 1996). Here, it is clear that the *evolution* of wealth and income in all forms, and consumption, "cause" production choices. Although this point is not new (e.g., Wright and Hewitt 1994), it has not been formally modeled and estimated.

With the results in table 5.2, one can trace the effects of any policy altering the distribution of agricultural crop income on the choices that restore equilibrium. More specifically, it means that significant responses may be outside of agriculture by changing nonagricultural investment and consumption. These responses likely will alter the normative and positive conclusions of the effects of policies substantially.

Indeed, returning to the example of crop insurance discussed in the introduction, the social value of public insurance will likely be reduced as more margins for adjustment (arbitrage conditions) are included in the analysis. In contrast, an increase in uncertainty (the covariance term) in nonagricultural investments as witnessed recently could increase the demand for risk-reducing agricultural instruments. The key point is that unless one has a model that provides for these interactions, one will not obtain reasonable policy conclusions.

The second general policy insight that can be obtained here is a distinction between long-run and shorter-run effects, which has been one of the foundations of agricultural policy analyses, preceding the seminal work of Nerlove (1958). Yet models in current vogue can only be interpreted as long-run analyses where adjustment costs are zero. This means that one has a natural structural way in the current model to distinguish short-run and long-run elasticities. For example, this implies that policies that raise the return to insurance (e.g., through public subsidies) have larger responses in the long run than in the short run.

Appendix A

Let $x \in \mathcal{X} \subseteq \mathbb{R}_{++}^{n_x}$ be an n_x-vector of variable inputs; let $w \in \mathcal{W} \subseteq \mathbb{R}_{++}^{n_x}$ be an n_x-vector of variable input prices; let $y \in \mathcal{Y} \subseteq \mathbb{R}_{++}^{n_x}$ be an n_y-vector of outputs; let $z \in \mathcal{Z} \subseteq \mathbb{R}_{++}^{n_x}$ be an n_z-vector of quasi-fixed inputs; let $F : \mathcal{X} \times \mathcal{Y} \times \mathcal{Z} \rightarrow \mathbb{R}$ be a transformation function that defines the boundary of a closed, convex production possibilities set with free disposal in inputs and outputs; let $X :$ $\mathcal{W} \times \mathcal{Y} \times \mathcal{Z} \rightarrow \mathcal{X}$, be an n_x-vector of variable input demand functions; and let $C : \mathcal{W} \times \mathcal{Y} \times \mathcal{Z} \rightarrow \mathbb{R}_{++}$ be a variable cost function,

$$(A1) \qquad c = C(w,y,z) \equiv \min_{x}\{w^{\mathsf{T}}x : F(x,y,z) \le 0, x \ge 0\} \equiv w^{\mathsf{T}}X(w,y,z),$$

where the symbol ^{T} denotes vector and matrix transposition. The purpose of this appendix is to prove that short-run cost-minimizing variable input demands, $x = X(w,y,z)$, can be written in the form $x = \tilde{X}(w,c,z)$ if and only if $c = C[w,z,\theta(y,z)] \Leftrightarrow F[x,z,\theta(y,z)]$.

The neoclassical model of conditional demands for variable inputs with joint production, quasi-fixed inputs, and production uncertainty is

$$(A2) \qquad X(w,y,z) = \arg\min \{w^{\mathsf{T}}x : F(x,y,z) \le 0, x \ge 0\},$$

where x is an n_x-vector of positive variable inputs with corresponding positive prices, w; y is an n_y-vector of planned outputs; z is an n_z-vector of quasi-fixed inputs; F is the real valued transformation function that defines the boundary of a closed, convex production possibilities set with free disposal in the inputs and the outputs; X maps variable input prices, planned outputs, and quasi-fixed inputs into variable input demand functions; and $C(w,y,z)$ $\equiv w^{\mathsf{T}}X(w,y,z)$, is the positive-valued variable cost function. By Shephard's lemma, we have

$$(A3) \qquad X(w,y,z) = \nabla_w C(w,y,z) \equiv \left(\frac{\partial C}{\partial w_1}, \ldots, \frac{\partial C}{\partial w_{n_x}}\right)^{\mathsf{T}}.$$

X is homogeneous of degree zero in w by the derivative property of homogeneous functions. Integrating with respect to w to recover the variable cost function, we obtain

$$(A4) \qquad c = C(w,y,z) \equiv \tilde{C}[w,y,z,\theta(y,z)],$$

where $\theta : \mathcal{Y} \times \mathcal{Z} \rightarrow \mathbb{R}$ is the *constant of integration*. In the present case, this means that θ is constant with respect to w. In general, θ is a function of y and z, and its structure cannot be identified from the variable input demands because it captures that part of the joint production process relating to quasi-fixed inputs and outputs that is separable from the variable inputs.

Under standard conditions, the variable cost function is strictly decreasing in z, strictly increasing in y, jointly convex in (y,z), increasing, concave and homogeneous of degree one in w. We are free to choose the *sign* of θ so that, with no loss of generality, $\partial \tilde{C}/\partial \theta > 0$.

Because $\tilde{C}$ is strictly increasing in θ, a unique inverse exists such that $\theta = \gamma(w,y,z,c)$, where $\gamma : \mathcal{W} \times \mathcal{Y} \times \mathcal{Z} \times \mathbb{R}_+ \to \mathbb{R}$, is the inverse of $\tilde{C}$ with respect to θ. $\gamma(w,y,z,c)$ is called the *quasi-indirect production transformation function,* analogous to the quasi-indirect utility function of consumer theory (Hausman 1981; Epstein 1982; LaFrance 1985, 1986, 1990, 2004; LaFrance and Hanemann 1989). For all interior and feasible (y,z), the function γ is strictly increasing in c, strictly decreasing and quasi-convex in w, and positively homogeneous of degree zero in (w,c).

The following two identities are simple implications of the inverse function theorem:

(A5) $$c \equiv \tilde{C}[w,y,z,\gamma(w,y,z,c)];$$

and

(A6) $$\theta \equiv \gamma[w,y,z,\tilde{C}(w,y,z,\theta)].$$

This lets one write the conditional demands for the variable inputs as

(A7) $$x = \nabla_w \tilde{C} \equiv G(w,y,z,c).$$

Equation (A7) gives the rationale for writing the factor demands as a function of c as well as (w,y,z). Thus, given the preceding regularity conditions for F and C, one can always write the system of factor demands as functions of cost.

Now define the quasi-production transformation function by

(A8) $$\upsilon(w,y,z) \equiv \min_{w \geq 0}\{\gamma(w,y,z,w^\mathsf{T}x)\}.$$

The terminology *quasi-production transformation function* indicates that $\upsilon(x, y, z)$ only reveals part of the structure of the joint production process. It cannot, and does not, reveal $\theta(y,z)$. This is analogous to the situation where one only recovers part of a direct utility function when analyzing the market demands for a subset of consumption goods.

The identity $\theta(y,z) \equiv \gamma\{w,y,z,\tilde{C}[w,y,z,\theta(y,z)]\}$ implies

(A9) $$\theta(y,z) \equiv \gamma\{w,y,z,\tilde{C}[w,y,z,\theta(y,z)]\} \geq \min_{w \geq 0}\{\gamma(w,y,z,w^\mathsf{T}x)\}$$
            $$\equiv \upsilon(x, y, z),$$

for all interior and feasible (x,y,z). This inequality follows from the fact that $\theta(y,z)$ is feasible but not necessarily optimal in the minimization problem.

The part of $F(x,y,z)$ not contained in $\upsilon(x,y,z)$ is given by (Diewert 1978; Epstein 1975; Hausman 1981; LaFrance and Hanemmann 1989),

(A10) $$F(x,y,z) \equiv \tilde{F}[x,y,z,\theta(y,z)].$$

The quasi-production transformation function is the unique solution, $\theta = \upsilon(x,y,z)$, to the implicit function, $\tilde{F}(x,y,z,\theta) = 0$; in other words, $\tilde{F}[x,y,z,\upsilon(x,y,z)] \equiv 0$.

The function $\upsilon(x,y,z)$ in equation (A9) conveys full information about the marginal rates of substitution between variable inputs but only partially so for outputs and quasi-fixed inputs. This is again analogous to the situation in consumption theory when one analyzes only a subset of the goods purchased and consumed. This can be shown by applying the implicit function theorem to $\tilde{F}$, which gives

(A11) $$\nabla_x \upsilon(x,y,z) = -\frac{\nabla_x \tilde{F}[x,y,z,\upsilon(x,y,z)]}{\nabla_\theta \tilde{F}[x,y,z,\upsilon(x,y,z)]},$$

$$\nabla_y \upsilon(x,y,z) = -\frac{\nabla_y \tilde{F}[x,y,z,\upsilon(x,y,z)]}{\nabla_\theta \tilde{F}[x,y,z,\upsilon(x,y,z)]},$$

$$\nabla_z \upsilon(x,y,z) = -\frac{\nabla_z \tilde{F}[x,y,z,\upsilon(x,y,z)]}{\nabla_\theta \tilde{F}[x,y,z,\upsilon(x,y,z)]}.$$

This demonstrates that υ conveys full information on marginal rates of substitution between variable inputs,

(A12) $$\frac{\partial \upsilon(x,y,z)/\partial x_i}{\partial \upsilon(x,y,z)/\partial x_j} = \frac{\partial \tilde{F}[x,y,z,\upsilon(x,y,z)]/\partial x_i}{\partial \tilde{F}[x,y,z,\upsilon(x,y,z)]/\partial x_j}$$
$$= \frac{\partial F(x,y,z)/\partial x_i}{\partial F(x,y,z)/\partial x_j}, \forall\, i,j = 1, \ldots, n_x,$$

but only partial information on marginal rates of product transformation between outputs,

(A13) $$\frac{\partial F(x,y,z)/\partial y_i}{\partial F(x,y,z)/\partial y_j}$$

$$= \frac{\partial \tilde{F}[x,y,z,\theta(y,z)]/\partial y_i + \partial \tilde{F}[x,y,z,\theta(y,z)]/\partial \theta \cdot \partial \theta(y,z)/\partial y_i}{\partial \tilde{F}[x,y,z,\theta(y,z)]/\partial y_j + \partial \tilde{F}[x,y,z,\theta(y,z)]/\partial \theta \cdot \partial \theta(y,z)/\partial y_j}$$

$$\neq \frac{\partial \upsilon(x,y,z)/\partial y_i}{\partial \upsilon(x,y,z)/\partial y_j}, \forall\, i,j = 1, \ldots, n_y,$$

and marginal rates of substitution between quasi-fixed inputs,

(A14) $\dfrac{\partial F(x,y,z)/\partial z_i}{\partial F(x,y,z)/\partial z_j}$

$$= \frac{\partial \tilde{F}[x,y,z,\theta(y,z)]/\partial z_i + \partial \tilde{F}[x,y,z,\theta(y,z)]/\partial \theta \cdot \partial \theta(y,z)/\partial z_i}{\partial \tilde{F}[x,y,z,\theta(y,z)]/\partial z_j + \partial \tilde{F}[x,y,z,\theta(y,z)]/\partial \theta \cdot \partial \theta(y,z)/\partial z_j}$$

$$\neq \frac{\partial \upsilon(x,y,z)/\partial z_i}{\partial \upsilon(x,y,z)/\partial z_j}, \forall\, i,j = 1, \ldots, n_z,$$

This background leads directly to the following result.

PROPOSITION 1. *The following functional structures are equivalent:*

(A15) $x = X(w,y,z) \equiv \tilde{X}(w,c,z);$

(A16) $c = C(w,y,z) \equiv \tilde{C}(w,z,\theta(y,z));$

and

(A17) $0 = F(w,y,z) \equiv \tilde{F}(x,z,\theta(y,z)).$

PROOF. (A16) $\Rightarrow$ (A15). Differentiating equation (A16) with respect to w, Shephard's lemma implies,

(A18) $x = \nabla_w \tilde{C}.$

$\tilde{C}$ is strictly monotonic in and has a unique inverse with respect to θ, say $\theta = \tilde{\gamma}(w,z,c)$. Substituting this into equation (A18) obtains

(A19) $x = \nabla_w \tilde{C}[w,z,\tilde{\gamma}(w,z,c)] \equiv \tilde{X}(w,c,z).$

(A17) $\Rightarrow$ (A15) $\Rightarrow$ (A16). If the representation of technology has the separable structure in equation (A17), then

(A20) $\arg\min \{w^\top x : \tilde{F}[x,z\theta(y,z)] \leq 0, x \geq 0\} \equiv \tilde{X}[w,z,\theta(y,z)].$

This implies that the variable cost function has the separable structure

(A21) $w^\top \tilde{X}[w,z,\theta(y,z)] \equiv \tilde{C}[w,z,\theta(y,z)].$

(A16) $\Rightarrow$ (A17). Given equation (A16), the quasi-production transformation function satisfies

(A22) $\upsilon(x,z) \equiv \min_{w \geq 0}\{\tilde{\gamma}(w,z,w^\top x)\}.$

This implies that

(A23) $\theta(y,z) \equiv \gamma\{x,z,\tilde{C}[x,z,\theta(y,z)]\} \geq \upsilon(x,z),$

for all interior, feasible $(x,y,z) \in \mathcal{X} \times \mathcal{Y} \times \mathcal{Z}$, with the boundary of the closed and convex production possibilities set defined by equality on the far right. Because υ is independent of y, equations (A11) and (A13) imply

(A24) $$\frac{\partial F(x,y,z)/\partial y_i}{\partial F(x,y,z)/\partial y_j} = \frac{\partial \theta(y,z)/\partial y_i}{\partial \theta(y,z)/\partial y_j}, \forall\, i,j = 1, \ldots, n_y.$$

Hence, the marginal rates of transformation between outputs are independent of variable inputs,

(A25) $$\frac{\partial}{\partial x_k}\left[\frac{\partial F(x,y,z)/\partial y_i}{\partial F(x,y,z)/\partial y_j}\right] = \frac{\partial}{\partial x_k}\left[\frac{\partial \theta(y,z)/\partial y_i}{\partial \theta(y,z)/\partial y_j}\right] = 0,$$

$$\forall\, i,j = 1, \ldots, n_y, \forall\, k = 1, \ldots, n_x,$$

Thus, y is separable from x in the joint production transformation function (Goldman and Uzawa 1964, Lemma 1); that is, $F(x,y,z) = \tilde{F}[x,z,\theta(y,z)]$. ∎

Appendix B
Sufficiency Algebra for Proposition 2

Define the function $\omega : \mathbb{R}_+ \times \mathbb{R} \to \mathbb{R}$ by

(B1) $$\omega(x,y) = y + \int_0^x [\lambda(s) + \omega(s,y)^2]ds,$$

where $\lambda : \mathbb{R} \to \mathbb{R}$ is an arbitrary smooth function, and w is subject to the pair of initial conditions, $w(0,y) = y$ and $\partial w(0,y)/\partial x = \lambda(0) + y^2$, to ensure that the definition is unique and smooth. Given two arbitrary smooth functions $\eta : \mathbb{R}_{++}^{n_x} \to \mathbb{R}_+$ and $\theta : \mathbb{R}^{n_z} \times \mathbb{R}^{n_y} \to \mathbb{R}$, by the Leibniz rule of differentiation, we have

(B2) $$\frac{\partial \omega[\eta(w), \theta(z,\bar{y})]}{\partial w} = \{\lambda[\eta(w)] + \omega[\eta(w), \theta(z,\bar{y})]^2\}\frac{\partial \eta(w)}{\partial w}.$$

Given a monotonic, smooth function $f : \mathbb{R}_{++} \to \mathbb{C}, f' \neq 0$, define the relationship between f and ω by $f = (\alpha\omega + \beta)/(\gamma\omega + \delta)$, $\alpha,\beta,\gamma,\delta : \mathbb{R}_{++}^n \to \mathbb{C}$, and $\alpha\delta - \beta\gamma \equiv 1$. Let the cost function be $c : \mathbb{R}_{++}^{n_x} \times \mathbb{R}_+^{n_z} \times \mathbb{R}_{++}^{n_y} \to \mathbb{R}_{++}$ and denote an arbitrary positive-valued, 1° homogeneous, increasing, and concave deflator by $\pi : \mathbb{R}_{++}^{n_x} \to \mathbb{R}_{++}$. The projective transformation group representation of any exactly aggregable ex ante cost function is

(B3) $$f\left[\frac{c(w,z,\bar{y})}{\pi(w)}\right] = \frac{\alpha(w)\omega[\eta(w), \theta(z,\bar{y})] + \beta(w)}{\gamma(w)\omega[\eta(w), \theta(z,\bar{y})] + \delta(w)}.$$

Hereafter, suppress all arguments of all functions and use bold italics subscripts to denote vector-valued partial derivatives. For example, rewrite equation (B2) compactly as $\omega_w = (\lambda + \omega^2)\eta_w$.

The inverse of equation (B3) with respect to ω is $\omega = (\delta f - \beta)/(-\gamma f + \alpha)$. Combine this with the identification normalization $\alpha\delta - \beta\gamma \equiv 1$ to obtain the following:

(B4) $\gamma\omega + \delta = \gamma\left(\dfrac{\delta f - \beta}{-\gamma f + \alpha}\right) + \delta = \dfrac{\gamma\delta f - \beta\gamma - \gamma\delta f + \alpha\delta}{-\gamma f + \alpha} = \dfrac{1}{-\gamma f + \alpha},$

or equivalently, $-\gamma f + \alpha = 1/(\gamma\omega + \delta)$. Multiply each side of this by the corresponding side of equation (B3) to obtain $(-\gamma f + \alpha)f = (\alpha\omega + \beta)/(\gamma\omega + \delta)^2$. These relationships are used in what follows to simplify expressions.

Our task is to differentiate equation (B3) with respect to w, combine terms, and rewrite the expression that results so that the elements of $\{1, f, f^2\}$ appear on the right. Differentiating gives

(B5) $f' \cdot \left(\dfrac{c_w}{\pi} - \dfrac{c\pi_w}{\pi^2}\right) = \dfrac{(\alpha_w\omega + \alpha\omega_w + \beta_w)}{(\gamma w + \delta)} - \dfrac{(\alpha\omega + \beta)(\gamma_w\omega + \gamma\omega_w + \delta_w)}{(\gamma w + \delta)^2}$

$= (-\gamma f + \alpha)[\alpha_w\omega + \alpha(\lambda + \omega^2)\eta_w + \beta_w]$

$- (-\gamma f + \alpha)f[\gamma_w\omega + \gamma(\lambda + \omega^2)\eta_w + \delta_w].$

The second line follows from $1/(\gamma\omega + \delta) = -\gamma f + \alpha$, $(\alpha\omega + \beta)/(\gamma\omega + \delta)^2 = (-\gamma f + \alpha)f$, and $\omega_w = (\lambda + \omega^2)\eta_w$. Group terms in ω on the second line of equation (B5) to obtain

(B6) $f' \cdot \left(\dfrac{c_w}{\pi} - \dfrac{c\pi_w}{\pi^2}\right) = (-\gamma f + \alpha)[\beta_w + \alpha\lambda\eta_w - (\delta_w + \gamma\lambda\eta_w)f]$

$+ (-\gamma f + \alpha)(\alpha_w - \gamma_w f)\omega + (-\gamma f + \alpha)^2\eta_w\omega^2.$

Substituting $\omega = (\delta f - \beta)/(-\gamma f + \alpha)$ into the second line of equation (B6) now leads to

(B7) $f' \cdot \left(\dfrac{c_w}{\pi} - \dfrac{c\pi_w}{\pi^2}\right) = (-\gamma f + \alpha)[\beta_w + \alpha\lambda\eta_w - (\delta_w + \gamma\lambda\eta_w)f]$

$+ (-\gamma f + \alpha)(\alpha_w - \gamma_w f)\left(\dfrac{\delta f - \beta}{-\gamma f + \alpha}\right)$

$+ (-\gamma f + \alpha)^2\eta_w\left(\dfrac{\delta f - \beta}{-\gamma f + \alpha}\right)^2$

$= (-\gamma f + \alpha)[\beta_w + \alpha\lambda\eta_w - (\delta_w + \gamma\lambda\eta_w)f]$

$+ (\alpha_w - \gamma_w f)(\delta f - \beta) + \eta_w(\delta f - \beta)^2.$

Expanding the quadratic forms and grouping terms in f in the last line of equation (B7) gives

(B8) $\quad f' \cdot \left(\dfrac{c_w}{\pi} - \dfrac{c\pi_w}{\pi^2} \right) = -\gamma f[\beta_w + \alpha\lambda\eta_w - (\delta_w + \gamma\lambda\eta_w)f]$

$\qquad\qquad + \alpha[\beta_w + \alpha\lambda\eta_w - (\delta_w + \gamma\lambda\eta_w)f]$

$\qquad\qquad\qquad - \alpha_w\beta + (\alpha_w\delta + \gamma_w\beta)f - \gamma_w\delta f^2$

$\qquad\qquad\qquad + \eta_w (\beta^2 - 2\beta\delta f + \delta^2 f^2)$

$\qquad\qquad = \alpha(\beta_w + \alpha\lambda\eta_w) - (\gamma\beta_w + \alpha\delta_w + 2\alpha\gamma\lambda\eta_w)f$

$\qquad\qquad\qquad + \gamma(\delta_w + \gamma\lambda\eta_v)f^2 - \alpha_w\beta + \eta_w\beta^2$

$\qquad\qquad\qquad + (\alpha_w\delta + \gamma w\beta - 2\beta\delta\eta_w)f + (-\gamma_w\delta + \eta_w\delta^2)f^2$

$\qquad\qquad = \alpha\beta_w - \beta\alpha_w + (\alpha^2\lambda + \beta^2)\eta_w$

$\qquad\qquad\qquad - [\alpha\delta_w - \delta\alpha_w + \gamma\beta_w - \beta\gamma_w + 2(\alpha\gamma\lambda + \beta\delta)\eta_w]f$

$\qquad\qquad\qquad + [\gamma\delta_w - \delta\gamma_w + (\gamma^2\lambda + \delta^2)\eta_w]f^2.$

Grouping terms in η_w as well gives

(B9) $\quad f' \cdot \left(\dfrac{c_w}{\pi} - \dfrac{c\pi_w}{\pi^2} \right) = \alpha\beta_w - \beta\alpha_w - (\alpha\delta_w - \delta\alpha_w + \gamma\beta_w - \beta\gamma_w)f$

$\qquad\qquad\qquad + (\gamma\delta_w - \delta\gamma_w)f^2 + [(\delta f - \beta)^2 + \lambda(-\gamma f + \alpha)^2]\eta_w.$

Finally, solving for $c_w = x$ gives

(B10) $\quad x = \dfrac{\pi_w}{\pi}c + \pi\left\{ [\alpha\beta_w - \beta\alpha_w + (\alpha^2\lambda + \beta^2)\eta_w]\dfrac{1}{f'} \right.$

$\qquad\qquad\qquad - [\alpha\delta_w - \delta\alpha_w + \gamma\beta_w - \beta\gamma_w + 2(\alpha\gamma\lambda + \beta\delta)\eta_w]\dfrac{f}{f'}$

$\qquad\qquad\qquad \left. + [\gamma\delta_w - \delta\gamma_w + (\gamma^2\lambda + \delta^2)\eta_w]\dfrac{f^2}{f'} \right\}$

$\qquad\qquad = \dfrac{\pi_w}{\pi}c + \dfrac{\pi}{f'}\{\alpha\beta_w - \beta\alpha_w - (\alpha\delta_w - \delta\alpha_w + \gamma\beta_w - \beta\gamma_w)f$

$\qquad\qquad\qquad + (\gamma\delta_w - \delta\gamma_w)f^2 + [(\delta f - \beta)^2 + (-\gamma f + \alpha)^2 \lambda]\eta_w\}.$

Appendix C
Specifying the Cost Function

The first $n_x - 1$ variable input prices, w, and total variable cost, c, are normalized by the average wage rate for hired farm labor, w_{n_x}. We consider the following transformation of normalized variable cost, which nests the price independent generalized logarithmic (PIGLOG) and price independent generalized linear (PIGL) class of models,

$$f(c) = \frac{(c^\kappa + \kappa - 1)}{\kappa}, f'(c) = c^\kappa, f''(c) = (\kappa - 1)c^{\kappa-2}, \kappa \in \mathbb{R}_+.$$

This includes all of the real-valued Gorman functional forms, with $f(c) = c$ when $\kappa = 1$, and $\lim_{\kappa \to 0} f(c) = 1 + \ln c$. Therefore, the highest rank that the variable input demands can achieve is three (Gorman 1981; Lewbel 1987; LaFrance and Pope 2009).

Previous empirical work considered translated Box-Cox functions of input prices, $(w_i^\lambda + \lambda - 1)/\lambda$, $\lambda \in [0,1]$, $i = 1, \ldots, n_x - 1$, to nest models with that have log prices, power functions of prices, and are linear prices. In the national model $\lambda = 1$ is optimal on this interval and for our data set. Hence, we restrict attention here to normalized input prices. Our previous empirical results using this data at state- and national-levels of aggregation and various levels of aggregation across inputs, suggests quite strongly that rank three overparameterizes this data set (Ball et al. 2010). Hence, we focus here on rank two:

(C1) $$f[c(w,A,K,a,\overline{Y})] = \alpha(w,A,K) + \beta(w)\theta(A,K,a,\overline{Y}),$$

$$\Leftrightarrow \tilde{\theta}(w,c,A,K) = \frac{f(c) - \alpha(w,A,K)}{\beta(w)},$$

$$\alpha(w,A,K) = (\alpha_{10} + \alpha_1^\top w)A + (\alpha_{20} + \alpha_2^\top w)K,$$

$$\beta(w) = \sqrt{w^\top Bw + 2\gamma^\top w + 1},$$

$$\overline{Y} = \overline{y} \cdot a_t = [\overline{y}_1 a_1 \ldots \overline{y}_{n_y} a_{n_y}]^\top,$$

where $\overline{y}_i$ is the expected (planned) yield for the ith crop, a_i is the acreage planted to this crop, and the symbol • denotes the Hadamard/Schur product for matrices and vectors. This appendix identifies restrictions on the parameters in equation (C1) that are necessary and sufficient for economic regularity of the variable cost function.

Monotonicity in w:

(C2) $$c^{\kappa-1}\frac{\partial c}{\partial w} = \alpha_1 A + \alpha_2 K + \frac{\theta}{\beta}(Bw + \gamma) \geq 0$$

$$\Leftrightarrow \tilde{x} = c^{1-\kappa}\left[\alpha_1 A + \alpha_2 K + \left(\frac{f - \alpha}{\beta^2}\right)(Bw + \gamma)\right] \geq 0,$$

where $\tilde{x} = [x_1 \ldots x_{n_x-1}]^\top$ is the (n_x-1)-vector of the first n_x-1 input quantities, excluding labor.

Concavity in w:

(C3) $$(\kappa - 1)c^{\kappa-2}\frac{\partial c}{\partial w}\frac{\partial c}{\partial w^\mathsf{T}} + c^{\kappa-1}\frac{\partial^2 c}{\partial w \partial w^\mathsf{T}} = \frac{\theta}{\beta}B - \frac{\theta}{\beta^3}(Bw + \gamma)(Bw + \gamma)^\mathsf{T}$$

$$\Leftrightarrow \frac{\partial^2 c}{\partial w \partial w^\mathsf{T}} = \left(\frac{1-\kappa}{c}\right)\tilde{x}\tilde{x}^\mathsf{T} + c^{1-\kappa}\left(\frac{f-\alpha}{\beta^2}\right)\left[B - \frac{(Bw + \gamma)(Bw + \gamma)^\mathsf{T}}{\beta^2}\right],$$

The first matrix on the right-hand side of the second line is rank one and is negative semidefinite if and only if $\kappa \geq 1$. The matrix in square brackets on the far right of the second line will be positive semidefinite if $B = LL^\mathsf{T} + \gamma\gamma^\mathsf{T}$, where L is a triangular matrix with nonzero main diagonal elements. This makes the following $n_x \times n_x$ matrix positive definite:

(C4) $$\begin{bmatrix} B & \gamma \\ \gamma^\mathsf{T} & 1 \end{bmatrix} = \begin{bmatrix} L & \gamma \\ 0^\mathsf{T} & 1 \end{bmatrix}\begin{bmatrix} L^\mathsf{T} & 0^\mathsf{T} \\ \gamma^\mathsf{T} & 1 \end{bmatrix} = \begin{bmatrix} LL^\mathsf{T} + \gamma\gamma^\mathsf{T} & \gamma \\ \gamma^\mathsf{T} & 1 \end{bmatrix},$$

since it gives a Choleski factorization of the matrix on the left. It follows from this that

$$w^\mathsf{T}BW + 2\gamma^\mathsf{T}w + 1 = \begin{bmatrix} w^\mathsf{T} & 1 \end{bmatrix}\begin{bmatrix} B & \gamma \\ \gamma^\mathsf{T} & 1 \end{bmatrix}\begin{bmatrix} w \\ 1 \end{bmatrix} > 0 \;\forall w \in \mathbb{R}^{n_x-1}_{++},$$

and

$$\left[B - \frac{(Bw + \gamma)(Bw + \gamma)^\mathsf{T}}{w^\mathsf{T}Bw + 2\gamma^\mathsf{T}w + 1}\right]$$

is positive semidefinite, by the Cauchy-Schwartz inequality in n_x-dimensional Euclidean space. Given this, the second term on the right-hand side of the second line of equation (12) will be negative semidefinite if and only if $f \leq \alpha$.

Constant returns to scale (CRS):

(C5) $$\theta \equiv \frac{\partial\theta}{\partial A}A + \frac{\partial\theta}{\partial K}K + \frac{\partial\theta}{\partial a^\mathsf{T}}a + \frac{\partial\theta}{\partial \overline{Y}^\mathsf{T}}\overline{Y}.$$

We believe that we have a much more accurate measure of capital than we do of land. Hence, we normalize θ by the value of capital rather than land in farms.

Monotonicity in $(A, K, a, \overline{Y})$:

(C6) $$c^{\kappa-1}\frac{\partial c}{\partial A} = \alpha_{10} + \alpha_1^\mathsf{T}w + \beta\frac{\partial\theta}{\partial A} \leq 0, c^{\kappa-1}\frac{\partial c}{\partial K} = \alpha_{20} + \alpha_2^\mathsf{T}w + \beta\frac{\partial\theta}{\partial K} \leq 0,$$

$$c^{\kappa-1}\frac{\partial c}{\partial a} = \beta\frac{\partial\theta}{\partial a} \leq 0, c^{\kappa-1}\frac{\partial c}{\partial \overline{Y}} = \beta\frac{\partial\theta}{\partial \overline{Y}} \geq 0.$$

Joint Convexity in $(A,K,a,\overline{Y})$:

$$(C7)\quad
\begin{bmatrix}
\dfrac{\partial^2 c}{\partial A^2} & \dfrac{\partial^2 c}{\partial A \partial K} & \dfrac{\partial^2 c}{\partial A \partial a^{\mathsf T}} & \dfrac{\partial^2 c}{\partial A \partial \overline{Y}^{\mathsf T}} \\[2mm]
\dfrac{\partial^2 c}{\partial K \partial A} & \dfrac{\partial^2 c}{\partial K^2} & \dfrac{\partial^2 c}{\partial K \partial a^{\mathsf T}} & \dfrac{\partial^2 c}{\partial K \partial \overline{Y}^{\mathsf T}} \\[2mm]
\dfrac{\partial^2 c}{\partial a \partial A} & \dfrac{\partial^2 c}{\partial a \partial K} & \dfrac{\partial^2 c}{\partial a \partial a^{\mathsf T}} & \dfrac{\partial^2 c}{\partial a \partial \overline{Y}^{\mathsf T}} \\[2mm]
\dfrac{\partial^2 c}{\partial \overline{Y} \partial A} & \dfrac{\partial^2 c}{\partial K \partial \overline{Y}} & \dfrac{\partial^2 c}{\partial a \partial \overline{Y}} & \dfrac{\partial^2 c}{\partial \overline{Y} \partial \overline{Y}^{\mathsf T}}
\end{bmatrix}
$$

$$
= \frac{1-\kappa}{c}
\begin{bmatrix}
\dfrac{\partial c}{\partial A} \\[2mm] \dfrac{\partial c}{\partial K} \\[2mm] \dfrac{\partial c}{\partial a} \\[2mm] \dfrac{\partial c}{\partial \overline{Y}}
\end{bmatrix}
\begin{bmatrix}
\dfrac{\partial c}{\partial A} \\[2mm] \dfrac{\partial c}{\partial K} \\[2mm] \dfrac{\partial c}{\partial a} \\[2mm] \dfrac{\partial c}{\partial \overline{Y}}
\end{bmatrix}^{\mathsf T}
+ c^{1-\kappa}\beta
\begin{bmatrix}
\dfrac{\partial^2 \theta}{\partial A^2} & \dfrac{\partial^2 \theta}{\partial A \partial K} & \dfrac{\partial^2 \theta}{\partial A \partial a^{\mathsf T}} & \dfrac{\partial^2 \theta}{\partial A \partial \overline{Y}^{\mathsf T}} \\[2mm]
\dfrac{\partial^2 \theta}{\partial K \partial A} & \dfrac{\partial^2 \theta}{\partial K^2} & \dfrac{\partial^2 \theta}{\partial K \partial a^{\mathsf T}} & \dfrac{\partial^2 \theta}{\partial K \partial \overline{Y}^{\mathsf T}} \\[2mm]
\dfrac{\partial^2 \theta}{\partial a \partial A} & \dfrac{\partial^2 \theta}{\partial a \partial K} & \dfrac{\partial^2 \theta}{\partial a \partial a^{\mathsf T}} & \dfrac{\partial^2 \theta}{\partial a \partial \overline{Y}^{\mathsf T}} \\[2mm]
\dfrac{\partial^2 \theta}{\partial \overline{Y} \partial A} & \dfrac{\partial^2 \theta}{\partial K \partial \overline{Y}} & \dfrac{\partial^2 \theta}{\partial a \partial \overline{Y}} & \dfrac{\partial^2 \theta}{\partial \overline{Y} \partial \overline{Y}^{\mathsf T}}
\end{bmatrix}.
$$

The first matrix on the right is rank one and will be positive semidefinite if and only if $\kappa \le 1$. Therefore, $c(w,A,K,a,\overline{Y})$ will be concave in w and jointly convex in $(A,K,a,\overline{Y})$ more than locally if and only if $\kappa = 1$. We estimated the rank two model using the Box-Cox transformation on cost. The NL3SLS/GMM point estimate for κ is 1.124 with a classical (Gaussian) asymptotic standard error of .152 and a White/Huber heteroskedasticity consistent standard error of .111. We cannot reject a null hypothesis of $\kappa = 1$ in either case at the 25 percent significance level. Hence, in this chapter, we restrict our attention to $\kappa = 1$.

Given this restriction, the cost function will be jointly convex in $(A,K,a,\overline{Y})$ if and only if the Hessian matrix for θ,

$$(C8)\quad
\begin{bmatrix}
\dfrac{\partial^2 \theta}{\partial A^2} & \dfrac{\partial^2 \theta}{\partial A \partial K} & \dfrac{\partial^2 \theta}{\partial A \partial a^{\mathsf T}} & \dfrac{\partial^2 \theta}{\partial A \partial \overline{Y}^{\mathsf T}} \\[2mm]
\dfrac{\partial^2 \theta}{\partial K \partial A} & \dfrac{\partial^2 \theta}{\partial K^2} & \dfrac{\partial^2 \theta}{\partial K \partial a^{\mathsf T}} & \dfrac{\partial^2 \theta}{\partial K \partial \overline{Y}^{\mathsf T}} \\[2mm]
\dfrac{\partial^2 \theta}{\partial a \partial A} & \dfrac{\partial^2 \theta}{\partial a \partial K} & \dfrac{\partial^2 \theta}{\partial a \partial a^{\mathsf T}} & \dfrac{\partial^2 \theta}{\partial a \partial \overline{Y}^{\mathsf T}} \\[2mm]
\dfrac{\partial^2 \theta}{\partial \overline{Y} \partial A} & \dfrac{\partial^2 \theta}{\partial K \partial \overline{Y}} & \dfrac{\partial^2 \theta}{\partial \overline{Y} \partial a^{\mathsf T}} & \dfrac{\partial^2 \theta}{\partial \overline{Y} \partial \overline{Y}^{\mathsf T}}
\end{bmatrix}.
$$

is positive semidefinite. Given these considerations, the specification for θ employed in the chapter is

(C9) $\theta(A_t, K_t, a_t, \overline{Y}_t) =$

$$-\theta_1 A_t - \theta_2 K_t - \theta_3^\top a_t + \theta_4^\top \overline{Y}_t + \frac{1}{2}\left(\frac{\theta_5 A_t^2 + a_t^\top \Theta_6 a_t + \overline{Y}_t^\top \Theta_7 \overline{Y}_t}{K_t} \right),$$

where $\theta_1, \theta_2, \theta_5 > 0$, $\theta_3, \theta_4 > 0$, and Θ_6, Θ_7 are symmetric and positive semidefinite. The implied constraints for monotonicity can be written as

(C10) $\dfrac{\partial c_t}{\partial A_t} < 0 \,\forall\, t \Leftrightarrow \min_t\left(\theta_1 - \theta_5 \dfrac{A_t}{K_t} \right) > \max_t\left(\dfrac{\alpha_0 + \alpha_1^\top w_t}{\sqrt{\beta(w)}} \right),$

$\dfrac{\partial c_t}{\partial K_t} < 0 \,\forall\, t \Leftrightarrow \min_t\left[\theta_2 + \dfrac{1}{2}\left(\dfrac{\theta_5 A_t^2 + a_t^\top \Theta_6 a_6 + \overline{Y}_t^\top \Theta_7 Y}{K_t^2} \right) \right] > \max_t\left(\dfrac{\alpha_2 + \alpha_3^\top w_t}{\sqrt{\beta(w)}} \right),$

$\dfrac{\partial c_t}{\partial a_t} < 0 \,\forall\, t \Leftrightarrow \theta_3 > \max_t\left(\dfrac{\Theta_6 a_6}{K_t} \right),$

$\dfrac{\partial c_t}{\partial \overline{Y}_t} > 0 \,\forall\, t \Leftrightarrow \theta_4 + \min_t\left(\dfrac{\Theta_7 \overline{Y}_t}{K_t} \right) > 0.$

These can be imposed iteratively in estimation if necessary (LaFrance 1991). In this chapter, we checked for the monotonicity conditions at each data point given the parameter estimates obtained without imposing monotonicity.

Also, given that $K_t > 0$, the implied curvature conditions are that the matrix

(C11) $$\begin{bmatrix} \theta_5 K_t^2 & -\theta_5 A_t K_t & \mathbf{0}_{n_y}^\top & \mathbf{0}_{n_y}^\top \\ -\theta_5 A_t K_t & (\theta_5 A_t^2 + a_t^\top \Theta_6 a_6 + \overline{Y}_t^\top \Theta_7 \overline{Y}_t) & -K_t a_t^\top \Theta_6 & -K_t \overline{Y}_t^\top \Theta_7 \\ \mathbf{0}_{n_y} & -K_t \Theta_6 a_6 & K_t^2 \Theta_6 & \mathbf{0}_{n_y \times n_y} \\ \mathbf{0}_{n_y} & -K_t \Theta_7 \overline{Y}_7 & \mathbf{0}_{n_y \times n_y} & K_t^2 \Theta_7 \end{bmatrix}$$

is positive semidefinite. This can be imposed during estimation with the Choleski factors, $\Theta_6 = L_6 L_6^\top$ and $\Theta_7 = L_7 L_7^\top$ with L_6 and L_7 lower triangular Choleski factors for Θ_6 and Θ_7, respectively, and the inequality $\theta_5 > 0$. In this chapter, only the matrix $\Theta_7 = L_7 L_7^\top$ is estimated as part of the arbitrage conditions.

Appendix D
Specification Errors and Parameter Stability Tests

Many diagnostic procedures for testing parameter stability and model specification errors have been developed. Few are designed for large systems of nonlinear simultaneous equations in small samples. These properties preclude using recursive-forecast residuals or Chow tests based on sequential sample splits to analyze specification errors or nonconstant parameters (Brown, Durbin, and Evans 1975; Harvey 1990, 1993; Hendry 1995). It is desirable to test whether the data are consistent with the model specification and constant parameters. LaFrance (2008) derived a set of specification and parameter stability diagnostics for this class of problems. These test statistics rely on the estimated in-sample residuals and have power against a range of alternatives, including nonconstant parameters and specification errors. The purpose of this section is to discuss briefly the main ideas that underpin this class of test statistics.

If the model is stationary and the errors are innovations, then consistent estimates of the model parameters can be found in any number of ways. Given consistent parameter estimates, the estimated errors converge in probability (and, therefore, in distribution) to the true errors, $\hat{\varepsilon}_t \xrightarrow{P} \varepsilon_t$. Therefore, for each $i = 1, \ldots, n_x - 1$, by the central limit theorem for stationary Martingale differences, we have

$$(D1) \qquad \frac{1}{\sqrt{T}\sigma_i} \sum_{t=1}^{T} \varepsilon_{it} \xrightarrow{D} N(0,1),$$

where $\sigma_i^2 = E(\varepsilon_{it}^2)$ is the variance of the residual for the ith demand equation. Moreover, for any given proportion of the sample, uniformly in $z \in [0, 1]$,

$$(D2) \qquad \frac{1}{\sqrt{T}\sigma_i} \sum_{t=1}^{[zT]} \varepsilon_{it} \xrightarrow{D} N(0,z),$$

where $[zT]$ is the largest integer that does not exceed zT. The variance is z because we sum $[zT]$ independent terms each with variance $1/T$. Multiplying equation (D1) by z and subtracting from equation (D2) then gives

$$(D3) \qquad \frac{1}{\sqrt{T}\sigma_i} \sum_{t=1}^{[zT]} \left(\varepsilon_{it} - \overline{\varepsilon}_i \right) \xrightarrow{D} W(z) - z\,W(1) \equiv B(z),$$

where $W(z)$ is a standard Brownian motion on the unit interval, with $W(z) \sim N(0, z)$, and $B(z)$ is a standard Brownian bridge, or tied Brownian motion. For all $z \in [0,1]$, $B(z)$ has an asymptotic Gaussian distribution, with mean zero and standard deviation $[z(1-z)]^{1/2}$ (Bhattacharya and Waymire 1990). For a given z—that is, to test for a break point in the model at a fixed and known date—an asymptotic 95 percent confidence interval for $B(z)$ is $\pm 1.96[z(1 - z)]^{1/2}$. To check for an unknown break point, a statistic based on the supremum norm,

(D4)
$$Q_T = \sup_{z \in [0,1]} \left| B_t(z) \right|$$

has an asymptotic 5 percent critical value of 1.36 (Ploberger and Krämer 1992).

We can use consistently estimated residuals and consistently estimated standard errors to obtain sample analogues to these asymptotic Brownian bridges. This gives

(D5)
$$B_{iT}(z) \equiv \frac{1}{\sqrt{T}\hat{\sigma}_i} \sum_{t=1}^{[zT]} \left(\hat{\varepsilon}_{it} - \bar{\hat{\varepsilon}}_i \right) \xrightarrow{D} B(z),$$

also uniformly in $z \in [0, 1]$ so long as the model specification is correct and the parameters are constant across time periods. This statistic is a single equation first-order specification/parameter stability statistic because it is based on the first-order moment conditions, $E(\varepsilon_{it}) = 0 \; \forall \; i, t$. A systemwide first-order specification/parameter stability statistic can be defined by

(D6)
$$B_T(z) \equiv \frac{1}{\sqrt{T}} \sum_{t-1}^{[zT]} \left[\frac{1}{\sqrt{n_x}} \sum_{i=1}^{n_q} \left(\hat{\xi}_{it} - \bar{\hat{\xi}} \right) \right] \xrightarrow{D} B(z),$$

where $\hat{\xi}_t = \hat{\Sigma}^{-1/2}\hat{\varepsilon}_t$ is the tth estimated standardized error vector and $\bar{\hat{\xi}} \equiv \sum_{t=1}^{T}\sum_{i=1}^{n_x} \hat{\xi}_{it}/n_x T$.

Similar methods apply to second-order stationarity and parameter stability. We focus on systemwide statistics. Let Σ be factored into LL^T, where L is lower triangular and nonsingular. Define the random vector ξ_t by $\varepsilon_t = L\xi_t$. In addition to the preceding assumptions, add $\sup_{i,t} E(\varepsilon_{it}^4) < \infty$. Estimate the within-period average sum of squared standardized residuals by

(D7)
$$\hat{\upsilon}_t = \frac{1}{n_x}\hat{\xi}_t^T\hat{\xi}_t = \frac{1}{n_x}\hat{\varepsilon}_t^T\hat{\Sigma}^{-1}\hat{\varepsilon}_t,$$

where $\hat{\varepsilon}_t$ is the vector of consistently estimated residuals in period t, and $\hat{\Sigma} = \sum_{t=1}^{T}\hat{\varepsilon}_t\hat{\varepsilon}_t^T/T$ is the associated consistently estimated error covariance matrix. The mean of the true υ_t is one for each t, and the martingale difference property of ε_t is inherited by $\upsilon_t - 1$. A consistent estimator of the asymptotic variance of υ_t is

(D8)
$$\hat{\sigma}_\upsilon^2 = \frac{1}{T}\sum_{t=1}^{T}(\hat{\upsilon}_t^2 - 1).$$

A systemwide second-order specification/parameter stability test statistic is obtained by calculating centered and standardized partial sums of $\hat{\upsilon}_t$,

(D9)
$$B_T(z) = \frac{1}{\sqrt{T}\hat{\sigma}_\upsilon} \cdot \sum_{t=1}^{[zT]} (\hat{\upsilon}_t - 1) \xrightarrow[T\to\infty]{D} B(z),$$

uniformly in $z \in [0, 1]$, where the limiting distribution on the far right follows from the identity $\bar{\hat{\upsilon}} \equiv \sum_{t=1}^{T}\hat{\upsilon}_t/T \equiv 1$.

References

Akridge, J. T., and T. W. Hertel. 1986. "Multiproduct Cost Relationships for Retail Fertilizer Plants." *American Journal of Agricultural Economics* 68:928–38.

Babcock, B. A., and D. Hennessy. 1996. "Input Demand under Yield and Revenue Insurance." *American Journal of Agricultural Economics* 78:416–27.

Ball, E., R. Cavazos, J. T. LaFrance, R. Pope, and J. Tack. 2010. "Aggregation and Arbitrage in Joint Production." In *The Economic Impact of Public Support to Agriculture,* edited by E. Ball, R. Fanfani, and L. Gutierrez, 309–28. New York: Springer.

Ball, V. E., C. Hallahan, and R. Nehring. 2004. "Convergence of Productivity: An Analysis of the Catch-up Hypothesis within a Panel of States." *American Journal of Agricultural Economics* 86:1315–21.

Banks, J., R. Blundell, and A. Lewbel. 1997. "Quadratic Engel Curves and Consumer Demand." *Review of Economics and Statistics* 79:527–39.

Bellman, R., and S. Dreyfus. 1962. *Applied Dynamic Programming.* Princeton, NJ: Princeton University Press.

Bhattacharya, R. N., and E. C. Waymire. 1990. *Stochastic Processes with Applications.* New York: Wiley.

Blackorby, C., D. Primont, and R. Russell. 1977. "Dual Price and Quantity Aggregation." *Journal of Economic Theory* 14:130–48.

———. 1978. *Separability and Functional Structure: Theory and Economic Applications.* New York: American Elsevier/North-Holland.

Blundell, R. 1988. "Consumer Behavior: Theory and Empirical Evidence—A Survey." *Economic Journal* 98:16–65.

Brown, R. L., J. Durbin, and J. M. Evans. 1975. Techniques for Testing the Constancy of Regression Relationships over Time. *Journal of the Royal Statistical Society* Series B 37:149–192.

Chambers, R. G. 1984. "A Note on Separability of the Indirect Production Function and Measures of Substitution." *Southern Economic Journal* 4:1189–91.

———. 1989. "Insurability and Moral Hazard in Agricultural Insurance Markets." *American Journal of Agricultural Economics* 71:604–16.

Chambers, R. G., and R. D. Pope. 1991. "Testing for Consistent Aggregation." *American Journal of Agricultural Economics* 73:808–18.

———. 1994. "A Virtually Ideal Production System: Specifying and Estimating the VIPS Model." *American Journal of Agricultural Economics* 76:105–13.

Chambers, R. G., and J. Quiggin. 2000. *Uncertainty, Production, Choice and Agency: The State-Contingent Approach.* Cambridge, UK: Cambridge University Press.

Chavas, J. P. 2008. "A Cost Approach to Economic Analysis under State-Contingent Production Uncertainty." *American Journal of Agricultural Economics* 90: 435–66.

Chavas J. P., and M. Holt. 1996. "Economic Behavior under Uncertainty: A Joint

Analysis of Risk Preferences and Technology." *Review of Economics and Statistics* 78:329–35.

Deaton, A., and J. Muellbauer. 1980. "An Almost Ideal Demand System." *American Economic Review* 70:312–26.

Diewert, W. 1978. "Hicks' Aggregation Theorem and the Existence of a Real Value-Added Function." In *Production Economics: A Dual Approach to Theory and Applications.* Vol. 2, edited by M. Fuss and D. McFadden, 17–51. New York: North-Holland.

Diewert, W. E., and T. J. Wales. 1987. "Flexible Functional Forms and Global Curvature Conditions." *Econometrica* 55:43–68.

———. 1988. "Normalized Quadratic Systems of Consumer Demand Functions." *Journal of Business and Economic Statistics* 6:303–12.

Epstein, L. G. 1975. "A Disaggregate Analysis of Consumer Choice under Uncertainty." *Econometrica* 47:877–92.

———. 1982. "Integrability of Incomplete Systems of Demand Functions." *Review of Economic Studies* 49:411–25.

Färe, R., and D. Primont. 1995. *Multi-Output Production and Duality: Theory and Applications.* Boston: Kluwer-Nijhoff.

Gardner, B. L., and R. A. Kramer. 1986. "Experience with Crop Insurance Programs in the United States." In *Crop Insurance for Agricultural Development: Issues and Experience,* edited by P. Hazell, C. Pomereda, and A. Valdes, 195–222. Baltimore: Johns Hopkins University Press.

Goldman, S. M., and H. Uzawa. 1964. "A Note on Separability in Demand Analysis." *Econometrica* 32:387–98.

Goodwin, B. K., and V. H. Smith. 1999. "An Ex-Post Evaluation of the Conservation Reserve, Federal Crop Insurance, and other Government Programs: Program Participation and Soil Erosion." *Journal of Agricultural and Resource Economics* 28 (2): 201–16.

Gorman, W. M. 1953. "Community Preference Fields." *Econometrica* 21:63–80.

———. 1961. "On a Class of Preference Fields." *Metroeconomica* 13:53–6.

———. 1981. "Some Engel Curves." In *Essays in Honour of Sir Richard Stone,* edited by A. Deaton, 7–29. Cambridge, UK: Cambridge University Press.

Hall, R. E. 1973. "The Specification of Technology with Several Kinds of Output." *Journal of Political Economy* 81:878–92.

Hansen, L. P., and K. J. Singleton. 1983. "Stochastic Consumption, Risk Aversion, and the Temporal Behavior of Asset Returns. *Journal of Political Economy* 91: 249–65.

Harvey, A. C. 1990. *The Econometric Analysis of Time Series.* Cambridge, MA: MIT Press.

———. 1993. *Time Series Models.* 2nd ed. Cambridge, MA: MIT Press.

Hausman, J. 1981. "Exact Consumer's Surplus and Deadweight Loss." *American Economic Review* 71:662–76.

Hendry, D. F. 1995. *Dynamic Econometrics.* Oxford, UK: Oxford University Press.

Hoppe, R. A., and D. E. Banker. 2006. *Structure and Finances of U.S. Farms: 2005 Family Farm Report.* Economic Information Bulletin no. EIB-12. Washington, DC: Economic Research Service, U.S. Department of Agriculture.

Horowitz, J., and E. Lichtenberg. 1994. "Risk Reducing and Risk Increasing Effects of Pesticides." *Journal of Agricultural Economics* 45:82–9.

Howe, H., R. A. Pollak, and T. J. Wales. 1979. "Theory and Time Series Estimation of the Quadratic Expenditure System." *Econometrica* 47:1231–47.

Jerison, M. 1993. "Russell on Gorman's Engel Curves: A Correction." *Economics Letters* 23:171–75.

Jorgenson, D. W. 1990. "Consumer Behavior and the Measurement of Social Welfare." *Econometrica* 58:1007–40.

Jorgenson, D. W., L. J. Lau, and T. M. Stoker. 1980. "Welfare Comparisons under Exact Aggregation." *American Economic Review* 70:268–72.

———. 1982. "The Transendental Logarithmic Model of Aggregate Consumer Behavior." In *Advances in Econometrics,* edited by R. L. Basmann and G. F. Rhodes, Jr., 97–238. Greenwich, CT: JAI Press.

Jorgenson, D. W., and D. T. Slesnick. 1984. "Aggregate Consumer Behavior and the Measurement of Inequality." *Review of Economic Studies* 51:369–92.

———. 1987. "Aggregate Consumer Behavior and Household Equivalence Scales." *Journal of Business and Economic Statistics* 5:219–32.

Just, R., D. Zilberman, and E. Hochman. 1983. "Estimation of Multicrop Production Functions." *American Journal of Agricultural Economics* 65:770–80.

Keeton, K., J. Skees, and J. Long. 1999. "The Potential Influence of Risk Management Programs on Cropping Decisions." Paper presented at the annual meeting of the American Agricultural Economics Association, Nashville, Tennessee.

Kohli, U. 1983. "Non-Joint Technologies." *Review of Economic Studies* 50:209–19.

LaFrance, J. T. 1985. "Linear Demand Functions in Theory and Practice." *Journal of Economic Theory* 37:147–66.

———. 1986. "The Structure of Constant Elasticity Demand Models." *American Journal of Agricultural Economics* 68:543–52.

———. 1990. "Incomplete Demand Systems and Semilogarithmic Demand Models." *Australian Journal of Agricultural Economics* 34:118–31.

———. 1991. "Consumer's Surplus versus Compensating Variation Revisited." *American Journal of Agricultural Economics* 73 (5): 1495–1507.

———. 2004. "Integrability of the Linear Approximate Almost Ideal Demand System." *Economic Letters* 84:297–303.

———. 2008. "The Structure of U.S. Food Demand." *Journal of Econometrics* 147:336–49.

LaFrance, J. T., T. K. M. Beatty, and R. D. Pope. 2006. "Gorman Engel Curves for Incomplete Demand Systems." In *Exploring Frontiers in Applied Economics: Essays in Honor of Stanley R. Johnson,* edited by M. T. Holt and J.-P. Chavas, 1–15. Berkeley, CA: Berkeley Electronic Press.

LaFrance, J. T., T. K. M. Beatty, R. D. Pope, and G. K. Agnew. 2002. "Information Theoretic Measures of the U.S. Variable Cost Distribution in Food Demand." *Journal of Econometrics* 107:235–57.

LaFrance, J. T., and W. M. Hanemann. 1989. "The Dual Structure of Incomplete Demand Systems." *American Journal of Agricultural Economics* 71:262–74.

LaFrance, J. T., and R. D. Pope. 2008. "Homogeneity and Supply." *American Journal of Agricultural Economics* 92:606–12.

———. 2009. "The Generalized Quadratic Expenditure System." In *Contributions to Economic Analysis: Quantifying Consumer Preferences,* edited by Daniel Slottje, 84–116. New York: Elsevier-Science.

———. 2010. "Duality Theory for Variable Costs in Joint Production." *American Journal of Agricultural Economics* 92:755–62.

LaFrance, J. T., J. P. Shimshack, and S. Y. Wu. 2000. "Subsidized Crop Insurance and the Extensive Margin." Washington DC: USDA Economists Group, September.

———. 2001. "The Environmental Impacts of Subsidized Crop Insurance." Proceedings of California Workshop on Environmental and Resource Economics, Santa Barbara, California.

———. 2002. "Crop Insurance and the Extensive Margin." Paper presented at World

Congress of Environmental and Resource Economists (AERE/EAERE), Monterey, California.

————. 2004. "Subsidized Crop Insurance and the Extensive Margin." In *Risk Management in Agriculture.* Rome, Italy: Instituto di Servizi per il Mercato Agricolo Alimentare.

Lau, L. 1972. "Profit Functions of Technologies with Multiple Input and Output." *Review of Economics and Statistics* 54:281–9.

————. 1978. "Applications of Profit Functions." In *Production Economics: A Dual Approach to Theory and Applications,* edited by M. Fuss and D. McFadden, 134–216. Amsterdam: North-Holland.

————. 1982. "A Note on the Fundamental Theorem of Exact Aggregation." *Economics Letters* 9:119–26.

Lewbel, A. 1987. "Characterizing Some Gorman Systems that Satisfy Consistent Aggregation." *Econometrica* 55:1451–59.

————. 1988. "An Exactly Aggregable Trigonometric Engel Curve Demand System." *Econometric Reviews* 2:97–102.

————. 1989. "A Demand System Rank Theorem." *Econometrica* 57:701–5.

————. 1990. "Full Rank Demand Systems." *International Economic Review* 31:289–300.

————. 1991. "The Rank of Demand Systems: Theory and Nonparametric Estimation." *Econometrica* 59:711–30.

————. 2003. "A Rational Rank Four Demand System." *Journal of Applied Econometrics* 18:127–35.

Lopez, R. 1983. "Structural Implications of a Class of Flexible Functional Forms for Profit Functions." *International Economic Review* 26:593–601.

Muellbauer, J. 1975. "Aggregation, Variable Cost Distribution and Consumer Demand." *Review of Economic Studies* 42:525–43.

————. 1976. "Community Preferences and the Representative Consumer." *Econometrica* 44:979–99.

Nelson, C. H., and E. T. Loehman. 1987. "Further toward a Theory of Agricultural Insurance." *American Journal of Agricultural Economics* 69:523–31.

Nerlove, M. 1958. The Dynamics of Supply: Estimation of Farmers' Response to Price. Baltimore: Johns Hopkins University Press.

Pope, R. D., and J.-P. Chavas. 1994. "Cost Functions under Production Uncertainty." *American Journal of Agricultural Economics* 76:196–204.

Pope, R. D., J. T. LaFrance, and R. E. Just. 2007. "Imperfect Price Deflation in Production Systems." *American Journal of Agricultural Economics* 89:738–54.

Ploberger, W., and W. Krämer. 1992. *"The CUSUM Test with OLS Residuals."* Econometrica 60:271–86.

Quiggin, J. 1992. "Some Observations on Insurance, Bankruptcy, and Input Demand." *Journal of Economic Behavior and Organization* 18:101–10.

Russell, T. 1983. "On a Theorem of Gorman." *Economic Letters* 11:223–24.

————. 1996. "Gorman Demand Systems and Lie Transformation Groups: A Reply." *Economic Letters* 51:201–4.

Russell, T., and F. Farris. 1993. "The Geometric Structure of Some Systems of Demand Functions." *Journal of Mathematical Economics* 22:309–25.

————. 1998. "Integrability, Gorman Systems, and the Lie Bracket Structure of the Real Line." *Journal of Mathematical Economics* 29:183–209.

Shumway, C. R. 1983. "Supply, Demand, and Technology in a Multiproduct Industry: Texas Field Crops." *American Journal of Agricultural Economics* 65:748–60.

Smith, V. H., and B. K. Goodwin. 1996. "Crop Insurance, Moral Hazard and Agricultural Chemical Use." *American Journal of Agricultural Economics* 78:428–38.

Soule, M., W. Nimon, and D. Mullarkey. 2000. "Risk Management and Environmental Outcomes: Framing the Issues." Paper presented at Crop Insurance, Land Use and the Environment Workshop (USDA/ERS), Washington D.C.

Turvey, C. G. 1992. "An Economic Analysis of Alternative Farm Revenue Insurance Policies." *Canadian Journal of Agricultural Economics* 40:403–26.

van Daal, J., and A. H. Q. M. Merkies. 1989. "A Note on the Quadratic Expenditure Model." *Econometrica* 57:1439–43.

Williams, J. R. 1988. "A Stochastic Dominance Analysis of Tillage and Crop Insurance Practices in a Semiarid Region." *American Journal of Agricultural Economics* 70:112–20.

Wright, B. D., and J. A. Hewitt. 1994. "All-Risk Crop Insurance: Lessons from Theory and Experience." In *Economics of Agricultural Crop Insurance: Theory and Evidence,* edited by D. L. Hueth and W. H. Furtan, 73–112. Natural Resource Management and Policy Series. Boston: Kluwer Academic.

Wu, J. 1999. "Crop Insurance, Acreage Decisions and Non-Point Source Pollution." *American Journal of Agricultural Economics* 81:305–20.

Young, C. E., R. D. Schnepf, J. R. Skees, and W. W. Lin. 2000. "Production and Price Impacts of U.S. Crop Insurance Subsidies: Some Preliminary Results." Paper presented at Crop Insurance, Land Use and the Environment Workshop (USDA/ERS), Washington D.C.

II

Biofuels and Biotechnology

6

Commodity Price Volatility in the Biofuel Era
An Examination of the Linkage between Energy and Agricultural Markets

Thomas W. Hertel and Jayson Beckman

6.1 Introduction

U.S. policymakers have responded to increased public interest in reducing greenhouse gas (GHG) emissions and lessening dependence on foreign supplies of energy with a Renewable Fuels Standard (RFS) that imposes aggressive mandates on biofuel use in domestic refining. These mandates are in addition to the longstanding price policies (blending subsidies and import tariffs) used to promote the domestic ethanol industry's growth. Recently, a number of authors have begun to explore the linkages between energy and agricultural markets in light of these new policies (McPhail and Babock 2008; Hochman, Sexton, and Zilberman 2008; Gohin and Chantret 2010; Tyner 2009). It is clear from this work that we are entering a new era in which energy prices will play a more important role in driving agricultural commodity prices. However, based on experience during the past year, it is also clear that the coordination between energy and agricultural markets is fundamentally different at high oil prices versus low oil prices, as well as in the presence of binding policy regimes.

Figure 6.1 illustrates how the linkage between energy and corn prices has

Thomas W. Hertel is Distinguished Professor of Agricultural Economics, adjunct professor of economics, and executive director of the Center for Global Trade Analysis Project at Purdue University. Jayson Beckman is an economist at the Economic Research Service of the U.S. Department of Agriculture.

The authors thank Wally Tyner for valuable discussions on this topic. V. Kerry Smith served as our National Bureau of Economic Research (NBER) discussant; he and members of the NBER workshop, as well as two anonymous reviewers, provided useful comments on this chapter. The views expressed are those of the authors and do not necessarily reflect those of the Economic Research Service (ERS) or the U.S. Department of Agriculture (USDA).

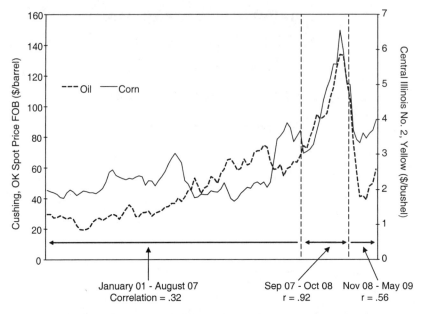

Fig. 6.1   Monthly oil (Cushing, OK Spot Price $/barrel) and corn (Central Illinois no. 2 Yellow $/bushel) prices, January 2001 to May 2009

varied over the 2001 to 2009 period. With oil prices below $75 a barrel from January 2001 to August 2007, the correlation between monthly oil and corn prices was just 0.32. During much of this period, the share of corn production going to ethanol was still modest, and ethanol capacity was still being constructed. Also, considerable excess profits appear to have been available to the industry over this period (figure 6.2)—a phenomenon that loosened any potential link between ethanol prices on the one hand and corn prices on the other. Indeed, Tyner (2009) reports a –0.08 correlation between ethanol and corn prices in the period 1988 to 2005. The year 2006 was a key turning point in the ethanol market, as this was when methyl tertiary butyl ether (MTBE) was banned as an additive and ethanol took over the entire market for oxygenator/octane enhancers in gasoline. In this use, the demand for ethanol was not very price-responsive and ethanol was priced at a premium when converted to an energy equivalent basis.

When oil prices reached and remained above $75 a barrel from September 2007 to October 2008, the correlation between crude oil and corn became much stronger (0.92, see figure 6.1 again), with per bushel corn prices remaining consistently at about 5 percent of crude oil prices per barrel. In this price range, the 2008 RFS appeared to be nonbinding. However, as oil prices subsequently fell, many ethanol plants were mothballed, and the RFS became binding at year's end in 2008. That is to say, without this mandate, even less ethanol would have been produced in December of that

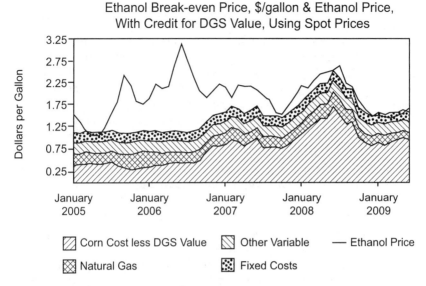

Fig. 6.2   Relationship between output and input prices in the ethanol industry over time: 2005–April 2009

Source: Iowa Ethanol Report, EIA. Compiled by Robert Wisner, Iowa State University.

year. Markets moved into a different price regime with the difference being made up in the value of the renewable fuel certificates required by blenders under the RFS.

While the RFS became temporarily nonbinding with the onset of a new year in 2009, a new phenomenon began to emerge, namely the presence of a blend wall (Tyner 2009). With refineries unable to blend more than 10 percent ethanol into gasoline for normal consumption at that time, an excess supply of ethanol began to emerge in many regional markets. (Due to infrastructure limitations and state regulations, there is not a single national market for ethanol.)[1] This led to a weakening of the link between ethanol and oil prices, with the crude oil price continuing to fall, while corn prices and, hence, ethanol prices remained at levels that no longer permit ethanol to compete with petroleum on an energy basis; therefore, the monthly corn-petroleum price correlation in the final period reported in figure 6.1 is much weaker (0.56).[2]

In this chapter, we develop a framework specifically designed for analyzing the linkages between energy and agricultural markets under different

1. See ASTM-D4814.
2. An output-based link still exists under the blend wall because changes in the liquid fuel price affect the demand for biofuels by altering the consumption of liquid fuels. However, this now works in the opposite direction as lower oil prices boost fuel consumption and, hence, ethanol demand.

policy regimes.[3] We employ a combination of theoretical analysis, econometrics, and stochastic simulation. Specifically, we are interested in examining how energy price volatility has been transmitted to commodity prices and how changes in energy policy regimes affect the inherent volatility of agricultural commodity prices in response to traditional supply-side shocks. We find that biofuels have played an important role in facilitating increased integration between energy and agricultural markets. In the absence of a binding RFS, and assuming that the blend wall is relaxed by expanding the maximum permissible ethanol content in petroleum as has recently been the case, we find that, by 2015, the contribution of energy price volatility to year-on-year corn price variation will be much greater—amounting to nearly two-thirds of the crop supply-induced volatility. However, if the RFS is binding in 2015, then the role of energy price volatility in crop price volatility is diminished. Meanwhile, the sensitivity of crop prices to traditional supply-side shocks is exacerbated due to the price inelastic nature of RFS demands. Indeed, the presence of a totally inelastic demand for corn in ethanol—stemming from the combination of a blend wall and a RFS both set in the range of fifteen billion gallons per year—would boost the sensitivity of corn prices to supply-side shocks by more than 50 percent.

6.2 Literature Review

Energy and energy intensive inputs play a large role in the production of agricultural products. Gellings and Parmenter (2004) estimate that energy accounts for 70 to 80 percent of the total costs used to manufacture fertilizers, which, in turn, represent a large component of corn production costs. Additional linkages come in the form of transportation of inputs and the final output as well as the use of diesel or gasoline on-farm. Overall, USDA/ERS *Cost of Production* estimates indicate that energy inputs accounted for almost 30 percent of the total cost of corn production for the United States in 2008.[4]

Another important linkage to energy markets is on the output side as agricultural commodities are increasingly being used as feedstocks for liquid biofuels. Hertel, Tyner, and Birur (2010) estimate that higher oil prices accounted for about two-thirds of the growth in U.S. ethanol output over the 2001 to 2006 period. The remainder of this growth is estimated to have

3. We ignore the nonmarket impacts of biofuels, which are important and have commanded much of the public's attention—particularly since the publication of Searchinger et al. (2008). Carbone and Smith (2008) point out how the presence of such considerations can introduce interactions that alter the market and welfare impacts of environmental policies.

4. Comparing the USDA numbers across time regimes further strengthens our argument that the link between energy and agricultural commodities has increased over time. From 1996 to 2000, the average share of energy inputs (fertilizer and fuel, lube, and electricity) in total corn producer costs was 19.6 percent. From 2001 to 2004, this average share was 20.9 percent. But for 2007 to 2008, the share increased to 31.5 percent.

been driven by the replacement of the banned gasoline additive, MTBE, with ethanol in petroleum refining. In the European Union (EU), those authors estimate that biodiesel growth over the same period was more heavily influenced by subsidies. Nonetheless, those authors estimate that oil price increases accounted for about two-fifths of the expansion in EU biofuel production over the 2001 to 2006 period.

These growing linkages between energy and agricultural commodities have received increasing attention by researchers. Tyner (2009) notes that, since 2006, the ethanol market has established a link between crude oil and corn prices that did not exist historically. He finds that the correlation between annual crude oil and corn prices was negative (−0.26) from 1988 to 2005; in contrast, it reached a value of 0.80 during the 2006 to 2008 period. And, as figure 6.1 shows, the correlation from September 2007 to October 2008 was 0.92.

Du, Yu, and Hayes (2009) investigate the spillover of crude oil price volatility to agricultural markets (specifically corn and wheat). They find that the spillover effects are not statistically significant from zero over the period from November 1998 to October 2006. However, when they look at the period October 2006 to January 2009, the results indicate significant volatility spillover from the crude oil market to the corn market.

In a pair of papers focusing on the cointegration of prices for oil, ethanol, and feedstocks, Serra and coauthors study the U.S. (Serra et al. 2010a) and Brazilian (Serra et al. 2010b) ethanol markets. In the case of the United States, they find the existence of a long-term equilibrium relationship between these prices, with ethanol deviating from this equilibrium in the short term (they work with daily data from 2005 to 2007 in the case of the United States and weekly data in the case of Brazil). For the United States, the authors find the prices of oil, ethanol, and corn to be positively correlated as might be expected, although they also find evidence of a structural break in this relationship in 2006 when the competing fuel oxygenator (MTBE) was banned, and ethanol demand surged to fill this need. The authors estimate that a 10 percent perturbation in corn prices boosts ethanol prices by 15 percent—a somewhat peculiar finding, given that corn represents only a portion of total ethanol costs.[5] From the other side, they find that a 10 percent rise in the price of oil leads to a 10 percent rise in ethanol, as one might expect of products that are perfect substitutes in use (perhaps an overly strong assumption in this case). In terms of temporal response time,

---

5. In an industry characterized by zero pure profits, a cost share-weighted sum of input price changes must equal the percentage change in output price. With corn comprising less than full costs, its price should change at a rate less than the output price, not more than the output price as reported in this study. For an industry starting in equilibrium to remain in equilibrium after corn prices rise by 10 percent and ethanol prices rise by 15 percent, returns to other inputs must also rise—likely by a very significant amount. Yet recent evidence suggests that higher corn prices reduce returns to capital in the U.S. ethanol industry. So this is a puzzling result.

they find that the response to corn prices is much quicker (1.25 months to full impact) than for an oil price shock (4.25 months).

In the Serra et al. (2010b) study of Brazil, the relevant feedstock is sugar cane. This presents a rather different commodity relationship because many of the sugarcane refining facilities can produce either ethanol or refined sugar, the latter of which sells into the food market, not the energy market. Brazil also has a much more mature ethanol market. Ethanol production and use has been actively promoted by the government since the 1973 oil crisis, and it now dominates petroleum in the domestic transportation market, with more than 70 percent of new car sales comprising flex-fuel vehicles accommodating either a 25 percent or 75 percent ethanol-gasoline blend or 100 percent ethanol-based fuel. Serra et al. (2010b) build on the long-run price parity relationships between ethanol and oil on the one hand (substitution in use) and ethanol and refined sugar on the other (substitution in production). They find that sugar and oil prices are exogenously determined and focus their attention on the response of ethanol prices to changes in these two exogenous drivers. The authors conclude that ethanol prices respond relatively quickly to sugar price changes but more slowly to oil prices. A shift in either of these prices has a very short-run impact on ethanol price volatility as well. Within one year, most of the adjustment to long-run equilibrium in both markets has occurred. However, it takes nearly two years for the full effect of an oil price shock to be reflected in ethanol prices. So overall, these commodity markets are not as quick to regain long-run equilibrium as those in the United States, based on the results in these two studies. The authors do not find evidence of ethanol prices or oil prices affecting long-run sugar prices over the period of their analysis, which spans the period July 2000 to February 2008.

Using similar time series econometric techniques, Ubilava and Holt (2010) investigate a different but related hypothesis regarding energy and feedstock prices in the United States. They test the hypothesis that including energy prices in a time series model of corn prices should improve the latter's ability to forecast corn prices. Recognizing that this relationship might well be regime-dependent (e.g., a closer linkage at high oil prices), they allow for such nonlinear responses. However, their findings, using weekly averages of daily futures data for the United States over the period October 2006 to June 2009, do not support these hypotheses; that is, the inclusion of energy prices in the time series model does not improve its forecast accuracy. While they are asking a different question (and using weekly instead of daily data), this finding appears to stand at odds with the findings of Serra et al. (2010a) and suggests the need for replication and further testing of these models.

Based on this evidence it appears that, where it exists, the close link between crude oil prices and corn prices in the United States is a relatively recent phenomenon; hence, econometric investigations of price transmission suffer from insufficient historical time series. For this reason, stochastic simulation has been an important vehicle to examine this topic in the United States. McPhail and Babcock (2008) developed a partial equilibrium model to simulate the outcomes for the 2008 to 2009 corn market based on stochastic shocks to planted acreage, corn yield, export demand, gasoline prices, and the ethanol industry capacity. They estimate that gasoline price volatility and corn price volatility are positively related, and, for example, gasoline price volatility of 25 percent standard deviation (i.e., if prices are normally distributed 68 percent of the time, the gasoline price will be within ±25 percent of the mean gasoline price) would lead to volatility in the corn price of 17.5 percent standard deviation.

Thompson, Meyer, and Westhoff (2009) also utilize a stochastic framework (based on the Food and Agricultural Policy Research Institute [FAPRI] model) to examine how shocks to the crude oil (and corn) markets can affect ethanol price and use. They note that the RFS introduces a discontinuity between crude oil and ethanol prices. As a consequence, they find that the implied elasticity of a change in oil price on corn price is 0.31 (i.e., a 1 percent increase in the price of oil leads to a 0.31 percent increase in the corn price) with no RFS and 0.17 with the RFS.[6] In subsequent work, Meyer and Thompson (2010) provide a more comprehensive analysis of the impact of biofuels and biofuel policies on corn price volatility using the FAPRI model baseline. They find (perhaps not surprisingly) that the presence of tariffs and credits does not alter corn price volatility significantly. However, the introduction of a mandate, in the form of the U.S. Renewable Fuels Standard, does cause some rise in volatility, although they do not provide information about how often the mandate is binding in their stochastic simulations.

A final paper in this line of partial equilibrium, stochastic simulation analyses of corn ethanol policies and corn prices is that of Gohin and Tréguer (2010) who find that biofuels policies destabilize corn prices by reducing the frequency with which farm policy instruments are binding. These authors also introduce producer risk aversion into their model. Inclusion of downside risk aversion dampens the supply response of producers to the biofuel policy. The presence of downside risk aversion also serves to contribute to additional welfare gains from biofuels policies, as producers are less exposed to low-end prices in the presence of these policies.

This review of the literature suggests the potential for some interesting hypotheses about potential linkages between agricultural and energy markets. The purpose of the next section of the chapter is to develop an analytical framework within which these can be clearly stated as a set of formal propositions.

---

6. These figures appear to be quite different from those offered by Serra et al. (2010a) for the United States, which appear to suggest a tighter relationship between oil and ethanol and between corn and ethanol. However, those authors do not offer a comparable number in their paper.

6.3   Analytical Framework

Consider an ethanol industry producing total output (Q_E) and selling it into two domestic market segments: in the first market, ethanol is used as a gasoline additive (QA_E), in strict proportion to total gasoline production.[7] As previously discussed, legal developments in the additive market (the banning of more economical MTBE as an oxygenator/octane enhancer) were an important component of the U.S. ethanol boom between 2001 and 2006. The second market segment is the market for ethanol as a price-sensitive energy substitute (QP_E). In contrast to the additive market, the demand in this market depends importantly on the relative prices of ethanol and petroleum. For ease of exposition, and to be consistent with the general equilibrium specification introduced later on, we will model the additive demand as a derived demand by the petroleum refinery sector and the energy substitution as being undertaken by consumers of liquid fuel. By assigning two different agents in the economy to these two functions, we can clearly specify the market shares governed by the two different types of behavior.[8]

Market clearing for ethanol, in the absence of exports, may then be written as:

$$(1) \qquad\qquad Q_E = QA_E + QP_E,$$

or, in percentage change form, where lowercase denotes the percentage change in the uppercase variable:

$$(2) \qquad\qquad q_E = (1 - \alpha)qa_E + \alpha qp_E$$

We denote the share of total ethanol output (Q_E) going to the price-sensitive side of the market with $\alpha = QP_E/Q_E$.

Now we formally characterize the behavior of each source of demand for ethanol as follows (again, lowercase variables denote percentage changes in their uppercase counterparts):

$$(3) \qquad\qquad qa_E = q_F,$$

where q_F is the percentage change in the total production of liquid fuel, for which the additive/oxygenator is demanded in fixed proportions. The price-sensitive portion of ethanol demand can be parsimoniously parameterized as follows:

$$(4) \qquad\qquad qp_E = qp_F - \sigma(p_E - p_F),$$

where qp_F is the percentage change in total liquid fuel consumption by the price-sensitive portion of the market (i.e., households), and σ is the

7. This may also be viewed as the "involuntary" demand for ethanol, in the words of Meyer and Thompson (2010). Those authors also include in this category additional state-level regulations such as the 10 percent ethanol blending requirement in the state of Minnesota.

8. This modeling of the two different ethanol uses gives rise to the "kinked-demand" curve referred to by some authors (e.g., McPhail and Babcock 2008).

constant elasticity of substitution among liquid fuel sources consumed by households. The price ratio P_E/P_F refers to the price of ethanol, relative to a composite price index of all liquid fuel products consumed by the household. The percentage change in this ratio is given by the difference in the percentage changes in the two prices: ($p_E - p_F$). When premultiplied by σ, this determines the price-sensitive component of households' change in demand for ethanol. Substituting equations (3) and (4) into equation (2), we obtain a revised expression for ethanol market clearing:

$$(5) \qquad q_E = (1 - \alpha)(qa_E) + \alpha[qp_F - \sigma(p_E - p_F)]$$

On the supply side, we assume constant returns to scale in ethanol production, which, along with entry/exit (a very common phenomenon in the ethanol industry since late 2007—indeed today plants shut down one month and start up the next), gives zero pure profits:

$$(6) \qquad p_E = \sum_j \theta_{jE} p_{jE}$$

Where p_E is the percentage change in the producer price for ethanol, p_{jE} is the percentage change in price of input j, used in ethanol production, and θ_{jE} is the share of that input in total ethanol costs (see figure 6.2 for evidence of the validity of equation [6] since 2007). Assuming noncorn inputs supplied to the ethanol sector in this partial equilibrium model (e.g., labor and capital) are in perfectly elastic supply, and abstracting from direct energy use in ethanol production (both assumptions will be relaxed in the following numerical general equilibrium model), we have $p_{jE} = 0$, $\forall j \neq C$, and we can solve equation (6) for the corn price in terms of ethanol price changes:

$$(7) \qquad p_{CE} = \theta_{CE}^{-1} p_E.$$

Assuming that corn is used in fixed proportion to ethanol output (i.e., Q_{CE}/Q_E is fixed), we can complete the supply-side specification for the ethanol market with the following equations governing the derived demand for and supply of corn in ethanol:

$$(8) \qquad q_{CE} = q_E$$

$$(9) \qquad q_{CE} = v_{CE} p_{CE},$$

where v_{CE} is the *net* supply elasticity of corn to the ethanol sector; that is, it is equal to the supply elasticity of corn, net of the price responsiveness in other demands for corn (outside of ethanol). This will be developed in more detail in the following when we turn to equilibrium in the corn market. Substituting equation (9) into equation (8) and then using equation (7) to eliminate the corn price, we obtain an equation for the *market supply of ethanol:*

$$(10) \qquad q_E = v_{CE} \theta_{CE}^{-1} p_E$$

Now turn to the corn market, where there are two sources of demand for corn output (Q_C): the ethanol industry, which buys Q_{CE}, and all other uses of corn, Q_{CO}. Letting β denote the share of total corn sales to ethanol, market clearing in the corn market may thus be written as:

$$(11) \qquad q_C = \beta q_{CE} + (1-\beta)q_{CO}.$$

We characterize nonethanol corn demands as consisting of two parts: a price-sensitive portion governed by a simple, constant elasticity of corn demand, η_{CD}, as well as a random demand shock (e.g., stemming from a shock to gross domestic product [GDP] in the home or foreign markets), Δ_{CD}. Ethanol demand for corn has already been specified in equation (8). We will shortly solve for q_{CE}, so we leave that in the equation, giving us the following market clearing condition for corn:

$$(12) \qquad q_C = \beta q_{CE} + (1-\beta)(\eta_{CD}p_C + \Delta_{CD})$$

As with demand, corn supply is specified via a price-responsive portion, governed by the constant elasticity of supply, η_{CS}, and a random supply shock (e.g., driven by weather volatility), Δ_{CS}, yielding:

$$(13) \qquad q_C = \eta_{CS}p_C + \Delta_{CS}.$$

At this point, we can derive an expression for the net corn supply to ethanol production by solving equation (12) for q_{CE} and using equation (13) to eliminate corn supply (q_C). This yields the following expression for net corn supply to the ethanol industry:

$$(14) \qquad q_{CE} = \left\{ \frac{[\eta_{CS} - (1-\beta)\eta_{CD}]}{\beta} \right\} p_C + \frac{[\Delta_{CS} - (1-\beta)\Delta_{CD}]}{\beta}$$

The term in brackets {.} is v_{CE}, the *net* supply elasticity of corn to the ethanol sector.[9] With $\beta < 1$ and $\eta_{CD} < 0$, this net supply elasticity is larger than the conventional corn supply elasticity, with the difference between the two diminishing as the share of corn sold to ethanol grows ($\beta \rightarrow 1$) and the price responsiveness of other corn demands falls ($\eta_{CD} \rightarrow 0$).

The second term in equation (14) translates random shocks to corn supply and other corn demands into random shocks to net corn supply to ethanol. The larger the shocks, the more volatile are the shocks to corn supply and demand (which we will assume to be independently distributed in the following empirical section) and the smaller the share of ethanol demand in total corn use. We denote the total effect of this random component (the second term in equation [14]) by the term Δ_{CE}, which we term the random shock to the net supply of corn to the ethanol industry.

We can now solve this simple model for equilibrium in the corn ethanol market. To do so, we make a number of additional assumptions. First, we assume that growth in the household portion of the liquid fuel market (qp_F) is equal to growth in total liquid fuel use (q_F) and that this aggregate liquid

9. This expression closely resembles the earlier work of de Gorter and Just (2008).

fuel demand may be characterized via a constant elasticity of demand for liquid fuels, η_{FD}. This permits us to write the aggregate demand for ethanol as follows:

(15) $$q_E = \eta_{FD}p_F - \alpha\sigma(p_E - p_F)$$

For purposes of this simple, partial equilibrium analytical exercise, we will assume that the share of ethanol in aggregate liquid fuel use is small so that we may ignore the impact of p_E on p_F. In so doing, we will consider the liquid fuels price to be synonymous with the price of petroleum. Thus, a 1 percent shock to the price of ethanol will reduce total ethanol demand by $\alpha\sigma$. Conversely, a 1 percent exogenous shock to the price of petroleum has two separate effects on the demand for ethanol, one negative (the expansion effect) and one positive (the substitution effect): $\eta_{FD} + \alpha\sigma$. Provided the share of total sales to the price-responsive portion of the market (α) is large enough, and assuming ethanol is a reasonably good substitute for petroleum, then the second (positive) term dominates, and we expect the rise in petroleum prices to lead to a rise in the demand for ethanol. However, if for some reason the second term is eliminated—for example, due to ethanol demand encountering a blend wall, as described by Tyner (2009)—then this relationship may be reversed; that is, a rise in petroleum prices will reduce the aggregate demand for liquid fuels, and, in so doing, it will reduce the demand for ethanol.

We solve the model by equating ethanol supply in equation (14) to ethanol demand in equation (15), noting that corn demand in ethanol changes proportionately with ethanol production in equation (8), and using equation (7) to translate the change in corn price into a change in ethanol price:

(16) $$q_E = v_{CE}\theta_{CE}^{-1}p_E + \Delta_{CE} = \eta_{FD}p_F - \alpha\sigma(p_E - p_F)$$

Equation (16) may be solved for the price of ethanol as a function of exogenous shocks to the corn market and to the liquid fuels market:

(17) $$(v_{CE}\theta_{CE}^{-1} + \alpha\sigma)p_E = (\eta_{FD} + \alpha\sigma)p_F - \Delta_{CE}$$

This gives rise to:

(18) $$p_E = \frac{(\eta_{FD} + \alpha\sigma)p_F - \Delta_{CE}}{v_{CE} + \theta_{CE}\alpha\sigma}.$$

This equilibrium outcome may be translated back into a change in corn prices, via equation (7):

(19) $$p_C = \frac{(\eta_{FD} + \alpha\sigma)p_F - \Delta_{CE}}{v_{CE} + \theta_{CE}\alpha\sigma}$$

It is now clear that a random shock to the nonethanol corn market, which in turn perturbs the net supply of corn to ethanol (Δ_{CE}), will result in a larger change in corn price, the more inelastic are corn supply and demand (as reflected by the v_{CE} term in the denominator of equation [19]) and the smaller the elasticity of substitution between ethanol and petroleum (σ), the smaller the share of ethanol going to the price responsive portion of the fuel market (α), and the smaller the cost share of corn in ethanol production (θ_{CE}). However, the role of the sales share of corn going to ethanol (β) is ambiguous and requires further analysis.

Consider first the impact only of a random shock to corn supply. Substitute into equation (19) the following relationships:

$$(20) \quad v_{CE} = \left\{ \frac{[\eta_{CS} - (1-\beta)\eta_{CD}]}{\beta} \right\}, \text{ and } \Delta_{CE} = \frac{[\Delta_{CS} - (1-\beta)\Delta_{CD}]}{\beta},$$

and ignore the demand-side shock to obtain:

$$(21) \quad p_C = \frac{[-\Delta_{CS}/\beta]}{(\{[\eta_{CS} - (1-\beta)\eta_{CD}]/\beta\} + \theta_{CE}\alpha\sigma)}.$$

Multiplying top and bottom by β and rearranging the denominator, we get:

$$(22) \quad p_C = \frac{[-\Delta_{CS}]}{[(\eta_{CS} - \eta_{CD}) + \beta(\theta_{CE}\alpha\sigma + \eta_{CD})]}.$$

Now, it is clear that, provided the derived demand elasticity for corn in ethanol use exceeds that in other uses, that is, $\theta_{CE}\alpha\sigma > -\eta_{CD}$, a rise in the share of corn sales to ethanol will dampen the volatility of corn prices in response to a corn supply shock. Of course, if something were to happen in the fuel market, for example ethanol use hits the blend wall, then the potential for substituting ethanol for petroleum would be eliminated. In this case, the opposite result will apply, namely, an increased reliance of corn producers on ethanol markets will actually destabilize corn market responses to corn supply shocks. As we will see in the following, this is a very important result.

Similarly in the case of a corn demand shock, substitution into equation (19) and reorganization yields the following expression:

$$(23) \quad p_C = \frac{[(1-\beta)\Delta_{CD}]}{[(\eta_{CS} - \eta_{CD}] + \beta(\theta_{CE}\alpha\sigma + \eta_{CD})]}$$

The presence of $(1-\beta)$ in the numerator means that higher values of β reduce the size of the numerator. Provided the derived demand for corn by ethanol is more price responsive than nonethanol demand, such that higher values of β increase the denominator in equation (23), we can say unambiguously that increased ethanol sales to corn results in more corn price stability in response to a given nonethanol demand shock. However, when the derived demand for corn by ethanol is less price responsive than nonethanol demand, the outcome is ambiguous.

Finally, consider the impact only of a random shock to fuel prices. Proceeding as before, we obtain the following expression:

$$(24) \qquad p_C = \frac{[(\eta_{FD} + \alpha\sigma)p_F]}{[(\eta_{CS} - \eta_{CD})/\beta + (\theta_{CE}\alpha\sigma + \eta_{CD})]}$$

Note that now the impact of higher values of β is unambiguous—resulting in smaller values for the denominator and, therefore, more volatile corn prices in response to fuel price shocks. This makes sense because a higher share of corn sold to ethanol boosts the importance of the liquid fuels market for corn producers. More generally, an increase in global fuel prices (p_F) will boost corn prices in all but extreme cases wherein the sales share-weighted elasticity of substitution between ethanol and petroleum in price-sensitive uses ($\alpha\sigma > 0$) is sufficiently dominated by the price elasticity of aggregate demand for liquid fuels ($\eta_{FD} < 0$). (Given the diminishing share of the additive market and the relatively inelastic demand for liquid fuels for transportation, this seems unlikely in the current economic environment.) The magnitude of this corn price change will be larger the more inelastic are corn supply and demand (as reflected in the denominator term v_{EC}), the larger the share of corn going to ethanol (β), and the smaller the cost share of corn in ethanol production (θ_{CE})

We are now able to state several important propositions that form the basis for the following empirical analysis:

PROPOSITION 1. *A random shock to the corn market—either to supply (Δ_{CS}) or to demand (Δ_{CD})—will result in a larger change in corn price, the more inelastic are corn supply and demand (as reflected in the numerator of v_{CE}), the smaller the elasticity of substitution between ethanol and petroleum (σ), the smaller the share of ethanol going to the price responsive portion of the fuel market (α), and the smaller the cost share of corn in ethanol production (θ_{CE}). The impact of the share of corn going to ethanol (β) depends on the relative responsiveness of corn demand in ethanol and nonethanol markets. If the ethanol market is more price responsive, then an increase in β dampens the corn price volatility in response to a corn demand or supply shock. However, if the ethanol market is less price responsive (e.g., due to the blend wall), then higher sales to ethanol serve to destabilize the corn price response to a random shock in the market for corn.*

PROPOSITION 2. *An increase in global fuel prices (p_F) will boost corn prices, provided the sales share-weighted elasticity of substitution between ethanol and petroleum in price sensitive uses ($\alpha\sigma > 0$) is not dominated by the price elasticity of aggregate demand for liquid fuels ($\eta_{FD} < 0$). The magnitude of this corn price change will be larger the more inelastic are corn supply and demand (as reflected in the denominator term v_{EC}), the larger the share of corn going to ethanol (β), and the smaller the cost share of corn in ethanol production (θ_{CE}).*

With a bit more information, we can also shed light on two important special cases in which policy regimes are binding. When oil prices are low, such that the RFS is binding, then the total sales of corn to the ethanol market are predetermined ($q_{CE} = 0$) so that the only price-responsive portion of corn demand is the nonethanol component. In this case, the equilibrium change in corn price simplifies to the following:

$$(25) \qquad p_C = \frac{-\Delta_{CE}}{v_{CE}}$$

Note that the price of liquid fuel does not appear in this expression at all. Because our partial equilibrium (PE) model abstracts from the impact of fuel prices on production costs of corn and ethanol, the RFS wholly eliminates the transmission of fuel prices through to the corn market by fixing the demand for ethanol in liquid fuels. The second point to note is that the responsiveness of corn prices to random shocks in the corn market is now magnified by the absence of the substitution-related term, $\theta_{CE}\alpha\sigma$, in the denominator. This leads to the third proposition.

PROPOSITION 3. *The binding RFS eliminates the output demand-driven link between liquid fuel prices and corn prices. Furthermore, with a binding RFS, the responsiveness of corn prices to a random shock in corn supply or demand is magnified. The extent of this magnification (relative to the nonbinding case) is larger, the larger the share of ethanol going to the price responsive portion of the market, the larger the elasticity of substitution between ethanol and petroleum, and the larger the cost share of corn in ethanol production.*

The other important special case considered in the following is that of a binding blend wall (BW). In this case, there is *no scope for altering the mix* of ethanol in liquid fuels. Therefore, the substitution effect in equation (15) drops out and the demand for ethanol simplifies to:

$$(26) \qquad q_E = \eta_{FD}p_F.$$

In this case, the equilibrium corn price expression simplifies to the following:

$$(27) \qquad p_C = \frac{(\eta_{FD}p_F - \Delta_{CE})}{v_{CE}}$$

Note that the price of liquid fuel has reappeared in the numerator, but the coefficient premultiplying this price is now negative. This gives rise to the fourth, and final, proposition.

PROPOSITION 4. *The presence of a binding blend wall changes the qualitative relationship between liquid fuel prices and corn prices. Now, a fall in liquid fuel prices, which induces additional fuel consumption, will stimulate the demand for corn and, hence, boost corn prices. As with the binding RFS, the responsiveness of corn prices to a random shock in corn supply or demand is again magnified. The extent of this magnification (relative to the nonbinding case) is larger, the larger the share of ethanol going to the price responsive portion of the market, the larger the elasticity of substitution between ethanol and petroleum, and the larger the cost share of corn in ethanol production.*

This simple, partial equilibrium analysis of the linkages between liquid fuel and corn markets has been useful in sharpening our thinking about key underlying relationships. However, it is necessarily quite simplified. As noted previously, we have ignored the role of energy input costs in corn and ethanol production—even though these are rather energy-intensive sectors. We have also ignored the important role of biofuel by-products. Yet sales of dried distillers grains with solubles (DDGS) account for about 16 percent of the industry's revenues, and their sale competes directly with corn and other feedstuffs in the livestock industry (Taheripour et al. 2010). And we have failed to distinguish feed demands for corn from processed food demands. Finally, we have abstracted from international trade, which has become an increasingly important dimension of the corn, ethanol, DDGS, and liquid fuel markets. For all these reasons, the empirical model introduced in the next section is more complex than that laid out in the preceding. Nonetheless, we will see that the fundamental insights offered by Propositions 1 to 4 continue to be reflected in our empirical results.

6.4 Empirical Framework

6.4.1 Overview of the Approach

Given the characteristic high price volatility in energy and agricultural markets; the complex interrelationships between petroleum, ethanol, ethanol by-products and livestock feed use, and agricultural commodity markets, as well as the constraining agricultural resource base; and the prominence of food and fuel in household budgets and real income determination, the economywide approach of an applied general equilibrium (AGE) analysis can offer a useful analytical framework for this chapter. The value of a global, AGE approach in analyzing the international trade and land use impacts of biofuel mandates has previously been demonstrated in the work of Banse et al. (2008), Gohin and Chantret (2010), and Keeney and Hertel (2009). The commodities in question are heavily traded and, by explicitly disaggregating the major producing and consuming regions of the world, we are better able to characterize the fundamental sources of volatility in these markets.

From Jorgenson's (1984) emphasis on the importance of utilizing econometric work in parameter estimation, to more recent calls for rigorous historical model testing (Hertel 1999; Kehoe 2003; Grassini 2004), it is clear that AGE models must be adequately tested against historical data to improve their performance and ensure reliability. The article by Valenzuela et al.

(2007) showed how patterns in the deviations between AGE model predictions and observed economic outcomes can be used to identify the weak points of a model and guide development of improved specifications for the modeling of specific commodity markets in a AGE framework. More recent work by Beckman, Hertel, and Tyner (2011) has focused on the validity of the Global Trade Analysis Project-Energy (GTAP-E) model for analysis of global energy markets.

Accordingly, we begin our work with a similar, historical validation exercise. In particular, we examine the model's ability to reproduce observed price volatility in global corn markets in the prebiofuels era (up to 2001). For the sake of completeness, as well as to permit us to analyze their relative importance, we augment the supply-side shocks (as derived from Valenzuela et al. 2007) by adding volatility in energy markets (specifically oil) and in aggregate demand (as proxied by volatility in national GDPs) following Beckman, Hertel, and Tyner (2011). With these historical distributions in hand, we are then in a position to explore the linkages between volatility in energy markets and volatility in agricultural markets.

6.4.2 Applied General Equilibrium Model

The impacts of biofuel mandates are far-reaching, affecting all sectors of the economy and trade, which creates potential market feedback effects. To capture these effects across production sectors and countries, we use the global AGE model, the biofuels-adapted version of the GTAP model ([GTAP-BIO] Taheripour et al. 2007), which incorporates biofuels and biofuel coproducts into the revised/validated GTAP-E model (Beckman, Hertel, and Tyner 2011). The GTAP-BIO model has been used to analyze the global economic and environmental implications of biofuels in Hertel et al. (2010), Taheripour et al. (2009), Keeney and Hertel (2009), and Hertel, Tyner, and Birur (2010).

6.4.3 Experimental Design

The GTAP database used here (v.6) is benchmarked to 2001; therefore, we undertake a historical update experiment to 2008 following the approach utilized by Beckman, Hertel, and Tyner (2011). Those authors show that by shocking population, labor supply, capital, investment and productivity changes (see table 6.1), along with the relevant energy price shocks, the resulting equilibrium offers a reasonable approximation to key features of the more recent economy.

This updating of the model also allows us the opportunity to test the model's ability to replicate the strengthened relationship between energy and agricultural prices. We do so by implementing the very same stochastic shocks used for the validation experiment in 2001, only now on our updated 2008 economy. As figure 6.1 illustrates, the observed correlation between oil and corn prices strengthened considerably over the 2001 to 2008 time

Table 6.1   **Exogenous shocks to update the database**

		Determinants of economic growth					
		Labor supply (% change)					
Region	Population (% change)	Unskilled	Skilled	Capital (% change)	Investment (% change)	TFP (% change)	Real GDP (% change)
USA	6.0	9.0	8.1	35.3	24.5	1.5	24.5
CAN	5.3	10.7	9.9	28.7	20.3	0.8	20.3
EU27	0.5	2.0	2.9	21.2	15.8	1.2	15.6
BRAZIL	8.5	1.8	28.1	24.0	22.7	0.6	22.7
JAPAN	0.1	1.4	−3.4	22.1	15.1	1.8	15.1
CHIHKG	4.7	6.6	29.0	96.6	66.7	2.9	65.5
INDIA	10.3	13.3	41.5	54.5	51.2	3.8	51.2
LAEEX	10.0	11.0	41.2	21.0	20.5	−0.6	20.6
RoLAC	11.6	13.6	43.2	34.7	25.2	−0.1	25.2
EEFSUEX	−1.2	3.6	7.9	22.7	41.3	3.7	40.0
RoE	8.6	8.0	26.7	16.7	24.1	2.2	25.4
MEASTNAEX	13.8	18.1	33.4	32.8	32.7	0.8	31.3
SSAEX	16.0	20.5	28.8	32.9	32.8	1.7	30.1
RoAFR	6.7	12.9	16.8	12.9	26.0	2.1	25.2
SASIAEEX	9.2	17.4	48.7	40.5	38.8	1.7	38.1
RoHIA	3.8	−2.1	27.7	42.8	38.6	2.7	38.2
RoASIA	12.9	15.2	36.1	33.4	39.9	2.8	40.6
Oceania	8.6	11.6	8.5	32.1	27.3	0.2	27.3

Source: GTAP-Dyn and Model Results (TFP).
Note: Regions are defined in table 6A.2. TFP = total factor productivity; GDP = gross domestic product.

period (note that before 2001, the correlation between the two was negative); therefore, our hypothesis (and, indeed, our model performance check) is that the transmission of energy price volatility will be higher than the pre-2001 period. Updating the model also allows us the chance to explore some of the empirical dimensions of Propositions 1 to 4, which emerged from the theoretical model.

All of this work sets the stage for an in-depth exploration of the role of biofuel policy regimes in governing the extent to which volatility in energy markets is transmitted to agricultural commodity markets and the extent to which increased sales of agricultural commodities to biofuels alters the sensitivity of these markets to agricultural supply-side shocks. For this part of the analysis, we focus on the year 2015, in which the RFS for U.S. corn ethanol reaches its target of fifteen billion gallons per year, and a blend wall could potentially be binding. In order to reach the target amount, we implement a quantity shock to the model that will increase U.S. ethanol production to fifteen billion gallons per year. We do not run a full update experiment as we did for the 2001 to 2008 time period because we do not know how the key exogenous variables will evolve over this future period.[10] We assume that the distributions of supply-side shocks in agriculture and energy markets, as well as the interannual volatility in regional GDPs, remain unchanged from their historical values; this has the virtue of allowing us to isolate the impact of the changing structure of the economy on corn price volatility.

Based on Proposition 4, we hypothesize that, at low oil prices, stochastic draws in the presence of a binding RFS will render corn markets more sensitive to agricultural supply-side shocks because a substantial portion of the corn market (the mandated ethanol use) will be insensitive to price, while at high corn prices, the opposite will be true, due to the highly elastic demand for ethanol as a substitute for corn. On the other hand, again, based on Proposition 4, we expect energy market volatility to have relatively little impact on corn markets at low oil prices.

At high oil prices, there are two possibilities—in the first case, the RFS is nonbinding, and the blend wall is not a factor (i.e., it has recently been increased from 10 percent to 15 percent for recent model vehicles). In this case, we expect to see the influence of a larger share of corn going to ethanol (β) and also a larger share of ethanol going to the price-responsive portion of the fuel market (α), translated into lesser sensitivity to random supply shocks emanating from the corn market (Proposition 1).

In the second case, high oil prices induce expansion of the ethanol industry to the point where the blend wall is binding so that Proposition 4 becomes relevant. In this case, the qualitative relationship between oil prices and corn prices is reversed; as with the binding RFS, the impact of random shocks to corn supply or demand will be magnified with a binding blend wall.

Before investigating these hypotheses empirically, we must first characterize the extent of volatility in agricultural and energy markets. In terms of the PE model developed in the preceding, we must estimate the parameters underlying the distributions of Δ_{CS}, Δ_{CD}, and p_F.

6.5 Characterizing Sources of Volatility in Energy and Agricultural Markets

The distributions of the stochastic shocks to corn production, corn demand, and oil prices are assumed to be normally and independently distributed. Given the great many uses of corn in the global economy, we prefer to shock the underlying determinant of economywide demand, namely GDP, allowing these shocks to vary by model region. Of course GDP shocks also result in oil price changes, and, in a separate line of work, we have focused on the ability of this model to reproduce observed oil price volatility

---

10. Obviously, we could use projections of key variables, but they would be uncertain, and we do not believe this would significantly alter our findings, which hinge primarily on the quantity and cost shares featured in equation (19).

Table 6.2   Time series residuals (mean-normalized standard deviations in annual percentage change: 1981–2008), used as inputs for the stochastic simulation analysis

Region	Corn production	Gross domestic product	Oil price
USA	19.05	3.18	24.91
CAN	14.84	4.27	24.91
EU27	11.91	2.04	24.91
BRAZIL	16.34	2.52	24.91
JAPAN	NA	1.81	24.91
CHIHKG	14.32	6.01	24.91
INDIA	16.54	3.55	24.91
LAEEX	13.54	3.27	24.91
RoLAC	8.64	4.36	24.91
EEFSUEX	NA	1.58	24.91
RoE	15.72	1.38	24.91
MEASTNAEX	9.66	5.27	24.91
SSAEX	11.87	4.65	24.91
RoAFR	NA	2.47	24.91
SASIAEEX	NA	4.90	24.91
RoHIA	19.93	3.65	24.91
RoASIA	6.71	4.84	24.91
Oceania	16.80	1.88	24.91

Note: Regions are defined in table 6A.2. NA = not available.

based on GDP shocks and oil supply shocks. However, in this chapter, we prefer to perturb oil prices directly so that we may separately identify the impact of energy price shocks and more general shocks to the economy.

To characterize the systematic component in corn production, time series models are fitted to National Agricultural Statistical Service (NASS) data on annual corn production (corn easily commands the largest share of coarse grains, the corresponding GTAP sector; hence, the focus on corn) over the time period of 1981 to 2008.[11] For crude oil prices, we use Energy Information Administration (EIA) data on U.S. average price and average import price (we take a simple average of the two series) over the same time periods. Here, we use the variation in regional GDP to capture changes in aggregate demand in each of the markets.

The summary statistic of interest from the time series regressions on both the supply and demand sides is the normalized standard deviation of the estimated residuals, reported in table 6.2.[12] This result summarizes variability of the nonsystematic aspect of annual production, prices and GDP

11. We use the 1981 to 2008 time period as the inputs for both the pre-2001 stochastic simulations and those of 2001 to 2008 in order to not influence the comparison across base periods with the higher volatility of the 2001 to 2008 time period.

12. Estimates for the time series models are available upon request.

in each region for the 1981 to 2008 time period (sectors and regions are defined in appendix tables 6A.1 and 6A.2). This is calculated as: √*variance* (of estimated residuals) divided by the mean value of production (or prices, or GDP), and multiplied by 100 percent. Not surprisingly, from table 6.2, we see that corn production and oil prices were much more volatile than GDP over the time period, with oil prices being somewhat more volatile than corn production. Note that we do not attempt to estimate region-specific variances for oil prices as we assume this to be a well-integrated global market.

6.6   Results for 2001 and 2008

6.6.1   Prebiofuel Era

Our first task is to examine the performance of the model with respect to the 2001 base period. The first pair of columns in table 6.3 reports the model-generated standard deviations in annual percentage change in coarse grains prices based on several alternative stochastic simulations undertaken using the Stroud Gaussian Quadrature as detailed in Arndt (1996) and Pearson and Arndt (2000). In the first column, we report the standard deviations in coarse grains prices when all three stochastic shocks from table 6.2 are simultaneously implemented. Focusing on the United States, the model with all three shocks predicts the standard deviation of annual percentage changes in corn prices to be 28.5, while the historical outcome (over the entire 1982 to 2008 period) revealed a standard deviation of just 20. So the model overpredicts volatility in corn markets. This is likely due to the fact that it treats producers and consumers as myopic agents who use only current information on planting and pricing to inform their production decisions. By incorporating forward-looking behavior as well as stockholding, we would expect the model to produce less price variation. Introducing more elastic consumer demand would be one way of mimicking such effects and inducing the model to more closely follow historical price volatility.

The second column under the 2001 heading reports the impact on coarse grains price volatility of oil price shocks only. From these results, it is clear that the energy price shocks have little impact on corn markets in the prebiofuel era. In the United States, the amount of coarse grains price variation generated by oil price-only shocks is just a standard deviation of 1.1 percent, whereas the variation from the three sources is 28.5 percent (resulting in oil's share of the total equaling 0.04, as reported in parentheses in table 6.3). This confirms the findings of Tyner (2009), who reports very little integration of crude oil and corn prices over the 1988 to 2005 period.

The third column in table 6.3 reports the observed variation in coarse grains prices from volatility in corn production. This indicates that the

Table 6.3   Model-generated coarse grain price variation in 2001, 2008, and 2015 economies

Region	2001 model volatility			2008 model volatility			2015 model volatility (Base–No RFS/BW)		
	All shocks	Oil price	Corn production	All shocks	Oil price	Corn production	All shocks	Oil price	Corn production
USA	28.5	1.1 (.04)	27.5 (.96)	30.7	10.0 (.32)	28.7 (.93)	29.8	15.6 (.53)	25.1 (.84)
CAN	16.7	1.1 (.07)	16.2 (.97)	18.8	4.4 (.23)	18.0 (.96)	18.6	5.5 (.29)	17.7 (.95)
EU27	18.3	1.0 (.05)	17.5 (.96)	20.4	3.1 (.15)	20.0 (.98)	20.2	3.2 (.16)	19.8 (.98)
BRAZIL	19.0	1.1 (.06)	18.8 (.99)	21.0	4.3 (.20)	20.7 (.99)	20.6	4.5 (.22)	20.3 (.99)
JAPAN	4.9	0.2 (.04)	3.8 (.77)	9.7	2.3 (.24)	8.9 (.92)	8.7	4.3 (.48)	7.6 (.88)
CHIHKG	34.0	0.1 (0)	32.4 (.95)	47.0	1.8 (.04)	46.4 (.99)	46.3	0.8 (.02)	46.0 (.99)
INDIA	31.4	1.5 (.05)	31.1 (.99)	37.6	5.1 (.14)	36.9 (.98)	37.5	3.9 (.10)	36.9 (.98)
LAEEX	18.7	1.0 (.05)	18.1 (.97)	20.4	5.0 (.25)	19.8 (.97)	20.2	5.8 (.29)	19.5 (.97)
RoLAC	11.7	0.4 (.03)	11.0 (.95)	13.4	2.2 (.16)	12.8 (.96)	13.0	3.4 (.26)	12.5 (.96)
EEFSUEX	2.4	1.1 (.49)	0.7 (.29)	2.9	1.8 (.65)	1.5 (.54)	2.9	1.3 (.46)	1.5 (.52)
RoE	20.7	1.9 (.09)	20.4 (.99)	22.2	3.8 (.17)	22.2 (1.00)	22.0	3.7 (.17)	22.0 (1.00)
MEASTNAEX	11.4	3.4 (.29)	10.8 (.94)	14.7	10.2 (.70)	12.9 (.88)	14.9	8.8 (.59)	12.7 (.85)
SSAEX	2.8	2.6 (.92)	0.7 (.26)	6.1	9.5 (1.56)	1.0 (.17)	6.1	7.6 (1.25)	1.0 (.17)
RoAFR	3.0	0.6 (.19)	1.9 (.64)	5.4	2.1 (.39)	4.7 (.88)	5.3	2.9 (.55)	4.2 (.79)
SASIAEEX	5.4	0.2 (.03)	4.0 (.74)	6.4	0.5 (.07)	5.6 (.87)	6.2	1.0 (.16)	5.4 (.88)
RoHIA	4.8	0.6 (.12)	3.7 (.77)	6.1	1.0 (.16)	5.6 (.91)	5.6	1.8 (.31)	4.9 (.88)
RoASIA	12.3	0.4 (.04)	11.7 (.95)	13.3	1.1 (.09)	12.9 (.96)	13.1	0.3 (.02)	12.7 (.97)
Oceania	18.9	0.5 (.03)	18.5 (.98)	19.8	3.0 (.15)	19.1 (.96)	19.2	4.2 (.22)	18.6 (.97)
World average	14.4			16.3			15.5		

Notes: Numbers in parentheses represent the share of volatility for oil price/corn production inputs in total volatility. Historical variation in corn prices for the United States was 21.6 standard deviations over the 1981–2008 time period. Regions are defined in table 6A.2

Table 6.4   Applied general equilibrium model parameters and data

Time period				Parameter			
	σ	η_{FD}	α	β	Θ_{CE}	ν_{EC}
2001	3.95	0.10	0.25	0.06	0.39	0.43
2008	3.95	0.10	0.44	0.26	0.67	0.31
2015	3.95	0.10	0.60	0.40	0.70	0.25

Source: Authors' calculations, based on the applied general equilibrium model parameter file and data bases.

majority of corn price variation in this historical period (a 0.96 share of the total) was due to volatility in corn production.

6.6.2 Biofuel Era

As discussed in the preceding section, we update the data base to 2008 in order to provide a reasonably current representation of the global economy in the context of the biofuel era. We then redo the same stochastic simulation experiments as 2001 to explore the energy or agricultural commodity price transmission in the biofuel era. The middle set of columns in table 6.3 present the results from this experiment.

The model estimates somewhat higher overall coarse grain price variation (standard deviation of 30.7 percent) in this case. Now, the ratio of the variation from energy price shocks to the total shocks is 0.32, versus the 0.04 for the 2001 database. This is hardly surprising in light of expression (19) and Proposition 3. Referring to table 6.4, which summarizes some of the key parameters or pieces of data from the three base years, we see that the shares of coarse grains going to ethanol production (β) rises fourfold over this period. In addition, the share of ethanol going to the price-sensitive side of the ethanol market (α) nearly doubles, and the net supply elasticity of corn to ethanol falls. Based on Proposition 3, all of these changes serve to boost the responsiveness of corn pries to liquid fuel prices. Meanwhile, the contribution of corn supply shocks to total volatility is somewhat reduced, as we would expect from the larger values for α, β, and θ_{CE}, although the smaller net supply elasticity of corn to ethanol works in the opposite direction.

6.7 The Future of Energy-Agriculture Interactions in the Presence of Alternative Policies

Having completed our analysis of energy and agricultural commodity interactions in the current environment, we now turn to the analysis of U.S. biofuel policies. U.S. policy, given current technologies, mandates that fifteen billion gallons of corn ethanol be produced by 2015 (this is known as the Renewable Fuel Standard [RFS]), up from roughly seven billion gallons produced in 2008.[13] We implement this mandate by increasing U.S. ethanol production through an exogenous quantity increase, following Hertel, Tyner, and Birur (2010).

Mathematically, the RFS effectively provides a lower bound on ethanol production and may be represented via the following complementary slackness conditions, where S is the per unit subsidy required to induce additional use of ethanol by the price sensitive agents in our model, and QR is the ratio of observed ethanol use to the quota as specified under the RFS:

$$S \geq 0 \perp (QR_{RFS} - 1) \geq 0 \qquad \text{which implies that either:}$$

$$S > 0, (QR_{RFS} - 1) = 0 \qquad \text{(RFS is binding) or:}$$

$$S = 0, (QR_{RFS} - 1) \geq 0 \qquad \text{(RFS is nonbinding)}$$

Because producers do not actually receive a subsidy for meeting the RFS, the additional cost of producing liquid fuels must be passed forward to consumers. We accomplish this by simultaneously taxing the combined liquid fuel product by the full amount of the subsidy.

The key point regarding the RFS is that it is asymmetric. Thus, when the RFS is just binding [$S = 0, (QR_{RFS} - 1) = 0$], any rise in the price of gasoline will increase ethanol production past the mandated amount because ethanol is now better able to compete with gasoline on an energy basis. In this case, corn demand (and price) will be responsive to changes in the oil price. In contrast, a decrease in the price of gasoline does nothing to ethanol production (i.e., it stays at the fifteen billion gallon mark) as this is the mandated amount; $S > 0$ ensures that the ethanol continues to be used at current levels. Of course, if the RFS is severely binding [$S \gg 0, (QR_{RFS} - 1) = 0$], then oil prices will have to rise considerably before reaching the point where $S = 0$ and the fuel price begins to translate through to the corn price. Because it is very difficult to predict whether the RFS will be binding in 2015, and if so, how severely binding it will be, we adopt the simple assumption that the RFS is just barely binding in the initial equilibrium. Therefore, any rise in oil prices will translate through to corn prices.

A blend wall works differently from the RFS; as pointed out by Tyner (2009), the blend wall is an effective constraint on demand.[14] Mathematically, the blend wall provides an upper bound on the ethanol intensity of liquid fuels and may be represented via the following complementary slack-

---

13. The RFS also mandates the production of advanced biofuels, which we do not consider here.

14. The Energy Information Agency estimates U.S. gasoline consumption at approximately 135 billion gallons; therefore, if the entire amount was blended with ethanol, we would fall short of the fifteen billion gallon mark. Several alternatives have been suggested, such as improving E85 demand and increasing the blending regulation (this is currently being investigated by the Environmental Protection Agency) to 12 or 15 percent.

ness conditions, where T is the per unit tax required to restrict additional use of ethanol, and QR is the ratio of observed ethanol intensity (Q_E / Q_F) to the blend wall.

$$T \geq 0 \perp (1 - QR_{BW}) \geq 0 \qquad \text{which implies that either:}$$

$$T > 0, (1 - QR_{BW}) = 0 \qquad \text{(BW is binding) or:}$$

$$T = 0, (1 - QR_{BW}) \geq 0 \qquad \text{(BW is nonbinding)}$$

For illustrative purposes, consider the case in which the blendwall is just barely binding so that $T = 0$, $(1 - QR_{BW}) = 0$, but the RFS is not binding. Then if the price of gasoline were to rise, the ethanol intensity of liquid fuel use would not change because it is up against the blend wall. Of course, the overall level of ethanol production may well fall as total liquid fuel consumption falls, thereby dragging down the maximum amount of ethanol that can be added. In this case, the tax adjusts to ensure the constraint remains binding. However, if the price of gasoline falls, the ethanol intensity of production will decline, thereby moving off this constraint such that the blend wall becomes nonbinding.

As with the RFS, it is difficult to predict the extent to which the blend wall will be binding in 2015. However, given the strong political interest in maintaining ethanol production, at the time of the NBER conference (Spring 2010), we viewed it as likely that the blend wall would be adjusted upward in the future in order to permit the industry to meet the RFS. At the time of our revision of this chapter, this has indeed been done by the U.S. Environmental Protection Agency (EPA), with the blend rate for recent model vehicles now raised to 15 percent. It seems unlikely that E85 (a fuel blend comprising 85 percent ethanol) use will expand greatly in the United States due to infrastructure limitations (the flex-fuel auto stock is limited, and for this reason, the number of fuel stations offering E85 is also quite limited); therefore, it is reasonable to consider the case wherein the blend wall is adjusted such that it is just becoming binding at the 2015 RFS level.

Given the many different combinations of RFS and blend wall policy regimes, we investigate the importance of energy price shocks on agricultural commodity prices under four different scenarios:

1. *Base case:* The RFS is not binding under any combination of commodity market shocks, and the blend wall is ignored. We expect that this base case will offer the largest scope for energy price shocks to influence agricultural commodity price volatility. Results from this case are reported in the last part of table 6.3.

2. *RFS is just binding:* That is, corn ethanol production is precisely fifteen billion gallons in 2015. In this case, if oil prices rise due to a random shock to the petroleum market, ethanol production will also rise as this fuel is substituted for the higher priced petroleum. However, the effect of declin-

ing petroleum prices will not be translated back to the corn market as the RFS will prevent a contraction of ethanol production. This has the effect of making corn demand more inelastic such that commodity price volatility is greater in the wake of the supply-side shocks. Results from this and the subsequent experiments are reported in table 6.5.

3. *RFS is not binding; however, the blend wall is binding:* In this case, we assume that the strength of the overall economy as well as the relative prices of petroleum and corn in 2015 are such that ethanol production is well above the level specified by the RFS so that the random shocks introduced in the following never threaten to push production below the fifteen billion gallon annual target. However, in this case, the blend wall is very likely to be binding, and we specify the initial conditions in the model such that $T = 0$, $(1 - QR_{BW}) = 0$; that is, the blend wall is on the verge of binding. In this case, we expect the impact of an oil price rise on corn price volatility to be very modest as it is not possible to increase the ethanol intensity of liquid fuels, so the only changes in ethanol use will be those emanating from changes in overall liquid fuel use.

4. *RFS and blend wall are both on the verge of binding:* This scenario could arise if the blend wall were continually adjusted upward, just reaching the point at which the RFS is met. In this case we have $T = 0$, $(1 - QR_{BW}) = 0$ and $S = 0$, $(QR_{RFS} - 1) = 0$.

Let us first consider the 2015 base case results presented in table 6.3. These indicate that, relative to the 2008 database, in the absence of any role for the RFS and blend wall (BW), energy price shocks contribute more to coarse grain price variation. Indeed, energy price volatility now contributes to a standard deviation of 15.6, which amounts to 0.53 of the total variation in corn prices (but still less than the independent variation induced by corn supply-side shocks). This result is expected as even more corn is going to ethanol production (table 6.4), and there is double the amount of ethanol produced, compared to the 2008 database. In addition, ethanol production is free to respond to both low and high oil price draws from the stochastic simulations because the RFS and BW are nonbinding. The contribution of corn supply-side volatility shocks to corn price variation is also lowest for this case.

For the second scenario, we follow the same process as before to stimulate ethanol production to the RFS amount, and we run the same stochastic simulations; however, as noted previously, we assume that the RFS is initially just binding, and we implement the requirement that U.S. ethanol production cannot fall below fifteen billion gallons. Results for this scenario indicate (refer to table 6.5) that the share of energy price volatility to total corn price variation is cut in half from the base case (from 0.53 to 0.26). This is due to the fact that we truncate consumers' response to low oil price draws by using less ethanol. Implementation of the RFS also leads to much higher

Table 6.5   Model-generated coarse grain price variation in the 2015 economy for the base case, renewable fuels standard (RFS), and a blend wall

Region	RFS binding			Blend wall binding			RFS and blend wall binding		
	All shocks	Oil price	Corn production	All shocks	Oil price	Corn production	All shocks	Oil price	Corn production
USA	37.1	9.5 (.26)	36.7 (.99)	31.6	6.7 (.21)	28.2 (.89)	40.4	1.2 (.03)	39.5 (.98)
CAN	19.6	3.8 (.19)	18.9 (.96)	18.7	3.1 (.17)	18.1 (.96)	19.9	1.8 (.09)	19.3 (.97)
EU27	20.6	2.5 (.12)	20.2 (.98)	20.2	2.3 (.11)	19.8 (.98)	20.6	1.8 (.09)	20.2 (.98)
BRAZIL	21.1	3.5 (.16)	20.7 (.99)	20.7	3.2 (.15)	20.3 (.99)	21.1	2.3 (.11)	20.9 (.99)
JAPAN	10.7	2.2 (.20)	10.2 (.95)	10.1	1.6 (.16)	8.9 (.88)	12.1	0.3 (.03)	11.4 (.94)
CHIHKG	46.7	1.3 (.03)	46.1 (.99)	46.6	1.4 (.03)	46.0 (.99)	46.7	1.8 (.04)	46.1 (.99)
INDIA	37.5	3.9 (.10)	36.9 (.98)	37.5	4.0 (.11)	36.9 (.98)	37.5	4.1 (.11)	36.9 (.98)
LAEEX	21.3	4.3 (.20)	20.7 (.97)	20.0	3.8 (.19)	19.7 (.98)	21.2	2.4 (.11)	20.9 (.99)
RoLAC	14.1	2.0 (.14)	13.6 (.97)	13.5	1.6 (.12)	12.8 (.95)	14.6	0.4 (.03)	14.0 (.96)
EEFSUEX	3.0	0.9 (.32)	1.8 (.60)	2.9	0.6 (.21)	1.6 (.55)	3.0	1.5 (.51)	1.9 (.62)
RoE	22.3	3.0 (.14)	22.3 (1.00)	22.0	2.9 (.13)	22.0 (1.00)	22.2	2.4 (.11)	22.3 (1.00)
MEASTNAEX	14.8	8.1 (.55)	13.0 (.88)	14.5	7.8 (.54)	12.8 (.88)	14.5	7.3 (.50)	13.0 (.00)
SSAEX	6.0	7.4 (1.24)	1.1 (.19)	6.0	7.4 (1.23)	1.0 (.17)	6.0	7.4 (1.22)	1.1 (.18)
RoAFR	6.3	1.9 (.30)	5.6 (.90)	5.7	1.6 (.27)	4.7 (.83)	6.8	0.7 (.11)	6.1 (.90)
SASIAEEX	6.7	0.4 (.06)	5.8 (.87)	6.4	0.3 (.05)	5.5 (.86)	6.9	0.2 (.02)	6.0 (.87)
RoHIA	6.4	0.8 (.13)	5.7 (.89)	6.2	0.7 (.11)	5.4 (.87)	7.0	0.2 (.04)	6.2 (.89)
RoASIA	13.4	0.8 (.06)	12.9 (.97)	13.3	0.9 (.07)	12.7 (.96)	13.5	1.3 (.10)	13.0 (.96)
Oceania	20.3	2.7 (.15)	19.3 (.95)	19.5	2.2 (.11)	18.9 (.97)	20.5	0.9 (.04)	19.7 (.96)
World average	17.8			16.5			19.3		

Note: See table 6.3 notes.

variation in corn prices. In Proposition 3, we demonstrated the cause of this; that is, the RFS severs the consumer demand-driven link between liquid fuels price and corn prices in the presence of low oil prices. The absence of price responsiveness in this important sector translates into a magnification of the responsiveness of corn prices to random shocks to corn supplies and nonethanol demand.

These results are similar to those from Yano, Blandford, and Surry (2010), who use Monte Carlo simulations of a PE model to show that the U.S. ethanol mandate reduces the impact that variations in petroleum prices have on corn prices (compared to a "no-mandate" scenario), while the impacts from variations in corn supply on corn prices are increased.

For the third scenario, we allow the RFS to be nonbinding, but we implement a blend wall, which itself is assumed to be just binding. The results from this case indicate that the share of energy price volatility in total corn price variation is even lower than when the RFS is just binding. This is substantiated by Tyner (2009), who notes that the blend wall effectively breaks the link between crude oil and corn prices as ethanol cannot react to high oil prices, but at low oil prices, the blend wall does little to reduce demand for ethanol.

The final scenario in table 6.5 is the case wherein both the RFS and the BW are on the verge of binding. This largely eliminates the demand-side feedback from energy prices to the corn market, which is what we see in the results, with oil price volatility accounting for just 0.03 of the total variation in corn prices. In contrast, the price responsiveness of corn to supply-side shocks is greatly increased. Indeed, when compared to the 2015 base case (no RFS, no BW), corn price volatility in the face of identical supply side shocks is 57 percent greater. If we look at the final row of table 6.5, we see that global price volatility is much increased under this scenario, rising by about one-quarter. Clearly the presence of biofuel mandates and associated fuel blending limits have the potential to greatly destabilize agricultural commodity markets in the future.

In addition to price volatility, it is useful to consider the mean price change from the 2015 base. Table 6.6 reports mean changes in both ethanol produc-

Table 6.6	Mean percentage changes in corn price and ethanol production in 2015 under the different stochastic scenarios	
Scenario	Mean percentage price change	Mean percentage change in ethanol production
Base case	8.9	3.7
Renewable fuels standard	18.7	22.8
Blend wall	2.1	−21.1
Renewable fuels standard/Blend wall	12.2	0

tion and corn prices in the United States under different policy regimes. Due to the nature of the demand relationships in the model, production shortfalls generate larger price changes than do symmetric instances of excess production, and the mean corn price change under the base case is greater than zero. When the RFS and blend wall are both binding, ethanol production is unchanged, and the mean change in corn price is even larger, at 12.2 percent. When only the RFS is binding, instances of high corn prices— potentially due to a production shortfall—are rewarded with persistent ethanol demand due to the mandate. This has a tendency to boost mean ethanol production as well as mean corn prices. On the other hand, when only the blend wall is binding, episodes of low corn prices—possibly due to a favorable draw from the coarse grains productivity distribution—no longer result in greater ethanol production as the blend wall prevents further expansion. However, high corn prices do result in lower ethanol use, which is why the mean change in ethanol production is –21 percent under the BW scenario. This results in lower expected corn prices as well.

6.8   Discussion

The relationship between agricultural and energy commodity markets has strengthened significantly with the recent increase in biofuel production. Energy has always played an important role in agricultural production inputs; however, the combination of recent high energy prices with policies aimed at promoting energy security and renewable fuel use have stimulated the use of crop feedstocks in biofuel production. With a mandate to further increase biofuel production in the United States, it is clear that the relationship among agricultural and energy commodities may grow even stronger.

Results from this work indicate that the era of rapid biofuel production strengthened the transmission of energy price volatility into agricultural commodity price variation. The additional mandated production has the potential to further strengthen this transmission. However, the outcome will depend critically on the policy regime in which ethanol markets find themselves. The presence of a Renewable Fuels Standard can hinder the ethanol's sector's ability to react to low oil prices, thereby destabilizing commodity markets. The presence of a liquid fuels blend wall causes a similar disconnection in the transmission of energy prices to agriculture—albeit at high oil prices—and, therefore, also serves to increase commodity price volatility.

Comparing all the scenarios considered here, the absence of all biofuel policies leads to the highest transmission of energy price volatility into commodity price variation and the lowest corn price volatility in response to traditional supply-side shocks. This is because consumers are able to respond to both high and low oil prices by changing their biofuel mix, and adjustment to corn supply shocks are absorbed by energy and nonenergy markets alike. When we implement biofuels policy (either the RFS or a blend wall), the impacts from energy price volatility are smaller than the base case, while the impacts from corn supply volatility are magnified. In the most extreme case, wherein the blend wall is expanded to the point where the RFS is just barely binding, U.S. coarse grains price volatility in response to corn supply shocks is 57 percent higher than in the nonbinding case, and world price volatility is boosted by 25 percent. This underscores the point made by Irwin and Good (2010), who highlight the risk introduced by sizable sales of corn for ethanol production in the United States, particularly in light of mandated minimum purchases. They suggest that this could lead to record price rises in the wake of an extreme weather event in the Corn Belt of the United States—something that has not been observed during recent years. This leads them to advocate introducing some type of safety valve for the biofuels program.

In summary, it seems likely we will experience a future in which agricultural price volatility—particularly for biofuel feedstocks—may rise. The extent of this volatility will depend critically on renewable energy policies. Indeed, in the future, these sources of uncertainty may become more important than traditional agricultural policies in many farm commodity markets.

Appendix

Table 6A.1   **Industries, commodities, and their corresponding Global Trade Analysis Project (GTAP) notation**

Industry name	Commodity name	Description	GTAP notation
CrGrains	CrGrains	Cereal grains	gro
OthGrains	OthGrains	Other grains	pdr, wht
Oilseeds	Oilseeds	Oilseeds	osd
Sugarcane	Sugarcane	Sugarcane and sugarbeet	c_b
Cattle	Cattle	Bovine cattle, sheep, and goats	ctl, wol
Nonrum	Nonrum	Nonruminants	oap
Milk	Milk	Raw milk	rmk
Forestry	Forestry	Forestry	frs
Ethanol2	Ethanol2	Ethanol produced from sugarcane	eth2
OthFoodPdts	OthFoodPdts	Other food products	b_t, ofdn
VegOil	VegOil	Vegetable oils	voln
ProcLivestoc	ProcLivestoc	Meat and dairy products	cmt, mil, omt
OthAgri	OthAgri	Other agriculture goods	ocr, pcr, pfb, sgr, v_f
OthPrimSect	OthPrimSect	Other primary products	fsh, omn
Coal	Coal	Coal	coa
Oil	Oil	Crude oil	oil
Gas	Gas	Natural gas	gas, gdt
Oil_Pcts	Oil_Pcts	Petroleum and coal products	p_c
Electricity	Electricity	Electricity	ely
En_Int_Ind	En_Int_Ind	Energy intensive industries	crpn, i_s, nfm, atp, cmn, cns, dwe, ele, fmp, isr, lea, lum, mvh
Oth_Ind_Se	Oth_Ind_Se	Other industry and services	nmn, obs, ofi, ome, omf, osg, otn, otp, ppp, ros, tex, trd, wap, wtp, wtr
EthanolC	Ethanol1	Ethanol produced from grains	ethl
	DDGS	Dried distillers grains with solubles	ddgs
Biodiesel	Biodiesel	Biodiesel	biod
	BDBP	Biodiesel by-products	bdbp

Table 6A.2   Regions and their members

Region	Corresponding countries in Global Trade Analysis Project
USA	United States
CAN	Canada
EU27	Austria; Belgium; Bulgaria; United Kingdom; Cyprus; Czech Republic; Germany; Denmark; Spain; Estonia; Finland; France; Greece; Hungary; Ireland; Italy; Lithuania; Luxembourg; Latvia; Malta; the Netherlands; Poland; Portugal; Romania; Slovakia; Slovenia; Sweden
BRAZIL	Brazil
JAPAN	Japan
CHIHKG	China; Hong Kong
INDIA	India
LAEEX	Argentina; Colombia; Mexico; Venezuela
RoLAC	Chile; Peru; Uruguay; rest of Andean Pact; Central America; rest of the Caribbean; rest of free trade area of the Americas; rest of North America; rest of South America
EEFSUEX	Russia; rest of European Free Trade Association; rest of former Soviet Union
RoE	Albania; Switzerland; Croatia; Turkey; rest of Europe
MEASTNAEX	Botswana; Tunisia; rest of Middle East; rest of North Africa
SSAEX	Madagascar; Mozambique; Malawi; Tanzania; Uganda; rest of South African Customs Union; rest of Southern African Development Community; rest of sub-Saharan Africa; Zimbabwe
RoAFR	Morocco; South Africa; Zambia
SASIAEEX	Indonesia; Malaysia; Vietnam; rest of Southeast Asia
RoHIA	Korea; Taiwan
RoASIA	Bangladesh; Sri Lanka; the Philippines; Singapore; Thailand; rest of East Asia; rest of South Asia
Oceania	Australia; New Zealand; rest of Oceania

References

Arndt, C. 1996. "An Introduction to Systematic Sensitivity Analysis via Gaussian Quadrature." GTAP Technical Paper no. 2. West Lafayette, IN: Purdue University, Global Trade Analysis Project.

Banse, M., H. van Meijl, A. Tabeau, and G. Woltjer. 2008. "Will EU Biofuel Policies Affect Global Agricultural Markets?" *European Review of Agricultural Economics* 35 (2): 117–41.

Beckman, J., T. Hertel, and W. Tyner. 2011. "Validating Energy-Oriented CGE Models." *Energy Economics,* forthcoming.

Carbone, J. C., and V. K. Smith. 2008. "Evaluating Policy Interventions with General Equilibrium Externalities." *Journal of Public Economics* 92 (5–6): 1254–74.

de Gorter, H., and D. R. Just. 2008. "'Water' in the U.S. Ethanol Tax Credit and Mandate: Implications for Rectangular Deadweight Costs and the Corn-Oil Price Relationship." *Review of Agricultural Economics* 30 (3): 397–410.

Du, X., C. Yu, and D. Hayes. 2009. "Speculation and Volatility Spillover in the Crude Oil and Agricultural Commodity Markets: A Bayesian Analysis." Paper presented at the Agricultural and Applied Economics Association, Milwaukee, Wisconsin.

Gellings, C., and K. E. Parmenter. 2004. "Energy Efficiency in Fertilizer Production and Use." In *Efficient Use and Conservation of Energy,* edited by C. W. Gellings and K. Blok, in *Encyclopedia of Life Support Systems (EOLSS).* Oxford, UK: Eolls. http://www.eolss.net.

Gohin, A., and F. Chantret. 2010. "The Long-Run Impact of Energy Prices on World Agricultural Markets: The Role of Macro-Economic Linkages." *Energy Policy* 38 (1): 333–39.

Gohin, A., and D. Tréguer. 2010. "On the (De) Stabilization Effects of Biofuels: Relative Contributions of Policy Instruments and Market Forces." *Journal of Agricultural and Resource Economics* 35 (1): 72–86.

Grassini, M. 2004. "Rowing along the Computable General Equilibrium Modeling Mainstream." Paper presented at conference, Input-Output and General Equilibrium Data, Modeling, and Policy Analysis, Brussels, Belgium.

Hertel, T. W. 1999. "Future Directions in Global Trade Analysis." GTAP Working Paper no. 4. West Lafayette, IN: Center for Global Trade Analysis. http://www.gtap.agecon.purdue.edu/resources/working_papers.asp.

Hertel, T. W., A. Golub, A. Jones, M. O'Hare, R. Plevin, and D. Kammen. 2010. "Global Land Use and Greenhouse Gas Emissions Impacts of U.S. Maize Ethanol: Estimating Market-Mediated Responses." *BioScience* 60 (3): 223–31. forthcoming.

Hertel, T. W., W. E. Tyner, and D. K. Birur. 2010. "Global Impacts of Biofuels." *Energy Journal* 31 (1): 75–100.

Hochman, G., S. Sexton, and D. Zilberman. 2008. "The Economics of Biofuel Policy and Biotechnology." *Journal of Agricultural and Food Industrial Organization* 6 (2), doi:10.2202/1542-0485.1237.

Irwin, S., and D. Good. 2010. "Alternative 2010 Corn Production Scenarios and Policy Implications." Marketing and Outlook Brief. University of Illinois, Department of Agricultural and Consumer Economics, March 11.

Jorgenson, D. 1984. "Econometric Methods for Applied General Equilibrium Analysis." In *Applied General Equilibrium Analysis,* edited by H. E. Scarf and J. B. Shoven, 139–203. New York: Cambridge University Press.

Keeney, R., and T. Hertel. 2009. "The Indirect Land Use Impacts of United States Biofuel Policies: The Importance of Acreage, Yield, and Bilateral Trade Response." *American Journal of Agricultural Economics* 91 (4): 895–909.

Kehoe, T. 2003. "An Evaluation of the Performance of Applied General Equilibrium Models of the Impact of NAFTA." Federal Reserve Bank of Minneapolis, Research Department Staff Report no. 320.

McPhail, L., and B. Babcock. 2008. "Ethanol, Mandates, and Drought: Insights from a Stochastic Equilibrium Model of the U.S. Corn Market." Working Paper no. 08-WP 464. Iowa State University, Center for Agricultural and Rural Development.

Meyer, S., and W. Thompson. 2010. "Demand Behavior and Commodity Price Volatility under Evolving Biofuel Markets and Policies." In *Handbook of Bioenergy Economics and Policy,* edited by M. Khanna, J. Scheffran, and D. Zilberman, 133–50. New York: Springer Science Verlag.

Pearson K., and C. Arndt. 2000. "Implementing Systematic Sensitivity Analysis Using GEMPACK." GTAP Technical Paper no. 3. West Lafayette, IN: Purdue University, Global Trade Analysis Project.

Searchinger, T., R. Heimlich, R. A. Houghton, F. Dong, A. Elobeid, J. Fabiosa, S. Tokgoz, D. Hayes, and T.-H. Yu. 2008. "Use of U.S. Croplands for Biofuels Increases Greenhouse Gases through Emissions from Land Use Change." *Science* 319:1238–40.

Serra, T., D. Zilberman, J. M. Gil, and B. K. Goodwin. 2010a. "Price Transmission in the U.S. Ethanol Market." In *Handbook of Bioenergy Economics and Policy,* edited by M. Khanna, J. Scheffran, and D. Zilberman, 55–72. New York: Springer Science Verlag.

———. 2010b. "Price Volatility in Ethanol Markets." Unpublished Manuscript. Energy Bioscience Institute, University of California, Berkeley.

Taheripour, F., D. Birur, T. Hertel, and W. Tyner. 2007. "Introducing Liquid Biofuels into the GTAP Data Base." GTAP Research Memorandum no. 11. West Lafayette, IN: Purdue University, Global Trade Analysis Project.

Taheripour, F., T. Hertel, W. Tyner, J. Beckman, and D. Birur. 2009. "Biofuels and Their By-products: Global Economic and Environmental Implications." *Biomass and Bioenergy* 33 (3): 278–89.

Thompson, W., S. Meyer, and P. Westhoff. 2009. "How Does Petroleum Price and Corn Yield Volatility Affect Ethanol Markets with and without an Ethanol Use Mandate?" *Energy Policy* 37 (2): 745–49.

Tyner, W. 2009. "The Integration of Energy and Agricultural Markets." Paper presented at the International Association of Agricultural Economists, Beijing, China.

Ubilava, D., and M. Holt. 2010. "Forecasting Corn Prices in the Ethanol Era." Unpublished Manuscript. Department of Agricultural Economics, Purdue University.

Valenzuela, E., T. W. Hertel, R. Keeney, and J. Reimer. 2007. "Assessing Global Computable General Equilibrium Model Validity Using Agricultural Price Volatility." *American Journal of Agricultural Economics* 89 (2): 383–97.

Yano, Y., D. Blandford, and Y. Surry. 2010. "The Impact of Feedstock Supply and Petroleum Price Variability on Domestic Biofuel and Feedstock Markets—The Case if the United States." Swedish University of Agricultural Sciences Working Paper no. 2010:3. Uppsala, Sweden: Swedish University of Agricultural Sciences.

Meeting the Mandate for Biofuels
Implications for Land Use, Food, and Fuel Prices

Xiaoguang Chen, Haixiao Huang,
Madhu Khanna, and Hayri Önal

7.1 Introduction

Concerns about energy security, high oil prices, and climate change mitigation have led to increasing policy support for the production of biofuels in the United States. In 2008, the production of U.S. corn ethanol more than tripled relative to 2001 with the production of nine billion gallons using one-third of U.S. corn production (U.S. Department of Agriculture [USDA] 2010). Prices of agricultural commodities doubled between 2001 and 2008, leading to a debate about the extent to which the price increase was caused by biofuels and the competition for land induced by them (U.S. Department of Agriculture/Economic Research Service [USDA/ERS] 2010). A number of studies have analyzed the impact of biofuel demand on the price of crops and obtained widely varying estimates depending on the choice of price index, the baseline, and the other contributing factors considered. Reviews of these studies by Pfuderer, Davies, and Mitchell (2010) and Abbott, Hurt, and Tyner (2008) show that biofuels did contribute to the spike in crop prices in 2008, but with the current relatively low levels of diversion of global corn

Xiaoguang Chen is a postdoctoral research associate at the Institute for Genomic Biology at the University of Illinois, Urbana-Champaign, and a research scientist at the Energy Biosciences Institute. Haixiao Huang is a research associate at the Institute for Genomic Biology at the University of Illinois, Urbana-Champaign, and a research scientist at the Energy Biosciences Institute. Madhu Khanna is professor of agricultural and consumer economics and a professor at the Institute of Genomic Biology at the University of Illinois, Urbana-Champaign. Hayri Önal is professor of agricultural and consumer economics at the University of Illinois, Urbana-Champaign.

Authorship is alphabetical. Funding from the Energy Biosciences Institute; the University of California, Berkeley; the U.S. Department of Energy; and the National Institute of Food and Agriculture (NIFA) at the U.S. Department of Agriculture (USDA) is gratefully acknowledged.

production to biofuels, they were not the key drivers of the price increase. The trade-offs between food and fuel production could, however, intensify in the future as the Renewable Fuel Standard (RFS) established by the Energy Independence and Security Act (EISA) of 2007 seeks a sixfold increase in biofuel production by 2022.

Recognition of these trade-offs and the limits to relying on corn-based ethanol to meaningfully reduce dependence on oil has led to growing interest in developing advanced biofuels from feedstocks other than cornstarch. A commercial technology to produce cellulosic biofuels is yet to be developed, but efforts are underway to produce them from several different feedstocks such as crop and forest residues and perennial grasses (such as miscanthus and switchgrass). The use of residues does not require diversion of land from food production, while perennial grasses are not only likely to be more productive in their biofuel yields per unit of land than corn ethanol but can also be grown on marginal lands. Cellulosic biofuels are expensive compared to corn ethanol and unlikely to be viable without biofuel support policies. The RFS mandates an increasing share of biofuel production from noncornstarch feedstocks; this increases to 58 percent in 2022. The Food, Conservation, and Energy Act (FCEA) of 2008 also provides a variety of volumetric tax credits for blending biofuels with gasoline, with higher tax credits for advanced biofuels ($0.27 per liter) than for corn ethanol ($0.12 per liter), with the intent of making them competitive with corn ethanol. These tax credits lower fuel prices and, to the extent that they shift the mix of biofuels toward cellulosic feedstocks relative to the mandate alone, they could also lower crop prices. The decrease in fuel prices could, however, lead to an increase in fuel consumption relative to the RFS alone.

This chapter examines the effects of the RFS and accompanying volumetric subsidies for land use, food, and fuel production and prices in the United States. We analyze the extent to which these policies lead to changes in cropping patterns on the intensive margin and to an expansion of cropland acreage. We also analyze the trade-off they pose between fuel and food production and the mix of cellulosic feedstocks that are economically viable under alternative policy scenarios.

Furthermore, we examine the welfare costs of these policies and the costs of these tax credits for domestic taxpayers. A recent report by the Congressional Budget Office [CBO] 2010) estimates that the volumetric tax credit costs tax payers $0.47 per liter of (gasoline energy equivalent) corn ethanol and $0.79 per liter of (gasoline energy equivalent) cellulosic biofuels. The study assumes that these tax credits lead to a 32 percent increase in corn ethanol production and a 47 percent increase in cellulosic biofuel production, over and above that otherwise. Metcalf (2008) attributes all of the corn ethanol consumption above the mandate in 2005 to the corn ethanol tax credits and estimates that tax credits increased consumption by 25 percent. McPhail and Babcock (2008) find a much smaller role for the effect of the corn ethanol tax credit in 2008 to 2009; they estimate that it increased domestic supply by about 3 percent compared to the mandate alone. With two types of biofuels, corn ethanol and cellulosic biofuels, receiving tax credits at differential rates, determining the incremental effect of these tax credits in the future is more challenging because they could affect not only the total volume of biofuels but could also create incentives to increase one type of biofuel at the expense of another. Moreover, the cost of these tax credits should include not only the direct effect on tax payers but also the indirect effect on consumers and producers of agricultural and fuel products. These policies will differ in their impacts on food and fuel consumers and producers and are likely to benefit agricultural producers and fuel consumers while adversely affecting gasoline producers and agricultural consumers. In an open economy with trade in agricultural products and gasoline, some of these costs are passed on to foreign producers and consumers by changing the terms of trade. We use the framework developed here to jointly determine the economic costs (in terms of domestic social welfare) of these tax credits as well as the extent to which they lead to incremental biofuel production above the mandated level and change the mix of biofuels. Finally, we analyze the sensitivity of the impact of the these biofuel policies on the mix of feedstocks used and on food and fuel prices to several supply-side factors, such as the costs of various feedstocks and biofuels, the growth in productivity of conventional crops, and the availability of land.

We develop a dynamic, multimarket equilibrium model, Biofuel and Environmental Policy Analysis Model (BEPAM), which analyzes the markets for fuel, biofuel, food or feed crops, and livestock for the period 2007 to 2022. We consider biofuels produced not only from corn but also from several cellulosic feedstocks and imported sugarcane ethanol while distinguishing between domestic gasoline supply and gasoline supply from the rest of the world. The BEPAM model treats each crop reporting district (CRD) as a decision-making unit where crop yields, costs of crop and livestock production, and land availability differ across CRDs. Food and fuel prices are endogenously determined annually and used to update price expectations, cropland acreage, and land use choices. The rest of the chapter is organized as follows. In section 7.2, we review the existing literature and the key contributions of our research. In section 7.3 we briefly describe the current legislations whose effects are being analyzed here. Section 7.4 describes the simulation model. Data used for the simulation model is described in section 7.5, followed by the results and conclusions in sections 7.6 and 7.7.

7.2   Previous Literature

A few studies have developed stylized models to analyze the economic and environmental effects of a biofuel mandate. De Gorter and Just (2008) examine the effects of a biofuel mandate with import tariffs, while de Gorter and Just (2009) examine the effect of a blend mandate with tax credits on fuel prices, assuming that biofuels and gasoline are perfect substitutes. With a blend mandate, the consumer price of the blended fuel is a weighted average of the price of gasoline and biofuel, with weights depending on the share of biofuels in the blend. The effect of the blend mandate on the price of the blended fuel is, therefore, theoretically ambiguous; the mandate increases the price of biofuel, but it lowers gasoline consumption and, thus, its price. Ando, Khanna, and Taheripour (2010) analyze the effects of a quantity mandate for biofuels on fuel prices and consumption, greenhouse gas (GHG) emissions, and social welfare and consider biofuels and gasoline to be imperfect substitutes. A quantity mandate imposes a fixed cost of blend-ing (the mandated quantity) on blenders. They show that if the mandate is small relative to the amount of gasoline consumed, and marginal cost pric-ing of the blended fuel is profitable, then the mandate unambiguously lowers the price of the blended fuel. It will, therefore, increase vehicle kilometers travelled (VKT) and have an ambiguous impact on GHG emissions. Our analysis here expands on the framework of Ando, Khanna, and Taheripour (2010) by analyzing the welfare effects of biofuel policies on both the fuel and agricultural sector and an open economy with trade in fuel and agri-cultural commodities.

A number of studies have examined the implications of biofuel produc-tion and policies for food or feed prices and land use in the long run. Using the partial equilibrium Food and Agricultural Policy Research Institute (FAPRI) model, Elobeid et al. (2007) analyze the long-run effects of crude oil price changes on demand for ethanol and corn, while Elobeid and Tok-goz (2008) expand that analysis to show the extent to which the effects of expansion in corn ethanol production on food or feed prices can be miti-gated by liberalizing import of biofuels from Brazil. More recently, Fabiosa et al. (2009) use the model to obtain acreage multiplier effects of corn etha-nol expansion. These studies (like Tyner and Taheripour 2008) consider an exogenously given price of gasoline and assume that ethanol and gasoline are perfectly substitutable. As a result, the price of ethanol is determined by the price of gasoline (based on its energy content relative to gasoline) and there is a one-directional link between gasoline prices and corn prices, resulting in a perfectly elastic demand for corn at the break-even price at which ethanol refineries can make normal profits. These studies also assume that crop yields are constant over time.

Ferris and Joshi (2009) use AGMOD (an econometric model of U.S. agri-culture) to examine the implications of the RFS for ethanol and biodiesel production (2008 to 2017), assuming perfect substitutability between gaso-line and ethanol and no cellulosic biofuel production. They find that the mandate could be met by potential crop yield increases and a decline in land under the Conservation Reserve Program (CRP) and cropland pasture.

Unlike the models used in the preceding studies that focus only on corn ethanol, the POLYSYS model (an agricultural policy simulation model of the U.S. agricultural sector) includes various bioenergy crops and investigates land use impacts of biofuel and climate policies (Ugarte et al. 2003). Walsh et al. (2003) apply POLYSYS to examine the potential for producing bioenergy crops at various exogenously set bioenergy prices. English et al. (2008) analyze the effects of the corn ethanol mandate (assuming that cellulosic biofuels are not feasible) and show that it will lead to major increases in corn production in the Corn Belt and in fertilizer use and soil erosion over the period 2007 to 2016. Most recently, Ugarte et al. (2009) apply POLYSYS to analyze the implications on agricultural income, over the 2010 to 2025 period, of various carbon prices and carbon offset scenarios under a GHG cap and trade policy assuming the RFS exists.

The impact of climate change policies on the agricultural sector and biofuel production has been examined by McCarl and Schneider (2001) using FASOM (Forest and Agricultural Sector Optimization Model), a multiperiod, price endogenous spatial market equilibrium model, with a focus on land allocation between agricultural crops and forests. Like the preceding studies, FASOM also assumes that gasoline and ethanol are perfectly substitutable, but determines the price of gasoline endogenously using an upward sloping supply curve for gasoline. The model includes an autonomous time trend in crop yields and considers various bioenergy feedstocks, such as crop and forest residues, switchgrass, and short-rotation woody crops. The FASOM model is used by the U.S. Environmental Protection Agency (EPA) to simulate the impacts of implementing the RFS relative to the 2007 Annual Energy Outlook (Energy Information Administration [EIA] 2007) reference case (EIA 2010a). Results show that the RFS would increase corn and soybeans prices in 2022 by 8 percent and 10 percent, respectively, and decrease gasoline price by 0.006 cents per liter relative to the Annual Energy Outlook 2007 reference case. Total social welfare in 2022 is $13 to 26 billion higher than the reference level.

In addition to these partial equilibrium studies, the general equilibrium Global Trade Analysis Project (GTAP) model has been used to examine the global land use effect of corn ethanol mandate in the United States and a biofuel blend mandate in the European Union in 2015, assuming no cellulosic biofuel production and imperfect substitutability between gasoline and ethanol (Hertel, Tyner, and Birur 2010). Reilly, Gurgel, and Paltsev (2009) use the general equilibrium Emissions Predictions and Policy Analysis (EPPA) model to examine the implications of GHG reduction targets over the 2015 to 2100 period for second-generation biomass production and changes in land use. Their simulations suggest that it is possible for significant biofuel production to be integrated with agricultural production in the long run without having dramatic effects on food and crop prices.

The model developed in this chapter has several key features. First, we allow imperfect substitutability between gasoline and ethanol because the extent to which biofuels can be substituted for gasoline in automobiles (at the aggregate level) depends on the current intensity of biofuels in the fuel mix and on the stock of flex-fuel vehicles in the national fleet. We consider a constant elasticity of substitution between gasoline and biofuels as in Ando, Khanna, and Taheripour (2010) and the GTAP model (Hertel, Tyner, and Birur 2010) because bottlenecks within the ethanol distribution infrastructure, the existing stock of vehicles, and constraints on the rate of turnover in vehicle fleet limit the substitutability between biofuels and gasoline. Empirical evidence shows that biofuel prices are not simply demand driven (based on energy equivalent gasoline prices and perfect substitutability); instead, they have been observed to be correlated with their costs of production as well.[1] Hayes et al. (2009) show that incorporating imperfect substitutability between ethanol and gasoline in the FAPRI model results in a substantially smaller impact of a change in crude oil prices on demand for ethanol and land use than in Tokgoz et al. (2007).

Additionally, we assume upward sloping supply functions for gasoline and for biofuels and distinguish between gasoline supply from domestic producers and the rest of world. The United States accounts for 23 percent of world petroleum consumption, and about 57 percent of the consumption is imported from the rest of the world (EIA 2010b); thus, the change in U.S. oil demand can significantly affect imports of gasoline and world gasoline prices. Our model allows biofuel production to have a feedback effect on gasoline prices and, thus, on the demand for biofuels (as in Hayes et al. 2009). It considers the effect of biofuel policy on imports and on domestic social welfare by separating the effect of price changes on domestic and foreign fuel providers. The welfare effects of biofuel policies, therefore, consider both the efficiency cost of these policies relative to a free market outcome and their terms of trade effects.

Crop yield changes over time influence the land needed to meet food and fuel needs to meet biofuel mandates. Dumortier et al. (2009) show that introduction of even a 1 percent increasing trend in corn yield in the FAPRI model can substantially reduce the corn acreage in response to changes in gasoline and biofuel prices. We allow for changes in crop yields over time from two sources, an endogenous price effect and an autonomous technology effect, using econometrically estimated elasticities and time trend.

Existing models such as FASOM rely on historically observed crop mixes to constrain the outcomes of linear programming models and generate results that are consistent with farmers' planting history. To accommodate new bioenergy crops and unprecedented changes in crop prices in the future, the FASOM model allows crop acreage to deviate 10 percent from observed historical mixes. In BEPAM, we use the estimated own and cross-price crop

---

1. See http://www.agmrc.org/renewable_energy/ethanol/the_relationship_of_ethanol _gasoline_and_oil_prices.cfm#.

elasticities to limit the flexibility of crop acreage changes instead of an arbitrary level of flexibility.

7.3 Policy Background

The EISA established the RFS in 2007 to provide an assurance of demand for biofuels beyond levels that might otherwise be supported by the market. It establishes a goal of 136 billion liters of biofuel production in 2022 that includes four separate categories of renewable fuels, each with a separate volume mandate. Of the 136 billion liters of the renewable fuel, the RFS requires that at least 80 billion liters should be advanced biofuels. Advanced biofuel specifically excludes ethanol derived from cornstarch. It includes ethanol made from cellulose, hemicelluloses, lignin, sugar, or any starch other than cornstarch as long as it achieves a GHG reduction of 50 percent compared to gasoline and is obtained from "renewable biomass." Renewable biomass limits the crops and crop residues used to produce renewable fuel to those grown on land cleared or cultivated at any time prior to enactment of EISA in December 2007. Crops used to produce renewable fuels that can meet the mandate must be harvested from agricultural land cleared or cultivated prior to December 2007. Land enrolled in the CRP is not allowed to be converted for the production of miscanthus and switchgrass (EIA 2010a).

Of the 80 billion liters of the advanced biofuels, at least 60 billion liters should be cellulosic biofuels derived from any cellulose, hemicelluloses, or lignin and achieve a life-cycle GHG emission displacement of 60 percent compared to gasoline, while the rest could be sugarcane ethanol from Brazil. The amount of conventional biofuels produced from cornstarch that can meet the RFS is capped at 56 billion liters in 2022; excess production can occur but cannot be considered for complying with the RFS. Cumulative production of biofuels over the 2007 to 2022 period mandated by the RFS requires 1,220 billion liters of renewable fuel and at least 420 billion liters of advanced biofuels, while the amount of conventional biofuels cannot exceed 800 billion liters during this period.

The FCEA of 2008 provides tax credits for blending biofuels with gasoline. The tax credits for corn ethanol peaked at $0.16 per liter in 1984, fell to $0.14 per liter in 1990, $0.13 per liter between 1998 and 2005, and is authorized at $0.12 cents per liter until December 2010.[2] The tax credit for cellulosic biofuels is $0.27 per liter and authorized until January 1, 2013. It also requires that cellulosic biofuels should be produced and consumed in the United States.

In addition to biofuel mandates and volumetric tax credits, the United States imposes trade barriers to restrict the imports of sugarcane ethanol

---

2. http://frwebgate.access.gpo.gov/cgi-bin/getdoc.cgi?dbname=110_cong_public_laws &docid=f:publ246.pdf.

from Brazil. The biofuel trade policy includes a 2.5 percent ad valorem tariff and a per unit tariff of \$0.14 per liter (authorized until January 2011). A key motivation for the establishment of the tariff is to offset a tax incentive for ethanol-blended gasoline. An exception to the tariff is the agreement of the Caribbean Basin Initiative (CBI) initiated by the 1983 Caribbean Basin Economic Recovery Act (CBERA). Under this agreement, ethanol produced from at least 50 percent agricultural feedstocks grown in CBI countries is admitted into the United States free of duty. If the local feedstock content is lower than the requirement, a tariff rate quota (TRQ) will be applied to the quantity of duty-free ethanol. Nevertheless, duty-free ethanol from CBI countries is restricted to no more than 0.2 billion liters or 7 percent of the U.S. ethanol consumption. To take advantage of this tariff-free policy, hydrous ethanol produced in other counties, like Brazil or European countries, can be imported to a CBI country and exported to the United States after dehydration. In 2007, total imports account for roughly 6 percent of U.S. consumption (25.7 billion liters), with about 40 percent of the import from Brazil and approximately 60 percent routed through CBI countries to avoid the import tariff. However, CBI countries have never reached the ceiling on their ethanol quota, partly due to insufficient capacity. Our analysis here assumes existing tariff policy remain in effect until 2022.

7.4   The Model

7.4.1   General Description

We develop a multimarket, multiperiod, price-endogenous, nonlinear mathematical programming model that simulates the U.S. agricultural and fuel sectors and formation of market equilibrium in the commodity markets including trade with the rest of the world. We refer to this model as the Biofuel and Environmental Policy Analysis Model (BEPAM). The BEPAM model is a dynamic, multimarket equilibrium model, which analyzes the markets for fuel, biofuel, food or feed crops, and livestock for an extendable future period (currently set for 2007 to 2022) in the United States. This model determines several endogenous variables simultaneously, including VKT, fuel and biofuel consumption, domestic production and imports of oil and imports of sugarcane ethanol, mix of biofuels and the allocation of land among different food and fuel crops, and livestock. This is done by maximizing the sum of consumers' and producers' surpluses in the fuel and agricultural sectors subject to various material balances and technological constraints underlying commodity production and consumption within a dynamic framework (Takayama and Judge 1971; McCarl and Spreen 1980). This model is designed specifically to analyze the implications of biofuel and climate policies on land use patterns, commodity markets, and the environment.

The agricultural sector in BEPAM includes several conventional crops, livestock, and bioenergy crops (crop residues from corn and wheat and perennial grasses, miscanthus, and switchgrass) and distinguishes between biofuels produced from corn, sugarcane, and cellulosic feedstocks. Crops can be produced using alternative tillage and rotation practices. The model incorporates spatial heterogeneity in crop and livestock production activity, where crop production costs, yields, and resource endowments are specified differently for each region and each crop assuming linear (Leontief) production functions. As the spatial decision unit, the model uses the CRDs in each state by assuming an aggregate representative producer who makes planting decisions to maximize the total net returns under the resource availability and production technologies (yields, costs, crop rotation possibilities, etc.) specified for that CRD. The model covers CRDs in forty-one of the contiguous U.S. states in five major regions.[3]

The model uses "historical" and "synthetic crop mixes" when modeling farms' planting decisions to avoid extreme specialization in regional land use and crop production. The use of historical crop mixes ensures that the model output is consistent with the historically observed planting behaviors (McCarl and Spreen 1980; Önal and McCarl 1991). This approach has been used in some existing models also, such as FASOM, to constrain feasible solutions of programming models and generate results that are consistent with farmers' planting history. To accommodate planting new bioenergy crops and unprecedented changes in crop prices in the future, FASOM allows crop acreage to deviate 10 percent from the observed historical mixes. In our model, we use synthetic (hypothetical) mixes to offer increased planting flexibility beyond the observed levels and allow land uses that might occur in response to the projected expansion in the biofuels industry and related increases in corn and cellulosic biomass production. Each synthetic mix represents a potential crop pattern generated by using the estimated own and cross-price crop acreage elasticities and considering a set of price vectors where crop prices are varied systematically. These elasticities are estimated econometrically using historical, county-specific data on individual crop acreages for the period 1970 to 2007 as described in Huang and Khanna (2010). Crop yields are assumed to grow over time at an exogenously given trend rate and to be responsive to crop prices.

The model includes five types of land (cropland, idle cropland, cropland pasture, pasture land, and forestland pasture) for each CRD. We obtain

3. Western region includes Arizona, California, Colorado, Idaho, Montana, Nevada, New Mexico, Oregon, Utah, Washington, and Wyoming; Plains includes Nebraska, North Dakota, Oklahoma, South Dakota, Texas, and Kansas; Midwest includes Illinois, Indiana, Iowa, Michigan, Minnesota, Missouri, Ohio, and Wisconsin; South includes Alabama, Arkansas, Florida, Georgia, Louisiana, Mississippi, and South Carolina; Atlantic includes Kentucky, Maryland, New Jersey, New York, North Carolina, Pennsylvania, Tennessee, Virginia, and West Virginia.

CRD-specific planted acres for fifteen row crops for the period 1977 to 2007 from the U.S. Department of Agriculture/National Agricultural Statistics Service ([USDA/NASS] 2009b) and use this to construct the historical and synthetic mixes of row crops. Cropland availability in each CRD is assumed to change in response to crop prices. The responsiveness of total cropland to crop prices as well as the own and cross-price acreage elasticities for individual crops is obtained from Huang and Khanna (2010). Data on idle cropland, cropland pasture, pasture, and forestland pasture for each CRD are also obtained from USDA/NASS (2009b). Idle cropland includes land use category for cropland in rotations for soil improvement and cropland on which no crops were planted for various physical and economic reasons. The estimates of idle land include land enrolled in the CRP that could be an additional source of land available for energy crops. Land in this program is farmland that is retired from crop production and converted to trees, grass, and areas for wildlife cover. We exclude land enrolled in CRP from our simulation model. Cropland pasture is considered as a long-term crop rotation between crops and pasture at varying intervals.

Pasture land consists of land with shrub, brush, all tame and native grasses, legumes, and other forage, while forestland pasture is stocked by trees of any size and includes a certain percentage of tree cover. Pasture land and forestland pasture are primarily for grazing uses. We keep the level of permanent pastureland and forestland pasture fixed at 2007 levels but allow idle land and cropland pasture to move into cropland and back into an idle state. It can also be used for perennial bioenergy crop production. A change in the composite crop price index triggers a change at the extensive margin and leads to a shift in land from idle cropland and cropland pasture to land available for crop production the following year. The responsiveness of aggregate cropland supply to a lagged composite price index is econometrically estimated, and the implications of expanding crop production to idle land and cropland acreage for average yields of conventional crops in each CRD are described in Huang and Khanna (2010). The remaining idle land or pasture land can be used for bioenergy crops. While yields of bioenergy crops are assumed to be the same on marginal land as on regular cropland, there is a conversion cost to the use of idle land or cropland pasture for bioenergy crop production. In the absence of an empirically based estimate of the ease of conversion of marginal land for perennial grass production, we assume a CRD-specific conversion cost equal to the returns the land would obtain from producing the least profitable annual crop in the CRD. This ensures consistency with the underlying assumption of equilibrium in the land market, in which all land with nonnegative profits from annual crop production is utilized for annual crop production. As annual crop prices increase, the cost of conversion increases; the "supply curve" for idle marginal land is, therefore, upward sloping. We impose a limit of 25 percent on the amount of land in a CRD that can be converted to perennial grasses due to concerns about the impact of monocultures of perennial grasses on biodiversity or subsurface water flows. We examine the sensitivity of model results to this assumption by lowering this limit to 10 percent.

The perennial nature of the energy crops included in the model requires a multiyear consideration when determining producers' land allocation decisions in any given year. For this, we use a rolling horizon approach where for each year of the period 2007 to 2022, the model determines production decisions and the corresponding dynamic market equilibrium for a planning period of ten years starting with the year under consideration. After each run, the first-year production decisions and the associated market equilibrium are used to update some of the model parameters (such as the composite crop price index, land supplies in each region, and crop yields per acre for major crops) based on previously generated endogenous prices, and the model is run again for another ten-year period starting with the subsequent year.

The behavior of agricultural consumers' behavior is characterized by linear demand functions that are specified for individual commodities, including crop and livestock products. In the crop and livestock markets, primary crop and livestock commodities are consumed either domestically or traded with the rest of the world (exported or imported), processed, or directly fed to various animal categories. Export demands and import supplies are incorporated by using linear demand or supply functions. The commodity demand functions and export demand functions for tradable row crops and processed commodities are shifted upward over time at exogenously specified rates. The crop and livestock sectors are linked to each other through the supply and use of feed items and also through the competition for land (because the grazing land needed by the livestock sector has alternative uses in crop production).

The biofuel sector distinguishes biofuels produced from corn, sugarcane ethanol, and cellulosic feedstock with all biofuels being perfect substitutes for each other. Biofuel from sugarcane is imported from Brazil and CBI countries subject to policies described in the preceding. Gasoline is produced domestically as well as imported from the rest of the world. The demand for gasoline and biofuels is derived from the demand for VKT. We assume a linear demand for VKT as a function of the cost per kilometer and that VKT is produced using a blend of gasoline and biofuels. At the individual consumer level (with a conventional vehicle), the two fuels are currently perfectly substitutable in energy equivalent units up to a 10 percent blend. For an individual consumer with a flex-fuel car, the two fuels are substitutable up to an 85 percent blend. At the aggregate level, we consider a representative consumer that owns a vehicle fleet that consists of a mix of the two types of vehicles; in 2007, only 2.9 percent of vehicles in 2007 were flex-fuel vehicles

(EIA 2010a). The ability to substitute gasoline for biofuels at the aggregate level is, therefore, limited by the mix of vehicles. It is also limited by the available ethanol distribution network and infrastructure for retail ethanol sales. We, therefore, consider gasoline and biofuel to be imperfectly substitutable at the aggregate level and use a constant elasticity of substitution (CES) function to model the aggregate blend of fuel produced. The VKT demand function and CES production function are calibrated for the base year assuming a specific value for the elasticity of substitution between gasoline and ethanol and observed base-year prices and quantities of these fuels and VKT. We examine the implications of varying the extent of substitutability on the consumption of the two types of fuels and on the agricultural and fuel sectors. The demand for VKT is shifted upward over time, and the VKT consumed is determined by the marginal cost of kilometers, which in turn depends on the marginal costs of gasoline and biofuels. The shares of various fuels are determined endogenously based on fuel prices.

In the presence of the RFS, the quantity mandate imposes a fixed cost of biofuel on blenders. The average cost of the blended fuel (gasoline and ethanol) will fall as the level of gasoline consumption increases, but the average cost will be greater than marginal costs for low levels of gasoline consumption. Thus, at low levels of fuel consumption, blenders can be expected to price fuel based on its average cost (if average cost is greater than the marginal cost) in order to avoid negative profits. In this case, VKT will be determined by the average cost of a kilometer rather than its marginal cost. If gasoline consumption is high enough (or if biofuel consumption is small), it could be profitable to use marginal cost pricing of the blended fuel. The model selects the appropriate rule for pricing the blended fuel depending on whether average cost of VKT is greater or smaller than its marginal cost.

The endogenous variables determined by the model include: (a) commodity prices; (b) production, consumption, export, and import quantities of crop and livestock commodities; (c) land allocations and choice of practices for producing row crops and perennial crops (namely, rotation, tillage, and irrigation options) for each year of the 2017 to 2022 planning horizon and for each CRD; and (d) the annual mix of feedstocks for biofuel production, domestic production, and imports of gasoline and consumption of VKT.

7.4.2   Algebraic Presentation

We describe the algebraic form of the numerical model using lowercase symbols to denote the exogenous parameters and uppercase symbols to represent endogenously determined variables. The objective function is the sum of discounted consumers' and producers' surpluses obtained from production, consumption, and trade of the crop and livestock products, plus the surplus generated in the fuels sector over the sixteen-year planning horizon 2007 to 2022 and the terminal values of standing perennial grasses in 2022. The algebraic expression is given explicitly in equation (1):

$$
\text{Max: } \sum_{0}^{T} e^{-rt} \left\{ \sum_{z} \int_{0}^{\text{DEM}_{t,z}} f^{z}(\cdot) d(\cdot) + \sum_{z} \int_{0}^{\text{EXP}_{t,z}} f^{z}(\cdot) d(\cdot) - \sum_{z} \int_{0}^{\text{IMP}_{t,z}} f^{z}(\cdot) d(\cdot) + \int_{0}^{\text{KIL}_{t}} f^{z}(\cdot) d(\cdot) \right.
$$

$$
- \sum_{r,q} rc_{r,q} \text{ACR}_{t,r,q} - \sum_{r,p} pc_{r,p} \text{ACR}_{t,r,p} - \sum_{r,q} rs_{r,q} \text{ACR}_{t,r,q} - \sum_{r,p} cc_{r} \Delta \text{ACR}_{t,r,p}
$$

$$
(1) \quad - \sum_{k} lc_{k} \text{LIV}_{t,k} - \sum_{i} sc_{i} \text{PRO}_{t,i}
$$

$$
\left. - \int_{0}^{\text{GAS}_{t}} f^{g}(\cdot) d(\cdot) - ec_{c} \text{ETH}_{t,c} - ec_{b} \text{ETH}_{t,b} \right\}
$$

$$
+ e^{-rT} \sum_{r,p} (v_{r,p} - w_{r}) \text{ACR}_{T,r,p}
$$

The first integral term in line of equation (1) represents the areas under the domestic demand functions from which consumers' surplus is derived. Each integral is associated with a crop, livestock, or processed commodity for which a domestic market demand is considered. ($\text{DEM}_{t,z}$ denotes the endogenous domestic demand variable in year t; $z = \{i,j,k\}$ denotes the index set for crop commodities (i), processed products from crops (j), and livestock commodities (k); $f^{z}(\cdot)$ denotes the inverse demand function for the commodity involved; and the $d(\cdot)$ denotes the integration variable). The next two integral terms account for the areas under the inverse demand functions for exports, $\text{EXP}_{t,z}$, and the areas under the import supply functions $\text{IMP}_{t,z}$ (such as sugar and sugarcane ethanol). The last integral term represents the area under the inverse demand function for kilometers traveled (denoted by KIL_{t}). The demand functions for crop products, livestock products, and kilometers traveled are all characterized by linear demand functions in the current version, but other functional forms, such as constant elasticity demand functions, can be incorporated without difficulty.

The second line in equation (1) includes the production costs of row crops, perennial crops, and crop or forest residues collected for biofuel production, and land conversion costs for marginal lands converted to the production of perennial crops. The land allocated to row crops and perennial crops (acreage) in region r and year t, denoted by $\text{ACR}_{t,r,q}$ and $\text{ACR}_{t,r,p}$, respectively, may use one of the various production practices that differ by crop rotation, tillage, and irrigation. Fixed input-output coefficients (Leontief production functions) are assumed for both row crops and perennial crops production. The third term represents the cost of collected crop residues (biomass for cellulosic biofuel production) and involves the management options for row crops that produce biomass (specifically, corn stover and wheat straw). The amount of marginal lands converted for perennial grasses are denoted by $\Delta \text{ACR}_{t,r,p}$ and cc_{r} represents the cost per unit of marginal land conversion. The last term denotes the costs of converted marginal lands (such as idle land and crop pasture land) for perennial crops. The land conversion costs include costs for land clearing, wind rowing, and any necessary activities for seedbed preparation.

The third line in equation (1) includes the costs associated with livestock activities. The amount of livestock is represented by $LIV_{t,k}$, and lc_k denotes the cost per unit of livestock category k (again employing Leontief production functions) that is assumed to be the same across all regions. The second term represents the total cost of converting primary crops (corn, soybeans, and sugarcane) to secondary (processed) commodities (oils, soymeal, refined sugar, high-fructose corn syrup [HFCS], and Distiller's Dried Grains with Solubles [DDGS]). The amount of processed primary crop i in year t is denoted by $PRO_{t,i}$, and sc_i denotes the processing cost per unit of i.

The fourth line involves the costs accruing to the fuel sector. The first integral represents the area under the supply functions for gasoline from domestic producers and the rest of the world, whose consumption and price are to be determined endogenously. The next two terms represent the processing costs of corn and cellulosic ethanol in refinery, namely $ETH_{t,c}$, $ETH_{t,b}$. Finally, the last line reflects the value of the remaining economic life of standing perennial grasses beyond the planning period T, denoted by $v_{r,p}$, net of the return from the most profitable cropping alternative in region r, denoted by w_r. The latter is used to account for the opportunity costs of land.

In the model, we assume that the consumers obtain utility from VKT (KIL_t), which is produced by blending gasoline (GAS_t), corn ethanol ($ETH_{t,c}$), cellulosic ethanol ($ETH_{t,b}$) and sugarcane ethanol ($IMP_{t,s}$). Gasoline and ethanol are assumed to be imperfect substitutes in kilometers production, while corn ethanol and cellulosic ethanol are perfect substitutes. The total amount of kilometers generated by use of all sources of fuels is formulated using a constant elasticity production function as shown in equation (2):

(2) $KIL_t = \gamma_t[\alpha_t(ETH_{t,c} + ETH_{t,b} + IMP_{t,s})^\rho + (1 - \alpha_t)GAS_t^\rho]^{1/\rho}$ for all t

The regional material balance equations link the production and usage of primary crops, as shown in constraint (3) for primary crop product i produced and marketed by region r:

(3) $MKT_{t,r,i} + \left\{CE_{t,r}\right\}_{i=\text{corn}} \leq \sum_j y_{r,q,i}ACR_{t,r,q}$ for all $t, r, i,$

where $MKT_{t,r,i}$ denotes the amount of primary crop product i sold in the commodity markets, and $y_{r,q,i}$ is the yield of product i per unit of the land allocated to crop production activity q in region r. For corn, $MKT_{t,r,i}$ includes nonethanol uses, and $CE_{t,r}$ is the amount of corn converted to ethanol production (which appears only in the balance constraint for corn).

The amount of primary crop i available in the market (excluding the corn used for ethanol) comes from domestic regional supply ($MKT_{t,r,i}$). This total amount is either consumed domestically ($DEM_{t,i}$), exported ($EXP_{t,i}$).

processed to secondary commodities ($PRO_{t,i}$), or used for livestock feed ($FED_{t,i}$). This is expressed in constraint (4):

$$(4) \qquad DEM_{t,i} + PRO_{t,i} + FED_{t,i} + EXP_{t,i} \leq \sum_{r} MKT_{t,r,i} \qquad \text{for all } t,i$$

Similar to equation (4), a balance equation is specified for each processed commodity. Like primary commodities, processed commodities can also be consumed domestically, exported, or fed to animals, as shown in constraint (5):

$$(5) \quad DEM_{t,j} + FED_{t,j} + EXP_{t,j} \leq v_{i,j} PRO_{t,i} + \left\{ \sum_{r} v_{i,j} CE_{t,r} \right\}_{j=ddg,i=corn} \qquad \text{for all } t,j,$$

where $v_{i,j}$ denotes the conversion rate of raw product i to processed product j.

A particularly important component of the model that links the crop and fuel sectors is the conversion of corn and cellulosic biomass to ethanol. During the conversion of corn a secondary commodity, (DDGS), is produced as a byproduct. The amount of DDGS produced is proportional to the amount of corn used for ethanol, $CE_{t,r}$, through a fixed conversion rate $v_{corn,ddg}$, and it can either be fed to livestock as a substitute for soymeal or exported.

The relations between ethanol production and crop production activities are expressed in the following:

$$(6) \qquad E_{t,c} = \alpha \sum_{r} CE_{tr} \qquad \text{for all } t$$

$$(7) \qquad E_{t,b} = \beta \left(\sum_{r,p} by_{r,p} AC_{r,p} + \sum_{r,q} ry_{r,q} AC_{t,r,q} \right) \qquad \text{for all } t,$$

where α and β denote the amounts of ethanol produced per unit of corn and cellulosic feedstock, respectively, and $by_{r,p}$ and $ry_{r,q}$ are the biomass and crop residue yields in region r for respective perennial and crop production activities.

Land is the only primary production factor considered in the model. In each region, the total amount of land used for all agricultural production activities cannot exceed the available land ($al_{t,r}$), which is specified separately for irrigated and nonirrigated land. Due to the steady increase in ethanol consumption, the demand for agricultural land is expected to increase through the conversion of some marginal lands (not currently utilized) to cropland. The extent of conversion is assumed to depend on variations in crop prices over time. Therefore, in the model, we determine the agricultural land supply "endogenously." Specifically, for a given year t in the planning horizon 2007 to 2022, we solve the model assuming a fixed regional land availability for each year of the ten-year production planning period considered in that run. From the resulting multiyear equilibrium solution, we take the first-year values of the endogenous commodity prices and use them to construct a composite commodity price index, CPI. Based on the CPI generated thereby, we adjust the land availability for the subsequent run (which considers another ten-year planning period starting with year $t + 1$). The land constraint is shown in equation (8).

$$(8) \qquad \sum_q \text{ACR}_{t,r,q} + \sum_p \text{ACR}_{t,r,p} \leq \text{al}_{t,r} \quad \text{for all } t, r$$

To prevent unrealistic changes and extreme specialization in land use, which may be particularly serious at regional level, we restrict farmers' planting decisions to a convex combination (weighted average) of historically observed acreage patterns ($h_{r,ht,i}$), where subscript ht stands for the observed time periods prior to the base year. Historical land uses may be valid when simulating farmer's planting decisions under "normal" conditions. However, they may be too restrictive for future land uses given the increased demand for ethanol and unprecedented land use patterns that are likely to occur in the future to produce the required biomass crops. To address this issue, we introduce "hypothetical" acreage patterns ($h'_{r,n,i}$) for each row crop and each region. To generate hypothetical acreage patterns (crop mixes), we first use the historical data on prices and acreages of row crops in each region to estimate acreage elasticities for each row crop with respect to its own price and cross-price changes while controlling other factors, such as social-economic changes and time trend. Then we estimate a number of hypothetical acreages using these price elasticities and consider a systematically varied set of crop prices. The resulting set of actual and hypothetical crop mixes are used in constraint (9) to limit the flexibility in planting decisions, where $\theta_{i,q}$ represents the share of row crop i in production activity q, and $W_{t,r,*}$ represents the weight assigned to historical or hypothetical crop mixes. The latter are defined as variables to be endogenously determined by the model.

$$(9) \qquad \sum_q \theta_{i,q} \text{ACR}_{t,r,q} = \sum_{ht} h_{r,ht,i} W_{t,r,ht} + \sum_n h'_{r,n,i} W_{t,r,n} \quad \text{for all } t, r, i$$

The sum of the endogenous weights assigned to individual mixes must be less than or equal to 1 (convexity requirement), as shown in equation (10).

$$(10) \qquad \sum_\tau W_{t,r,\tau} + \sum_n W_{t,r,n} \leq 1 \quad \text{for all } t, r$$

A similar set of crop mix constraints is introduced for irrigated crops too, which we do not show here, using only the historically observed irrigated land use patterns (no hypothetical mixes for irrigated crops).

Large-scale monocultures of perennial grasses may have unforeseen impacts on biodiversity and subsurface water flows. To prevent extreme specialization in the production of perennial grasses in some regions, we restrict the land allocated to perennial grasses to less than 25 percent of total land available in each region ($al_{t,r}$). The constraint is shown in equation (11).

(11) $$\sum_p ACR_{t,r,p} \leq 0.25 * al_{t,r} \quad \text{for all } t, r$$

In the livestock sector, we define production activity variables (number of animals) at the national level for each category of livestock except the beef and dairy cattle. Cattle production is given special emphasis in the model for two reasons. First, cattle require grazing land; therefore, they compete with crop production activities on total land in each region. Second, besides requirements of feed crops directly fed to different types of livestock, DDGS (a byproduct of corn ethanol production) is also used as a feed item that may substitute soymeal (both supplying protein). The regional cattle production activities are aggregated in equation (12) to obtain the total cattle activity at national level:

(12) $$LIV_{t,\text{cattle}} = \sum_r CTL_{t,r} \quad \text{for all } t,$$

where $CTL_{t,r}$ is the number of cattle stock in region r and year t. Cattle supply is constrained by the grazing land availability. Therefore, for each region, we specify the grazing rates and the supply of grazing land, $GL_{t,r,g}$, where g denotes the type of grazing land (namely pasture land, forest land, and cropland that can be used for grazing—such as wheat and oats). The amounts of other livestock (chicken, turkey, lamb, pork, and eggs) are also constrained by historical numbers at the national level. Constraint (13) relates the usage of grazing land and cattle activity in each region:

(13) $$CTL_{t,r} \leq \sum_g GL_{t,r,g} / ga_{r,g} \quad \text{for all } t, r,$$

where $ga_{r,g}$ denotes the amount of grazing land required per unit of cattle.

Equations (14) and (15) establish the balances between nutrition needs of livestock activities, in terms of protein and calories, and the amounts of nutrients provided by primary feed crops (grains) and by-products of crops processing (i.e., soymeal and DDGS):

(14) $$nr_{K,\text{nu}} LIV_{t,k} = \sum_i nc_{i,\text{nu}} F_{t,i,k} + \sum_j nc_{j,\text{nu}} F_{t,j,k} \quad \text{for all } t, k$$

(15) $$FED_{t,z} = \sum_k F_{t,z,k} \quad \text{for all } t, k \text{ and } z = i, j \text{ used for feed,}$$

where $nc_{z,\text{nu}}$ denotes the nutrition content per unit of feed item z, and $nr_{k,\text{nu}}$ and $F_{t,z,k}$ are the required amount of nutrient nu per unit of livestock and the amount of feed item z used by livestock category k, respectively.

To avoid unrealistic changes in feed mixes, we impose historical feed mixes used by all livestock categories. Constraints (16) and (17) constrain the consumption of feed to be within a convex combination of historical feed uses.

$$(16) \qquad \mathrm{FED}_{t,z} = \sum_{ht} hf_{z,ht} WF_{t,ht}$$

$$(17) \qquad \sum_{ht} WF_{t,ht} \leq 1$$

Soybean meal and DDGS are substitutes in the provision of protein up to a certain share level. Because the share of DDGS in total feed consumption of each livestock category is restricted (Babcock et al. 2008), we impose appropriate upper bounds for DDGS to reflect this aspect of feeding practices. Livestock commodities can be consumed domestically or exported. The total supply of each livestock commodity is then related to the respective livestock production activity through a fixed yield coefficient, denoted by $ly_{k,s}$. Constraint (18) establishes this relationship:

$$(18) \qquad \mathrm{DEM}_{t,k} + \mathrm{EXP}_{t,k} \leq \sum_{s} ly_{k,s} \mathrm{LIV}_{t,s} \quad \text{for all } t, k$$

7.5   Data

The simulation model uses CRD-specific data on costs of producing crops, livestock, biofuel feedstocks, yields of conventional and bioenergy crops, and land availability. We estimate the rotation, tillage, and irrigation specific costs of production in 2007 prices for fifteen row crops (corn, soybeans, wheat, rice, sorghum, oats barley, cotton, peanuts, potatoes, sugarbeets, sugarcane, tobacco, rye, and corn silage) and three perennial grasses (alfalfa, switchgrass, and miscanthus) at county level. These are aggregated to the CRD level for computational ease. Production of dedicated energy crops is limited to the rainfed regions, which include the Plains, Midwest, South, and Atlantic, while conventional crops can be grown in the Western region as well. The primary livestock commodities considered are eggs and milk. The secondary (or processed) crop and livestock commodities consist of oils from corn; soybeans and peanuts; soybean meal; refined sugar; HFCS; wool: and meat products such as beef, pork, turkey, chicken, and lamb. Feedstocks used for biofuel production in the model include corn, corn stover, wheat straw, forest residues, miscanthus, and switchgrass.

7.5.1   Dedicated Bioenergy Crops

Miscanthus and switchgrass have been identified as among the best choices for high yield potential and adaptability to a wide range of growing conditions and environmental benefits in the United States and Europe (Gunderson, Davis, and Jager 2008; Lewandowski et al. 2003b; Heaton, Dohleman, and Long 2008). Both grasses have high efficiency of converting solar radiation to biomass and in using nutrients and water and have good pest and disease resistance (Clifton-Brown, Chiang, and Hodkinson 2008; Semere and Slater 2007).

Switchgrass is a warm-season perennial grass native to North America, while Miscanthus is a perennial rhizomatous grass nonnative to the United States. A key concern with a large-scale introduction of a nonnative grass, such as miscanthus, is its potential to be an invasive species. The miscanthus variety being evaluated in this study as a feedstock for biofuels is the sterile hybrid genotype *Miscanthus* × *giganteus* that has been studied extensively through field trials in several European countries. Switchgrass stands can have a life span of fifteen to twenty years in a native state, but in cultivated conditions, the U.S. Department of Energy estimates stand-life at ten years.[4] In the United States, miscanthus stands that are more than twenty years old have been observed in experimental fields in Illinois (Heaton, Dohleman, and Long 2008). This study assumes a life span of ten years for switchgrass and fifteen years for miscanthus.

In the absence of long-term observed yields for miscanthus and limited data for switchgrass, we use a crop productivity model MISCANMOD to simulate their yields. The MISCANMOD estimates yields of miscanthus and Cave-in-Rock variety of switchgrass using GIS (geographic information system) data, at a 1° by 1° scale, on climate, soil moisture, solar radiation, and growing degree days as model inputs, as described in Jain et al. (2010). The Cave-in-Rock switchgrass cultivar studied here is an upland variety that originated in Southern Illinois and is cold-tolerant and well-suited for the upper Midwest (Lemus and Parrish 2009; Lewandowski et al. 2003a). Lowland varieties of switchgrass, like Alamo, are most suited for the southern United States (Lemus and Parrish 2009). Recent analysis of data from field trials across the United States shows that frequency distributions of yield for the upland and lowland varieties were unimodal, with mean (±SD) biomass yields of 8.7 ± 4.2 and 12.9 ± 5.9 metric tons dry matter per hectare (MT DM/ha) for the two varieties, respectively (Wullschlegera et al. 2010). This is consistent with estimates provided by a review of literature that shows that annual yield of lowland variety of switchgrass ranges between 11 to 16 MT DM/ha (Lemus and Parrish 2009) and is about 50 percent higher than that of the upland variety. We, therefore, increase switchgrass yields from MISCANMOD by 50 percent for all regions other than the Midwest (excluding Missouri) to account for higher yields of the lowland varieties.

The simulated yields show that the postharvest (delivered) biomass yield of miscanthus is about two times the yield of switchgrass at each location. For each crop, these yields vary from north to south and from west to east

---

4. See http://southwestfarmpress.com/energy/121107-switchgrass-challenges/ and http://www.osti.gov/bridge/servlets/purl/771591-9J657S/webviewable/771591.pdf.

in the United States. Atlantic states have high yields for miscanthus and switchgrass, while western states have very low yields due to insufficient soil moisture. Furthermore, southern states have higher yields for miscanthus and switchgrass as compared to northern states. The average delivered yield of miscanthus is the highest in the Atlantic states at 31.6 MT DM/ha, followed by the South at 30.2 MT DM/ha, the Midwest at 23.8 MT DM/ha, and the Plains at 19.8 MT DM/ha. Corresponding estimates for average switchgrass yield are 16.4, 15.2, 10.7, and 11 MT DM/ha, respectively.[5]

The costs of producing miscanthus and switchgrass differ over their lifetime due to lags between time of planting and harvestable yields. Costs of production of miscanthus and switchgrass are developed for each year of their lifetime for each CRD and include the costs of inputs including fertilizer, seed, and chemicals; machinery required for establishment and harvest of bioenergy crops; and storage and transportation. Cost of land for these crops is implicitly included given a land constraint in the model. The cost of labor, building repair and depreciation, and overhead (such as farm insurance and utilities) are excluded from the costs of production because they are likely to be the same for all crops and would not affect the relative profitability of crops. Costs of bioenergy crops in the first year differ from those in subsequent years because it involves costs of seeding and land preparation to establish the crops. Existing studies vary in their assumptions about input requirements, preharvesting, harvesting, and storage costs of bioenergy crops. This study constructs low-cost and high-cost scenarios for the production of the bioenergy crops, and the simulation model will test the sensitivity of the results to these assumptions. The low-cost scenario considers a low fertilizer application rate, low replanting probability, high second-year yield, low harvest loss, and low harvesting costs, while the high cost scenario considers the opposite scenario of production. These are described in Jain et al. (2010). Analysis of the break-even annualized costs of producing these grasses shows that there is considerable spatial variation in the cost of cellulosic feedstocks in the United States and that the mix of bioenergy crops will differ across geographic locations. Switchgrass is likely to have relatively lower costs of production in some of the northern Midwestern states (Minnesota and Wisconsin) and southern states (Texas and Louisiana) that have relatively high switchgrass yields, while miscanthus has lower costs in the southern, Atlantic, and central Plains states.

---

5. Delivered yields incorporate losses during harvesting, storing, and transporting. Switchgrass yield is typically about one-half of that for miscanthus. Exceptions to this are some northern states and some southern states, where switchgrass yields are relatively higher than those for miscanthus because minimum temperature are too low in the north and not low enough in the south for miscanthus growth. Perlaack et al. (2005) assume switchgrass yields of 18 MT/ha^{-1} in a high yield scenario and 12 MT/ha^{-1} otherwise.

7.5.2   Conventional Crops and Crop Residues

For row crops, we use the historical five-year average (2003 to 2007) yield per hectare for each CRD as the representative yield for that CRD (USDA/NASS 2009b) under dryland and irrigated land. The yields of corn, soybeans, and wheat are assumed to grow over time at the trend rate estimated using historical data. These yields are also assumed to be price-elastic with the price elasticities estimated econometrically. The trend rates and elasticities used in the model and more details of the econometric estimation methods can be found in Huang and Khanna (2010). Some crops are grown in rotation with each other to increase soil productivity and reduce the need for fertilizers. We adjust crop yields per hectare based on crop rotations for each CRD. We obtain fifteen crop rotation possibilities for each region of the United States from USDA/ERS (1997), including corn-soybean rotation, continuous corn rotation, fallow-wheat rotation, and continuous rotations for other crops. In Midwestern states where a corn-soybean rotation is the dominant rotation practice, we assume observed corn yields to be those under a corn-soybean rotation. Corn yields per hectare under a continuous corn rotation are assumed to be 12 percent lower than under a corn-soybean rotation. The fallow-wheat rotation is primarily used to conserve soil moisture over a two-year period for one-year production, which leads to a reduction in wheat yields by 50 percent in this rotation. The fallow-wheat rotation is widely used in the Northern wheat-growing region (such as Washington, Oregon, Idaho, Montana, and Colorado) and in parts of the Northern Plains states (such as North Dakota, South Dakota, Nebraska, and Kansas). Some counties in Minnesota and Texas also use the fallow-wheat rotation.[6]

Corn stover and wheat straw yields for each CRD are obtained based on a 1:1 grain-to-residue ratio of dry matter of crop grain to dry matter of crop residues and 15 percent moisture content in the grain reported in Sheehan et al. (2003); Wilcke and Wyatt (2002); and Graham, Nelson, and Sheehan (2007). Similar to Malcolm (2008), we assume that 50 percent of the residue can be removed from fields if no-till or conservation tillage is practiced, and 30 percent can be removed if till or conventional tillage is used. Corn stover yield ranges from 0.16-5.07 MT DM/ha under no-till, while wheat straw yield ranges from 0.34 to 4.38 MT DM/ha in the United States. In contrast to miscanthus, the average delivered yields for corn stover are the highest in Midwestern and Plains states at 4.0 MT/ha, followed by the southern and western states at 3.3 and 3.2 MT/ha respectively. Atlantic states have the lowest corn stover yield at 2.8 MT/ha. Wheat straw delivered yield is highest in

---

6. Information on crop rotation for each state is obtained from ERS/USDA report *Production Practices for Major Crops in U.S. Agriculture, 1990–1997*.

the West at 3.1 MT/ha followed by the Midwestern states at 2.3 MT/ha and less than 2 MT/ha in other regions.

Costs of producing row crops and alfalfa are obtained from the crop budgets complied for each state by state extension services and used to construct the costs of production for each CRD. Crop budgets vary by rotation, tillage, and irrigation choices. The costs of crop production include costs of inputs such as fertilizer, chemicals and seeds, costs of drying and storage, interest payments on variable inputs, costs on machinery and fuels, and costs of crop insurance. The costs of labor, building repair and depreciation, and overhead (such as farm insurance and utilities) are excluded from these costs of production because they are likely to be the same for all crops and would not affect the relative profitability of crops. We determine the cost of production of corn silage by estimating the foregone revenue per hectare by growing corn silage instead of corn, the additional cost of fertilizer replacement that is needed for corn silage, and harvesting costs as reported in FBFM (Illinois Farm Business Farm Management Association).[7]

Application rates for nitrogen, phosphorous and potassium, and seeds for row crops and alfalfa vary with crop yields and differ across CRDs. Other costs of producing crops are assumed to be fixed irrespective of crop yields per hectare but differ across states. In addition, costs of fertilizer, chemicals, and machinery under conventional tillage differ from those under conservation tillage.

The costs of collecting corn stover and wheat straw include the additional cost of fertilizer that needs to be applied to replace the loss of nutrients and soil organic matter due to removal of the crop residues from the soil. The fertilizer application rates per dry metric ton of stover and straw removed are assumed to be constant across regions and are obtained from Sheehan et al. (2003) and Wortmann et al. (2008), respectively. In addition, the collection of crop residues involves the costs of harvesting stover and staw (i.e., mowing, raking, baling, staging, and storage) that are estimated based on the state-specific crop budgets on hay alfalfa harvesting. We find that the costs of production of crop residues are higher than those of bioenergy crops grown on marginal lands, except for corn stover in Plains states, such as North Dakota, South Dakota, and Nebraska, where corn yields are high due to irrigation. High wheat yields in western mountain states (such as in Oregon, Idaho, and Washington) can make wheat straw in those states competitive with other biomass produced in the rain-fed eastern United States.

7.5.3   Land Availability

For each of the five types of land (cropland, idle cropland, cropland pasture, pasture land, and forestland pasture) we obtain CRD-specific data on

---

7. See www.farmdoc.uiuc.edu.

land availability. The CRD-specific planted acres for fifteen row crops are used to obtain the cropland available in 2007 (estimated at 123 M ha for the 295 CRDs considered here) and to obtain the historical and synthetic mixes of row crops. Cropland availability in each CRD is assumed to change in response to crop prices. The responsiveness of total cropland to crop prices as well as the own and cross-price acreage elasticities for individual crops are obtained from Huang and Khanna (2010).

Data on idle cropland, cropland pasture, pasture, and forestland pasture for each CRD are obtained from USDA/NASS (2009a). In 2007, the availability of pastureland and forestland pasture is estimated to be 155 M ha and 10.5 M ha, respectively while that of idle cropland is 15 M ha and of cropland pasture is 13 M ha. Most of the idle cropland in 2007 was enrolled in the CRP. This size of the CRP decreased to 13 M ha from 2008 onward. The analysis here assumes that land enrolled in CRP is preserved at 2008 levels and not used for conventional crop or bioenergy crop production.

7.5.4 Crop and Livestock Sector

In the livestock sector, we consider demands for several types of meat (chicken, turkey, lamb, beef, and pork), wool, dairy, and eggs. The demand functions are calibrated using the observed quantities consumed and prices and demand elasticities. The latter are obtained from Adams et al. (2005). The supply of livestock (chicken, turkey, lamb, and pork) is constrained by their historical numbers at the national level. The supply of beef is restricted by the number of cattle, which, in turn, depends on the amount of grazing land available at regional level. The historical livestock data at the national level and production of meat, dairy, and eggs for 2003 to 2007 are used to obtain the average livestock productivity. The data on grazing land requirements for cattle, nutrition requirements (in terms of protein and grain) for each livestock category, and production and processing costs are obtained from Adams et al. (2005). We use the nutrient content of feed crops, soymeal, and DDGS to find the least cost feed rations for each type of livestock. The price of DDGS is determined by the lagged prices of corn and soymeal using the relationship estimated by Ellinger (2008). To prevent unrealistic feed mixes consumed by livestock, we constrain the consumption of different types of feed based on the historically observed levels obtained from USDA/NASS (2009b).

The crops sector consists of markets for primary and processed commodities. The demands for primary commodities, such as corn and soybeans are determined in part by the demands for processed commodities obtained from them and by other uses (such as seed). The conversion rates from primary crop commodities to processed commodities are obtained from USDA/NASS (2009b). Conversion costs are obtained from Adams et al. (2005) and inflated to 2007 prices using the respective gross domestic product (GDP) deflator. We use two-year (2006 to 2007) average prices, consumption, exports, and imports of crop and livestock commodities to calibrate the domestic demand, export demand, and import supply functions for all commodities.[8] The data on prices, consumption, exports, and imports are obtained from ERS/USDA. Elasticities are assembled from a number of sources including FASOM, the USDA, and existing literature as shown in table 7.1. Domestic demands, export demands, and import supplies are shifted upward over time at exogenously specified rates, listed in table 7.1. We obtain projected amounts of crop and livestock commodities for domestic consumption, exports, and imports for 2010 and 2020 from FAPRI and interpolate then for the intervening years assuming a uniform annual growth rate.[9]

7.5.6   Fuel Sector

We assume a linear demand function for VKT with a price elasticity of –0.2 that shifts out by 1 percent each year.[10] The elasticity of substitution between gasoline and ethanol is 3.95 (Hertel, Tyner, and Birur 2010). For the supply of gasoline, we consider two gasoline supply curves to distinguish domestic gasoline supply and gasoline supply from the rest of the world. The short-run supply of domestic gasoline is assumed to be linear with a slope of 0.9 (Greene and Tishchishyna 2000), implying a short-run supply elasticity of 0.049 when the oil price is $34/BBL (oil barrel) while the short-run gasoline supply to the United States from the rest of the world is assumed to have a constant elasticity form with a price elasticity of 2 (National Research Council 2002).

To calibrate the demand function of vehicle kilometers, production function of vehicle kilometers, and supply functions of gasoline, data on consumption of kilometers and fuel consumption and fuel prices in 2007 are assembled from several sources. The Federal Highway Administration (FHWA) reports that total vehicle-kilometers traveled in 2007 were 5,107 billion kilometers. The Energy Information Administration (EIA) reports that the consumption of gasoline and ethanol are 519.4 billion liters and 23.4 billion liters, respectively, in the United States in 2007. The EIA reports that average retail price of gasoline that year was $0.72 per liter. We calculate the retail price of ethanol as the wholesale rack price plus $0.10 per liter fuel taxes and a $0.05 per liter markup minus $0.13 per liter subsidy, yielding $0.61 per liter in 2007.[11] In the benchmark case, we assume the price elasticity of VKT demand is –0.2 and elasticity of substitution between gasoline and ethanol is 3.95 (Hertel, Tyner, and Birur 2010).

We assume linear supply functions for ethanol imports from Brazil and

---

8. An exception is the price of milk, which is kept fixed at its observed 2006 to 2007 level.
9. See http://www.fapri.iastate.edu/outlook/2010/text/Outlook_2010.pdf.
10. We obtain historical data on vehicle kilometers travelled (VKT) from Federal Highway Administration website (http://www.fhwa.dot.gov/policyinformation/statistics/2008/vm202.cfm) and use average growth rate of VMT from 2000 to 2008.
11. See www.neo.ne.gov/statshtml/66.html.

Table 7.1   Domestic demand, export demand, and import supply elasticities

Commodity	Use	Shift[a] (%)	Elasticity	Source
Barley	Domestic	0.0	−0.3	USDA/ERS (2009)
	Export	2.0	−0.2	Adams et al. (2005)
Corn	Domestic	0.8	−0.23	Adams et al. (2005)
	Export	2.0	−0.26	Fortenbery and Park (2008)
Cotton	Domestic	−2.0	−0.18	Adams et al. (2005)
	Export	0.3	−0.65	Bredahl, Meyers, and Collins (1979)
Oats	Domestic	−0.4	−0.21	Adams et al. (2005)
Sorghum	Domestic	−1.5	−0.2	Adams et al. (2005)
	Export	2.0	−2.36	Bredahl, Meyers, and Collins (1979)
Wheat	Domestic	1.0	−0.3	USDA/ERS (2009)
	Export	−2.0	−1.67	Bredahl, Meyers, and Collins (1979)
Soybean	Domestic	1.4	−0.29	Piggott and Wohlgenant (2002)
	Export	0.4	−0.63	Piggott and Wohlgenant (2002)
Soybean meal	Export	2.0	−1.41	Adams et al. (2005)
Vegetable oil[b]	Domestic	0.2	−0.18	Piggott and Wohlgenant (2002)
	Export	2.0	−2.24	Piggott and Wohlgenant (2002)
Rice	Domestic	2.0	−0.11	Gao, Wailes, and Cramer (1995)
	Export	−0.4	−1.63	Gao, Wailes, and Cramer (1995)
Peanut	Domestic	0.8	−0.25	Carley and Fletcher (1989)
Beef	Domestic	0.3	−0.75	FAPRI (2009)
	Export	2.0	−0.8	Adams et al. (2005)
Chicken	Domestic	1.4	−0.46	Adams et al. (2005)
	Export	1.4	−0.8	Adams et al. (2005)
Eggs	Domestic	0.8	−0.11	Adams et al. (2005)
	Export	NA	NA	
Pork	Domestic	1.0	−0.83	Adams et al. (2005)
	Export	2.0	−0.8	Adams et al. (2005)
Turkey	Domestic	0.8	−0.53	Adams et al. (2005)
	Export	1.4	−0.8	Adams et al. (2005)
Lamb	Domestic	0.0	−0.4	Adams et al. (2005)
	Import	NA	NA	
Wool	Domestic	0.0	0.4	Adams et al. (2005)
	Export	0.0	−0.8	Adams et al. (2005)
Refined sugar	Domestic	0.0	−0.368	Adams et al. (2005)
	Import	0.0	0.99	Adams et al. (2005)
High-fructose corn syrup	Domestic	0.5	−0.91	Adams et al. (2005)
	Export	2.0	−0.2	Adams et al. (2005)

Notes: Table shows the commodities that can be used for domestic consumption or traded with the rest of the world. Domestic demand for commodities excludes uses for feed and ethanol production, and prices are fixed at 2007 prices if the elasticities are zeros. NA = not applicable.
[a]Demand shifts are computed based on FAPRI 2010 *U.S. and World Agricultural Outlook.*
[b]Vegetable oil includes corn oil, soybean oil, and peanut oil.

CBI countries and use two-year (2006 to 2007) average prices and imports of ethanol imports to calibrate the ethanol import supply functions. The excess supply elasticity of imported ethanol from Brail and CBI counties is assumed to be 2.7 (as in de Gorter and Just 2008). We calculate the sugarcane ethanol price in Brazil and CBI countries as U.S. retail price minus $0.02 per liter transportation cost, fuel tax, and tariff and plus subsidy, yielding $0.49 and $0.62 per liter, respectively.[12]

Ethanol yield from corn grain is 417.3 liters of denatured ethanol per metric ton of corn, while cellulosic biofuel yield from an nth-generation stand alone plant is estimated as 330.5 liters per metric ton of dry matter of biomass (Wallace et al. 2005). The cost of conversion of corn grain to ethanol is estimated as $0.20 per liter in 2007 prices based on Environmental Protection Agency (EPA) estimates (EPA 2010), while the nonfeedstock costs of producing cellulosic ethanol are estimated as $0.37 per liter in 2007 prices (EPA 2010). We assume that the current unit cost of conversion of feedstock to biofuel, C_{cum}, is a declining function of cumulative production, that is, $C_{cum} = C_0 \text{Cum}^b$, where C_0 is the cost of the first unit of production, Cum is the cumulative production, b is the experience index. We assume b for corn ethanol is equal to -0.20 (Hettinga et al. 2009) and calibrate C_0 using data on the processing cost and cumulative corn ethanol production in 2007. To calibrate the function for cellulosic ethanol, we assume C_{cum} in 2022 is $0.18 per liter (EPA 2010) and use the production quantities specified in the RFS to obtain a value for b of -0.05.[13] The feedstock and refinery costs of sugarcane ethanol in Brazil and CBI countries are also assumed to be declining functions of cumulative production. We assume b for sugarcane ethanol is -0.32 (Van Den Wall Bake et al. 2009). Parameter C_0 is calibrated using data on the feedstock and refinery costs of sugarcane ethanol and cumulative sugarcane ethanol production in 2007. The growth rate of sugarcane ethanol production is assumed to be constant and equal to 8 percent (Van Den Wall Bake et al. 2009) and is used to compute the feedstock and refinery costs of sugarcane ethanol for 2007 to 2022.

7.6   Results

7.6.1   Effect of Biofuel Policies on the Agricultural and Fuel Sectors

We first validated the simulation model assuming existing fuel taxes and corn ethanol tax credits and compared the model results on land allocation, crop production, biofuel production, and commodity prices with the corresponding observed values in the base year (2007). The corn ethanol mandate was exceeded in the aggregate in 2007; it is, therefore, imposed as a lower limit to corn ethanol production. As shown in table 7.2, the differences

---

12. Transportation cost of ethanol is estimated to be $0.02 per liter in Crago et al. (2010). The difference in ethanol prices in Brazil and CBA countries can be attributed to additional processing cost in CBA countries because ethanol needs to be dehydrated before admitted to the United States.

13. These functions imply that the per liter conversion cost for corn ethanol declines by about 27 percent, while that for cellulosic ethanol declines by 50 percent by 2022.

Table 7.2   Model validation for 2007

	Observed	Model	Difference (%)
Land use (million hectares)			
Total land	123.05	121.76	−1.04
Corn	34.31	31.12	−9.30
Soybeans	28.15	28.41	0.94
Wheat	21.52	22.46	4.38
Sorghum	2.69	2.93	9.05
Commodity price ($/metric ton)			
Corn	142.51	133.22	−6.52
Soybeans	303.69	319.40	5.17
Wheat	197.31	220.33	11.67
Fuel sector			
Gas price ($/liter)	0.72	0.72	0.00
Ethanol price ($/liter)	0.61	0.61	−0.49
Gas consumption (billion liters)	519.94	519.34	−0.11
Ethanol consumption (billion liters)	23.51	24.22	3.02
Kilometers consumption (billion kilometers)	4863.29	4863.29	0.00

between model results and the observed land use allocations are less than 10 percent. Food prices are generally within 10 percent of the observed values except for the wheat price, which is 12 percent higher than the actual prices in 2007. The fuel prices and fuel consumption are also simulated well, within 5 percent deviation from the observed values. We consider these results as a fairly good sign of the model's validation capability.

We then examine the effects of two policy scenarios on the agricultural and fuel sectors: biofuel mandates under the RFS alone and biofuel mandates with volumetric tax credits. The RFS mandates are set as nested volumetric requirements for the production of biofuels at mandated levels for the period of 2007 to 2022. These mandates serve as the minimum quantity restrictions on biofuel production that can shift up if economically competitive with conventional fuels through policy support and technological improvements. We then compare model results under biofuel policies to those under a business-as-usual (BAU) scenario. The BAU scenario is defined as one without any biofuel policy, except for the tariff on biofuel imports, which is kept unchanged in all scenarios here. In all scenarios considered here, we also include a fuel tax on gasoline and biofuels, which is set at $0.10 per liter, and assume that the demands for crops and VKT increase over time. Results for cropland allocation are presented in table 7.3, while table 7.4 shows the results for production and prices of key crop and livestock commodities. The regional distribution of land for bioenergy feedstocks are presented in table 7.5. Tables 7.6 and 7.7 present the impact of biofuel policies on the fuel sector and on social welfare. Table 7.8 contains the results of the sensitivity analysis.

Table 7.3   Effect of biofuel policies on land use in 2022 (M ha)

	Baseline 2007	Baseline	Mandate	Mandate with tax credits
Total land	121.51	121.13	127.99	129.06
Corn	29.74	28.91	33.55	25.14
Soybeans	29.85	29.74	27.50	30.09
Wheat	23.02	24.24	22.25	23.35
Stover			3.45	10.10
Straw			1.01	1.99
Miscanthus[a]			4.43	8.70
Switchgrass[b]			3.03	4.16

[a]Of this, 0.32 million ha and 1.88 million ha are on regular cropland under the mandate and mandate and tax credits, respectively.

[b]Of this, 0.12 million ha and 0.43 million ha are on regular cropland under the mandate and mandate and tax credits, respectively.

Table 7.4   Effect of biofuel policies on commodity prices and production

	Baseline (2007)		Business as usual (2022)		Mandate (2022)		Mandate with tax credits (2022)	
	Price ($/MT)	Production (M MT)	Price ($/MT)	Production (M MT)	Price ($/MT)	Production (M MT)	Price ($/MT)	Production (M MT)
Corn	127.0	276.7	117.6	321.5	145.9	380.0	111.0	282.2
Soybean	283.4	81.4	287.0	89.5	343.6	82.9	288.0	92.6
Wheat	213.8	54.7	212.9	68.5	228.6	63.3	219.5	67.9
Beef	1298.1	16.6	1136.3	18.3	1230.2	17.8	1151.2	18.2

Note: M MT = million metric tons.

Business-As-Usual (BAU) Scenario

In the absence of any government intervention in the biofuel market, we find that total crop acreage decreases by 0.3 percent from 121.5 in 2007 to 121.1 M ha in 2022 with corresponding increases in idle or pasture land. Corn and soybean acreages would decrease by 0.8 M ha (2.8 percent) and 0.1 M ha (0.4 percent), while wheat acreage would increase by 1.2 M ha (5.3 percent) over the 2007 to 2022 period. Land under cotton in 2022 decreases by 0.3 M ha (7.8 percent) compared to 2007. Despite the reduction in corn and soybean acreages, their production would increase by 16 percent and 10 percent over the 2007 to 2022 period due to 19 percent and 10 percent increases in corn and soybean yields. The production of wheat also increases by 25 percent, which can be attributed to the increases in wheat acreage and yields from 2.4 metric tons per hectare to 2.8 metric tons per hectare over 2007 to 2022. In the livestock sector, beef production would increase by 10 percent between 2007 to 2022. Despite the increasing demand for corn for biofuel production, corn price decreases by 7 percent in 2022 due to the increase in corn yields. Because corn is a major source of

Table 7.5   **Regional distribution of cellulosic feedstocks in 2022 (M ha)**

	Stover	Straw	Switchgrass	Miscanthus
Mandate				
Midwest			0.47	1.25
South			0.44	0.79
Plains	3.44	0.22	1.91	1.36
Atlantic			0.21	1.03
West		0.79		
Mandate with subsidies				
Midwest	6.67		0.54	3.08
South		0.19	0.67	1.09
Plains	3.22	0.75	2.43	2.91
Atlantic			0.53	1.63
West	0.20	1.04		

Table 7.6   **Effect of biofuel policies on fuel sector**

	Baseline 2007	Baseline 2022	Mandate	Mandate with tax credits
Price in 2022 ($/km or $/liter)				
Vehicle kilometers	0.080	0.087	0.085	0.080
Corn ethanol	0.69	0.66	0.70	0.54
Cellulosic ethanol			0.70	0.46
Gasoline	0.73	0.78	0.72	0.73
Consumption in 2022 (billion liters or billion kilometers)				
Vehicle kilometers	4,863.29	5,513.13	5,531.19	5,595.92
Domestic gasoline	172.44	179.30	171.68	172.49
Gasoline from ROW	354.85	409.24	349.11	355.26
Total ethanol	15.24	27.70	136.27	136.27
Corn	13.79	24.82	53.35	0.00
Stover			5.74	17.72
Straw			1.02	1.81
Miscanthus			47.73	84.79
Switchgrass			13.01	17.25
Ethanol imports	1.45	2.88	3.23	2.24
Forest residues			12.19	12.46
Cumulative consumption (over 2007–2022; billion liters or billion kilometers)				
Vehicle kilometers		82,885.78	83,235.64	83,817.33
Domestic gasoline		2,815.63	2,747.38	2,748.44
Gasoline from ROW		6,107.17	5,586.40	5,589.91
Total ethanol		330.78	1,220.98	1,316.36
Corn		295.82	613.22	131.66
Stover			24.75	70.71
Straw			2.14	9.36
Miscanthus			299.76	674.18
Switchgrass			107.87	246.22
Ethanol imports		34.96	38.22	25.60
Forest residues			135.03	158.64

Note: ROW = rest of world.

feed for beef production, it leads to a reduction in beef price in 2022 by 12 percent compared to 2007. Soybean and wheat prices change only marginally between 2007 and 2022. There is a significant increase in exports of corn, soybean, and wheat by 31 percent, 6 percent, and 38 percent over the 2007 to 2022 period. Exports of beef would increase by 30 percent due to lower beef prices.

In the fuel sector, we find an 8 percent increase in the price of VKT and a 7 percent increase in gasoline price in 2022 compared to 2007. Ethanol consumption would be about 28 billion liters in 2022 or 4 percent of fuel consumed with no government intervention. Of the cumulative consumption of corn ethanol over the 2007 to 2022 period, a little over 10 percent is imported from Brazil.

Biofuels Mandate

With corn ethanol production at its maximum allowable level, or 56 billion liters from 2015 and beyond, it could constitute a maximum of two-thirds of the cumulative biofuel production between 2007 and 2022; the remaining mandate is met by advanced biofuels. With the nested volumetric provisions of the RFS, however, advanced biofuels can meet more of the mandate than the minimum level if they can compete with corn ethanol. Given the assumptions about the rate of decline in costs of producing advanced biofuels from cellulosic feedstocks in the United States (described in the preceding), we find that the RFS would lead to the production of about 613 billion liters of corn ethanol (instead of the maximum of 800 billion liters that can meet the mandate) and about 608 billion liters of advanced biofuels, including 38 billion liters of sugarcane ethanol imports over the 2007 to 2022 period. This would increase cumulative production of corn ethanol by 107 percent relative to the BAU over this period. The cumulative advanced biofuels (608 billion liters) are largely produced using miscanthus (49 percent) and forest residues (22 percent), with the rest produced using switchgrass, corn stover, and wheat straw.

The RFS leads to a 6 percent increase in total cropland (6.86 M ha); most of this is to enable an increase in corn production to produce the additional corn ethanol. There is a 16 percent increase (about 4.7 M ha) in land under corn in 2022 compared to the BAU. With a high yielding grass like miscanthus, only 4.4 M ha are required for miscanthus production and 3 M ha to switchgrass production to produce cellulosic biofuels. Of this 7.44 M ha under bioenergy crops, only 0.44 M ha is converted from cropland, and about 7 M ha is from currently idle cropland or cropland pasture. Thus, a total 12.14 M ha is required for biofuel production; of this, about 5 M ha of land is released by reducing acreage under other crops (including soybeans, wheat, rice, cotton, and pasture), representing 4 percent of the 121.5 M ha of cropland in 2007, and the rest is obtained by a change in land use at the extensive margin. Corn stover and wheat straw would be harvested from

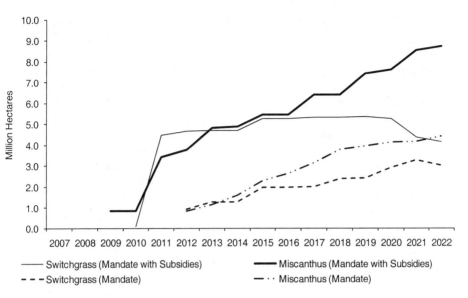

Fig. 7.1   Land under energy crops

10 percent and 5 percent of the land under corn and wheat, respectively, in 2022.

There is considerable variation in the mix of feedstocks produced across regions. Stover is harvested only in the Plains states, while wheat straw is harvested mainly in the Western states. More than half of the switchgrass acreage is in the Plains states, followed by the Midwest and the South. Miscanthus acreage is largely in the Plains and the Midwest, followed by the Atlantic and Southern states. This acreage also changes over time; it expands as the mandate requires more cellulosic biofuel production. Figure 7.1 shows the change in land under bioenergy crops over the 2007 to 2022 period under the mandate. Acreage under miscanthus expands from less than 1 M ha in 2012 to over 4 M ha in 2022. Initially, miscanthus and switchgrass acreage are similar as each is produced in areas where it has a comparative advantage; in latter years, miscanthus acreage expands much more rapidly, while switchgrass acreage levels off because of the relatively lower costs of producing a high-yielding crop like miscanthus.

The RFS would significantly affect production, exports, and prices of crop and livestock commodities. The increase in demand for corn results in an increase in corn production in 2022 by 18 percent relative to the BAU. However, corn price in 2022 is still 24 percent higher than under the BAU because 38 percent of corn production in 2022 is used for biofuel production. Soybean and wheat prices in 2022 are also 20 percent and 7 percent higher than the BAU due to 8 percent reduction in their production levels. The production of rice and cotton in 2022 would decrease by 8 percent and 2 percent, respectively, relative to the BAU due to the acreage shifts to the production of corn. This increases rice and cotton prices in 2022 by 5 percent and 2 percent relative to the BAU. Livestock prices also rise with beef price, increasing by 8 percent compared to the BAU due to the increases in feed prices and a 3 percent reduction in beef production. In response to higher prices of crop commodities, export of corn, soybean, and wheat would decrease by 4 percent, 11 percent, and 12 percent relative to the BAU, while the exports of rice would decrease by 42 percent. Higher livestock prices also lead to a reduction in beef exports by 2 percent relative to the BAU.

As a result of the mandate, the volumetric share of ethanol in total fuel consumption increases to 21 percent in 2022. The RFS results in a reduction in cumulative gasoline consumption over the 2007 to 2022 period by 7 percent and a reduction in gasoline price in 2022 by 8 percent compared to the BAU. While domestic gasoline production falls by 2.5 percent, gasoline imports from the rest of the world decrease by 8.5 percent relative to BAU. The overall cost of VKT falls from $0.087/km to $0.085/km; as a result, the VKT increases by 0.4 percent relative to the BAU scenario in 2022. This market-based feedback effect on gasoline prices tempers the extent to which biofuels replace gasoline. At a maximum, with perfect substitutability between gasoline and biofuels and a fixed price of gasoline, the additional 109 billion liters of biofuels produced in 2022 (over and above the 28 billion liters in the BAU) could have displaced an energy equivalent volume of 72 billion liters of gasoline. With imperfect substitutability and the reduction in gasoline price, the amount of gasoline reduced is 68 billion liters, implying a rebound effect on gasoline consumption of about 6 percent.

Biofuel Mandate and Volumetric Tax Credits

The provision of tax credits for biofuels leads to three significant impacts on total biofuel production and the mix of feedstocks used for biofuels. First, it increases total biofuel production over 2007 to 2022 from the minimum mandated level of 1,221 billion liters to 1,316 billion liters. Second, it makes cellulosic ethanol competitive with corn ethanol and sugarcane ethanol and reduces cumulative corn ethanol production from 613 billion liters under a mandate alone to 132 billion liters. Cumulative cellulosic ethanol production increases to twice the level under a mandate alone, from 570 billion liters to 1,159 billion liters over the 2007 to 2022 period. Third, it increases the share of miscanthus and switchgrass in cumulative advanced biofuels (cellulosic biofuels plus sugarcane ethanol) from 49 percent and 18 percent under a mandate alone to 57 percent and 21 percent. The corresponding shares of ethanol imports and biofuel produced from forest residues fall from 6 percent and 22 percent under a mandate alone to 2 percent and 13 percent. The reduction in production of corn ethanol (relative to the RFS) reduces the acreage under corn by 8.4 M ha. Of this, about 6 M ha is diverted to other conventional crops, while the rest is diverted to miscanthus and switchgrass.

In addition to this, 10.6 M ha of idle or cropland pasture is converted to produce these energy crops. The increase in biofuels produced from miscanthus leads to an increase in the land under miscanthus from 4.4 M ha under a mandate alone to 8.7 M ha under a mandate and volumetric tax credits and a corresponding increase in land under switchgrass from 3 M ha to 4.2 M ha. Switchgrass acreage expands in all rainfed regions as does miscanthus acreage. In particular, these tax credits enable miscanthus acreage to more than double in the Midwest and to expand by more than 50 percent in the Atlantic states. The biofuel tax credits also increase the acreage from which corn stover and wheat straw are harvested in 2022, to 40 percent of corn acres and 9 percent of wheat acres, respectively. With the tax credits, it is profitable to harvest corn stover in the Midwest and to harvest wheat straw in the Plains and Southern states. Switchgrass acreage expands in all rainfed regions as does miscanthus acreage. The expansion in acreage of energy crops over time is much more rapid for miscanthus than for switchgrass (figure 7.1). The volumetric tax credits also make the production of switchgrass and miscanthus viable earlier than otherwise. Moreover, they change the relative profitability of growing miscanthus and switchgrass. After 2016, miscanthus acreage continues to expand, while switchgrass acreage levels off and even declines in later years. This is because volumetric subsidies increase the relative profitability of biofuels with higher yields per hectare of land. After 2016, miscanthus and switchgrass compete for marginal land in the same locations, and the tax credits increase the relative profitability of miscanthus in those locations.

The change in the composition of biofuels due to the subsidy changes the total land under crop production and under various row crops. Total cropland increases by 1.1 M ha relative to that under the RFS alone, due to an expansion in acreage under energy crops. Acreage under corn and corn production in 2022 declines by 13 percent relative to the BAU scenario; corn production in 2022 is, however, still higher than that in 2007 under the BAU due to productivity increase. In comparison to BAU, acreage under soybeans and soybean production in 2022 would increase by 2 percent and 3 percent, respectively. The reduction in total cropland availability results in a decrease of 1.5 M ha in acreage under wheat, rice, cotton, and pasture compared to the BAU. However, the acreage under these crops in 2022 under a mandate and subsidy are still higher than those under a mandate alone.

The increase in the production of cellulosic biofuels due to biofuel subsidies alleviates the adverse impact of the mandate on the prices of crop and livestock commodities. Corn and soybean prices in 2022 would be 24 percent and 16 percent lower than under a mandate alone, while beef price in 2022 would be 6 percent lower. In comparison to the BAU, corn price in 2022 is 6 percent lower due to productivity increase and decrease in demand for corn ethanol. Prices of soybeans, wheat, rice, and cotton are similar to those under the BAU, deviating from –1 percent for rice to 3 percent for wheat. Beef price is about 1 percent higher relative to the BAU. In response to lower prices of corn, soybeans, and rice, exports demand for these commodities would increase by 0.7 percent, 0.1 percent, and 2 percent relative to the BAU. Lower beef price also leads to an increase in beef exports by 11 percent relative to the BAU.

The volumetric tax credits result in consumer prices of $0.54 per liter for corn ethanol and $0.46 per liter for cellulosic ethanol that are significantly lower than those under a mandate alone, while the gasoline price is marginally higher due to increased demand for fuel relative to the mandate alone. Relative to the RFS alone, cumulative VKT over the 2007 to 2022 period increases by 581 billion kilometers (0.7 percent), while gasoline consumption increases by 4.6 billion liters (0.05 percent), and biofuel consumption increases by 95.38 billion liters (8 percent). The tax credits lower the overall cost of fuel and, thus, the cost per kilometer by 6 percent.

7.6.2   Social Welfare Effects of Biofuel Policies

We use the modeling framework presented here to estimate the changes in consumer and producer surplus in each of the markets in the fuel and agricultural sector considered here and the change in government revenues due to fuel taxes or subsidies. As compared to a free market outcome, biofuel policies impose an efficiency cost by expanding biofuel production beyond free market levels and affecting food and fuel prices. However, they also have a terms-of-trade effect that benefits domestic agricultural producers and domestic fuel consumers. Moreover, the welfare costs of higher prices of agricultural commodities are partly borne by foreign consumers of agricultural goods, while the loss in surplus for gasoline producers due to lower fuel prices is partly borne by foreign gasoline producers. The terms-of-trade effect can offset a part or all of the efficiency costs of biofuel policies, and the net impact of biofuel policies on social welfare is, therefore, ambiguous.

We present the change in social welfare with the RFS compared to the BAU and the change in social welfare with the RFS and volumetric tax credits relative to the RFS alone in table 7.7. As described in the preceding, the RFS leads to lower gasoline price but higher costs of corn ethanol and cellulosic biofuels; nevertheless, it lowers cost per kilometer. Therefore, it increases the consumer surplus of the vehicle kilometer consumers. The RFS also raises conventional crop prices, and by increasing demand for

Table 7.7   Welfare costs of biofuel policies

	Mandate relative to business as usual	Mandate with tax credits relative to mandate
Change in social welfare ($ billions)	122.80	–78.93
Additional cumulative biofuel (billion liters)	890.21	95.38
Benefit/Cost per liter of additional biofuel ($/liter)	0.14	–0.83

residues and energy crops, it raises returns from existing land as well as from marginal land that was otherwise not used for agricultural production. It, therefore, benefits agricultural producers. This is at the expense of agricultural consumers; only a portion of these are, however, domestic. Thus, some of the loss in surplus is borne by foreign consumers. The RFS hurts gasoline producers by lowering demand for gasoline and its price. However, with two-thirds of the cumulative gasoline consumption over the 2007 to 2022 period being imported, the bulk of the loss in producer surplus is borne by foreign oil producers. As a result, the RFS leads to an increase in net present value of social welfare (in 2007 dollars) of $122 billion relative to the BAU. It also increases cumulative biofuel production relative to the BAU by 890 billion liters, implying a per liter benefit of $0.14.

As compared to the RFS, the provision of volumetric tax credits lowers crop prices and the cost per kilometer; therefore, they benefit agricultural consumers and vehicle kilometer consumers. Moreover, they benefit producers of cellulosic feedstocks by further increasing demand for crop residues and energy crops. However, producers of conventional crops are adversely affected as are gasoline producers. There is a significant government expenditure of $221 billion in present discounted value over the 2007 to 2022 period. As a result, aggregate social welfare is $79 billion lower than under the RFS alone. Focusing only on tax payer cost of these tax credits would significantly overestimate the cost of additional biofuel production. By estimating welfare cost, we consider not only the costs to tax payers but also the net costs to the economy after considering the gains and losses to fuel and crop consumers and producers. The tax credits do lead to additional biofuel production over and above the RFS alone (by 95 billion liters, that is, by about 8 percent) over the 2007 to 2022 period, implying a welfare cost of $0.83 per liter of biofuel. In gasoline energy equivalent terms, this implies a cost of about $1.25 per liter.

Although our estimate of the welfare cost of biofuel is lower than the direct cost to tax payers, our per liter cost of additional biofuel is higher than that obtained by the CBO (2010) because the incremental volume of biofuels attributable to the tax credits is lower than their assumption. The low volume of incremental biofuel attributed to the tax credits in this study is due to our assumption that the volume of biofuel mandated by the RFS will be achieved even in the absence of a tax credit. In the event that this is not the case, or if there are other constraints to increasing biofuel production, then the incremental biofuel production due to these tax credits could be smaller or larger than that estimated here. Moreover, the welfare cost of these tax credits cannot be disaggregated into those for corn ethanol and those for cellulosic biofuels because each of these tax credits not only has a direct effect on the particular type of biofuel toward which it is targeted but also indirectly affects the production of the other type of biofuel by changing their relative costs. Thus, it is the combined effect of both the volumetric tax credits for corn ethanol and cellulosic biofuels that together determines the effect on food and fuel prices and on social welfare.

7.6.3   Sensitivity Analysis

We examine the sensitivity of our results to changes in some key assumptions about technology and cost parameters in the agricultural sector (see table 7.8), such as the rate of yield increase of row crops, the costs of producing bioenergy crops, and land availability for bioenergy crops. Jain et al. (2010) describe two scenarios for the costs of production of miscanthus and switchgrass, a low-cost and a high-cost scenario. The benchmark case considered the low cost of miscanthus and switchgrass production described there. We now examine the implications of the costs of production being less optimistic for miscanthus than assumed in the benchmark case but the same for other feedstocks as in the benchmark case.[14] We also analyze the impacts of raising production costs of both miscanthus and switchgrass on the mix of biofuels and land use patterns. In addition, we examine the implications of constraining the amount of land in a CRD that can be used for bioenergy crops to 10 percent instead of 25 percent assumed in the benchmark case. In each case, only one parameter is changed at a time, while all other parameters remain the same. We report the results for the biofuel mandate alone (M) and biofuel mandates plus volumetric tax credits (MS) scenarios. We present the percentage variations due to the parameter changes relative to the same policy scenarios with the benchmark parameters.

We find that compared to the benchmark case, a 50 percent reduction in rate at which crop productivity reduces the acreage under corn under the RFS by about 5 percent, increases corn price by 2 percent, and decreases the production of corn ethanol by 25 percent. It increases cellulosic biofuel production by 27 percent, and acreage under miscanthus and switchgrass increases by 31 percent and 7 percent. This raises the marginal cost of feedstocks for cellulosic biofuel production and makes it profitable to increase the area from which corn stover and wheat straw are harvested by 164 percent and 53 percent, respectively. The volumetric subsidies now shift land even more toward miscanthus and switchgrass (because they are relatively higher yielding feedstocks) and lowers acreage under corn stover and wheat straw acreages by 9 percent and 7 percent, respectively. Corn and soybean prices are 7 percent and 4 percent higher than in the benchmark case. The welfare cost of the tax credits is lower than in the benchmark case by 12 percent, primarily because the producers of conventional crops and of bioenergy crops are better off in this case, the former due to higher crop prices and the latter due to greater demand for cellulosic biofuels. Incremental biofuel production due to the tax credits is higher due to greater imports

14. This scenario considers higher fertilizer application rates, lower yields in the second year, and higher yield losses during harvest as well as higher harvesting costs per ton.

Table 7.8   Sensitivity analysis to technology parameters

	Rate of yield increase reduced by 50%		High cost of production of miscanthus		Upper limit of 10% on energy crop acres in a crop reporting district		High cost of production of miscanthus and switchgrass	
	M	MS	M	MS	M	MS	M	MS
Land use in 2022 (%)								
Total land	1.0	0.1	0.4	0.9	-1.2	-2.0	1.1	0.4
Corn	-4.8	-0.1	1.3	-1.6	1.2	1.6	2.6	8.0
Soybeans	4.1	-0.2	0.4	-3.0	0.0	-0.6	0.5	-3.2
Wheat	1.7	0.6	-0.9	-1.6	0.8	1.6	2.7	2.2
Cellulosic feedstock acres in 2022 (%)								
Stover	164.3	-8.9	345.8	111.0	209.2	92.3	451.6	128.7
Straw	52.8	-6.5	122.0	822.2	52.8	591.6	1273.2	938.4
Miscanthus	31.3	3.6	-100.0	-99.0	-20.1	-5.7	-76.0	-35.7
Switchgrass	6.6	0.5	149.8	305.9	-42.6	-61.0	-18.0	22.7
Crop production and price in 2022 (%)								
Corn production	-13.3	-7.9	2.2	0.2	2.3	2.3	2.1	8.6
Corn price	2.4	6.6	0.0	-3.1	0	0	2.3	3.4
Soybeans production	-1.2	-5.5	-0.7	-2.2	0	-0.9	-1.0	-3.2
Soybeans price	3.5	4.4	1.7	4.4	-0.1	2.2	1.7	2.3
Wheat production	-6.6	-7.6	0	-1.8	0.7	0.3	1.6	0.5
Wheat price	7.0	6.1	0	1.4	0	-0.9	-1.6	-1.9

(continued)

Table 7.8 (continued)

	Rate of yield increase reduced by 50%		High cost of production of miscanthus		Upper limit of 10% on energy crop acres in a crop reporting district		High cost of production of miscanthus and switchgrass	
	M	MS	M	MS	M	MS	M	MS
Fuel price in 2022 and cumulative consumption of fuels and kilometers (%)								
Gasoline price	-0.04	-0.01	-0.01	-0.3	-0.1	-0.3	-0.1	-0.3
Corn ethanol price	1.3	3.2	-1.7	-1.4	-1.5	-0.3	-1.0	1.0
Cellulosic ethanol price	1.5	0.7	1.8	16.2	5.7	16.4	10.1	17.8
Gasoline consumption	-0.05	-0.03	-0.03	0.1	-0.03	0.3	-0.1	0.5
Corn ethanol	-25.3	-2.8	12.9	-3.8	8.9	1.7	26.7	26.8
Cellulosic ethanol	27.1	0.6	-14.1	-2.8	-9.8	-5.9	-29.5	-11.6
Ethanol imports	2.0	0.7	3.0	7.2	2.7	9.9	11.1	22.5
Total biofuels	0	0.25	0	-2.74	0	-4.84	0.00	-7.12
Kilometer consumption	-0.04	-0.01	-0.03	-0.1	-0.03	-0.1	-0.1	-0.2
Welfare cost of biofuels								
Welfare cost ($ billions)[a]	110.2	-69.6	132.6	-81.3	133.7	-83.2	142.4	-18.2
Additional biofuels (%)		8.1		4.9		2.6		0.1
Welfare cost ($/liter)		0.71		1.37		2.62		69.6

Notes: Percentage changes are calculated relative to the same policy in the benchmark scenario. M = biofuel mandate alone; MS = biofuel mandates plus volumetric tax credits.

[a] Welfare cost of mandate is change in welfare relative to business as usual; welfare cost of mandate and volumetric tax credits is change in welfare relative to mandate alone.

and the shift toward cellulosic biofuels. As a result, the welfare cost per liter of biofuels decreases to $0.7.

Raising the production cost of miscanthus relative to other feedstocks leads to a significant decline in the production of miscanthus and expansion in the use of crop residues and switchgrass to produce cellulosic biofuels. It increases the share of corn ethanol, ethanol from forest residues, and of ethanol imports in the cumulative biofuel production under the RFS and under the RFS and tax credit scenario. The price of cellulosic biofuels increases by 16 percent, but overall impact on VKT and on gasoline consumption is small. There is a 3 percent reduction in cumulative biofuel consumption in the MS scenario relative to the benchmark due to the absence of the high yielding feedstock, miscanthus; the same level of land under bioenergy crops now yields a lower volume of biofuels. The welfare cost of the tax credits is significantly lower in this case but so is the incremental biofuel production due to the tax credit, resulting in an increase in the per liter welfare cost to $1.4.

If the production costs of both miscanthus and switchgrass are high, there is a significant expansion in the acreage on which crop residues are harvested and a reduction in the production of miscanthus and switchgrass under the RFS scenario. Although switchgrass acreage increases under the MS scenario, cumulative cellulosic biofuel production reduces by 12 percent. Despite the increase in the use of crop residues, forest residues, corn, and sugarcane ethanol imports to meet the RFS, total biofuel production under the MS scenario is 7 percent lower than that under the same policy scenario with the benchmark parameters, only 0.1 percent (1.7 billion liters) higher than the RFS mandates. The price of cellulosic biofuels increases by 10 percent and 17 percent, respectively, in these two scenarios due to high costs of production of bioenergy crops. We find the overall impact on VKT and on gasoline consumption is modest. The welfare cost of the tax credits, relative to the mandate alone, is now lower ($18.2 billion instead of $78.9 billion), but the per liter welfare cost of the incremental biofuel production due to the tax credit is very high ($70 per liter) because of the small volume of additional biofuel production induced by the tax credits.

A reduction in land available for bioenergy crops to a maximum of 10 percent of the CRD reduces the share of cellulosic biofuels to meet the RFS by 10 percent, while increasing the price of cellulosic biofuels by 5 percent. Biomass feedstock producers are better off as are row crop producers. The welfare costs of the subsidies are similar to those in the benchmark case, but cumulative biofuel production is 5 percent lower than in the benchmark case (by 63 billion liters). As a result, the welfare cost of biofuels is substantially higher.

In general, we find that changes in technology and cost parameters that limit the potential to expand production of high yielding biofuels reduce the ability of the volumetric tax credits to significantly increase biofuel production. The tax credits then primarily support biofuel production that occurs anyway to meet the RFS, provided the RFS is binding, resulting in high welfare costs per liter of biofuel production.

7.7   Conclusions and Discussion

Biofuel mandates and subsidy policies have been enacted with the intention of promoting renewable alternatives to reduce dependence on gasoline. Concerns about the competition they pose for land and its implications for food prices have led to a shift in policy incentives toward second-generation biofuels from nonfood-based feedstocks. This chapter develops a framework to examine the economic viability of these feedstocks and the extent to which biofuel expansion will imply a trade-off between food and fuel production. It analyzes the differential incentives provided by alternative policies for biofuel production and the mix of biofuels and the welfare costs of biofuel policies.

Even with the option of high yielding energy crops, we find that a biofuel mandate (without any subsidies) would rely on corn ethanol to meet 50 percent of the RFS mandate over 2007 to 2022; miscanthus and forest residues would produce 49 percent and 22 percent of the cumulative advanced biofuels over 2007 to 2022, with switchgrass, crop residues, and ethanol imports meeting the rest. In the benchmark case, the mandate leads to a 16 percent increase in corn acreage, which is largely met by reducing acreage under soybean and other crops. Despite gains in corn productivity over 2007 to 2022, the corn price in 2022 is 24 percent higher than in the BAU. In response to higher crop and livestock prices, exports of corn, soybeans, wheat, and beef decline relative to the BAU. The mandate lowers the price of gasoline by 8 percent in 2022 relative to the BAU, which results in a reduction in the cost per kilometer and increases cumulative VKT by 0.4 percent over the 2007 to 2022 period. The benefits to fuel consumers and agricultural producers more than offsets the costs to domestic agricultural consumers and gasoline producers; consequently, the RFS raises net present value of cumulative social welfare relative to the BAU by $122 billion. This ranges between $110 to $132 billion across the scenarios considered here.

Volumetric tax credits for corn ethanol and cellulosic biofuels significantly enhances the competitiveness of cellulosic biofuels relative to corn ethanol and shifts the mix of biofuels such that 88 percent of the cumulative biofuels over the 2007 to 2022 period would now be produced from cellulosic feedstocks. This mitigates the competition for land and reduces corn, soybean, wheat, rice, cotton, and beef prices relative to those with a mandate alone. Corn price in 2022 would now be 6 percent lower than in the BAU. These tax credits lead to substantial reduction in the consumer price of biofuels and in the cost per kilometer, despite marginal increases in the gasoline price. As a result, these tax credits benefit fuel consumers, agricultural consumers, gasoline producers, and biomass producers. However, they impose significant costs on tax payers and on conventional crop producers (by eventually leading to a transition from corn ethanol to cellulosic biofuels). As a result, they lower social welfare relative to the RFS alone. The discounted present value of the welfare costs of these tax credits range between $79 billion and $118 billion over the 2007 to 2022 period. The incremental gain in total biofuel production beyond the RFS alone ranges between 32 billion liters and 99 billion liters across the scenarios considered here. Thus, the welfare cost per liter varies between $0.7 per liter and $2.6 per liter. These welfare costs are based on the premise that the mandated volume of biofuel production is achieved even in the absence of these tax credits. Moreover, these cost estimates are sensitive to assumptions about the costs of producing cellulosic feedstocks and the extent to which there might be constraints to the expansion of bioenergy crop production on marginal land.

Our analysis also shows the role of productivity enhancing technologies both in the traditional crop sector and the bioenergy sector. Yield increases for major crops like, corn, and soybeans and the use of high yielding, long-lived energy crops like miscanthus contribute to mitigating the competition for land and the impact of biofuel production on food prices. Corn price in 2022 would be 2 to 7 percent higher if the rate of productivity growth of row crops is 50 percent of that assumed in the benchmark case. High relative costs of miscanthus production result in 14 percent lower cumulative cellulosic biofuel production under the RFS and 3 percent lower with the RFS and tax credits compared to the corresponding benchmark case.

Our analysis abstracted from considerations of the external benefits of biofuel production in the form of energy security and reduced greenhouse gas emissions relative to gasoline as well as other benefits of ethanol, such as its additive value as an oxygenate for gasoline. It does, however, show how high these benefits would need to be to offset the economic welfare costs of tax credits estimated here.

References

Abbott, P. C., C. Hurt, and W. E. Tyner. 2008. *What's Driving Food Prices?* Farm Foundation Issue Report. Oak Brook, IL: Farm Foundation.

Adams, D., R. Alig, B. McCarl, and B. C. Murray. 2005. "FASOMGHG Conceptual Structure, and Specification: Documentation." http://agecon2.tamu.edu/people/faculty/mccarl-bruce/FASOM.html.

Ando, A., M. Khanna, and F. Taheripour. 2010. "Market and Social Welfare Effect of the Renewable Fuels Standard." In *Handbook of Bioenergy Economics and Policy,* edited by M. Khanna, J. Scheffran, and D. Zilberman, 233–50. New York: Springer.

Babcock, B. A., D. J. Hayes, and J. D. Lawrence. 2008. "Using Distillers Grains in the U.S. and International Livestock and Poultry Industries." Working Paper. Iowa State University: Midwest Agribusiness Trade Research and Information Center.

Bredahl, M. E., W. H. Meyers, and K. J. Collins. 1979. "The Elasticity of Foreign Demand for U.S. Agricultural Products: The Importance of the Price Transmission Elasticity." *American Journal of Agricultural Economics* 61: 58–63.

Carley, D. H., and S. M. Fletcher. 1989. "Analysis of the Impact of Alternative Government Policies on Peanut Farmers." Experiment Station Special Publication no. 62, University of Georgia.

Clifton-Brown, J., Y.-C. Chiang, and T. R. Hodkinson. 2008. "*Miscanthus:* Genetic Resources and Breeding Potential to Enhance Bioenergy Production." In *Genetic Improvement of Bioenergy Crops,* edited by W. Vermerris, 273–94. New York: Springer Science+Business Media.

Congressional Budget Office. 2010. *Using Biofuels Tax Credits to Achieve Energy and Environmental Policy Goals.* Washington, DC: Congressional Budget Office.

Crago, C. L., M. Khanna, J. Barton, E. Giuliani, and W. Amaral. 2010. "Competitiveness of Brazilian Sugarcane Ethanol Compared to U.S. Corn Ethanol." *Energy Policy* 38 (11): 7404–15.

de Gorter, H., and D. R. Just. 2008. "The Economics of the U.S. Ethanol Import Tariff with a Blend Mandate and Tax Credit." *Journal of Agricultural and Food Industrial Organization* 6:1–21.

———. 2009. "The Economics of a Blend Mandate for Biofuels." *American Journal of Agricultural Economics* 91:738–50.

Dumortier, J., D. J. Hayes, M. Carriquiry, F. Dong, X. Du, A. Elobeid, J. F. Fabiosa, and S. Tokgoz. 2009. "Sensitivity of Carbon Emission Estimates from Indirect Land-Use Change." Working Paper. Center for Agricultural and Rural Development, Iowa State University, July.

Ellinger, P. 2008. *Ethanol Plant Simulator.* Urbana, Illinois: Department of Agricultural and Consumer Economics, University of Illinois, Urbana-Champaign. http://www.farmdoc.illinois.edu/pubs/FASTtool.asp?section=FAST.

Elobeid, A., and S. Tokgoz. 2008. "Removing Distortions in the U.S. Ethanol Market: What Does It Imply for the United States and Brazil?" *American Journal of Agricultural Economics* 90:918–32.

Elobeid, A. E., S. Tokgoz, D. J. Hayes, B. A. Babcock, and C. E. Hart. 2007. "The Long-Run Impact of Corn-Based Ethanol on Grain, Oilseed, and Livestock Sectors with Implications for Biotech Crops." *AgBioForum* 10:11–18.

Energy Information Administration (EIA). 2007. *Annual Energy Outlook 2007 with Projections to 2030.* Washington, DC: U.S. Energy Information Administration. http://tonto.eia.doe.gov/ftproot/forecasting/0383%282007%29.pdf.

———. 2010a. *Annual Energy Outlook 2010.* Washington, DC: U.S. Energy Information Administration. http://www.eia.doe.gov/oiaf/aeo/.

———. 2010b. *How Dependent Are We on Foreign Oil?* Washington, DC: U.S. Energy Information Administration. http://tonto.eia.doe.gov/energy_in_brief/foreign_oil_dependence.cfm.

English, B., D. G. Ugarte, R. Menard, and T. West. 2008. "Economic and Environmental Impacts of Biofuel Expansion: The Role of Cellulosic Ethanol." Paper presented at the Integration of Agricultural and Energy Systems Conference, Atlanta, Georgia.

Environmental Protection Agency (EPA). 2010. "Renewable Fuel Standard Program (RFS2) Regulatory Impact Analysis." Washington, DC: U.S. Environmental Protection Agency. http://www.epa.gov/otaq/fuels/renewablefuels/index.htm.

Fabiosa, J. F., J. C. Beghin, F. Dong, A. Elobeid, S. Tokgoz, and T.-H. Yu. 2009. "Land Allocation Effects of the Global Ethanol Surge: Predictions from the International FAPRI Model." Working Paper no. 09-WP 488. Ames, IA: Center for Agricultural and Rural Development, Iowa State University.

Ferris, J., and S. Joshi. 2009. "Prospects for Ethanol and Biodiesel, 2008 to 2017 and Impacts on Agriculture and Food." In *Handbook of Bioenergy Economics and Policy,* edited by M. Khanna, J. Scheffran, and D. Zilberman, 91–111. New York: Springer.

Food and Agricultural Policy Research Institute (FAPRI). 2009. *Elasticity Database.* Ames, IA: Iowa State University. http://www.fapri.iastate.edu/tools/elasticity.aspx.

Fortenbery, T. R., and H. Park. 2008. "The Effect of Ethanol Production on the U.S. National Corn Price." Staff Paper no. 523. Department of Agricultural and Applied Economics, University of Wisconsin-Madison.

Gao, X. M., E. J. Wailes, and G. L. Cramer. 1995. "Double-Hurdle Model with Bivariate Normal Errors: An Application to U.S. Rice Demand." *Journal of Agricultural and Applied Economics* 27:363–76.

Graham, R. L., R. Nelson, and J. Sheehan. 2007. "Current and Potential U.S. Corn Stover Supplies." *Agronomy Journal* 99:1–11.

Greene, D. L., and N. I. Tishchishyna. 2000. *Costs of Oil Dependence: A 2000 Update.* Oak Ridge, TN: Oak Ridge National Laboratory.

Gunderson, A. C., E. B. Davis, and I. H. Jager. 2008. *Exploring Potential U.S. Switchgrass Production for Lignocellulosic Ethanol.* ORNL/TM-2007/183. Oak Ridge, TN: Oak Ridge National Laboratory.

Hayes, D., B. Babcock, J. Fabiosa, S. Tokgoz, A. Elobeid, and T.-H. Yu. 2009. "Biofuels: Potential Production Capacity, Effects on Grain and Livestock Sectors, and Implications for Food Prices and Consumers." *Journal of Agricultural and Applied Economics* 41:1–27.

Heaton, E., F. Dohleman, and S. Long. 2008. "Meeting U.S. Biofuel Goals with Less Land: The Potential of Miscanthus." *Global Change Biology* 14:2000–14.

Hertel, T. W., W. E. Tyner, and D. K. Birur. 2010. "The Global Impacts of Biofuel Mandates." *The Energy Journal* 31:75–100.

Hettinga, W. G., H. M. Junginger, S. C. Dekker, M. Hoogwijk, A. J. Mcaloon, and K. B. Hicks. 2009. "Understanding the Reductions in U.S. Corn Ethanol Production Costs: An Experience Curve Approach." *Energy Policy* 37:190–203.

Huang, H., and M. Khanna. 2010. "An Econometric Analysis of U.S. Crop Yields and Cropland Acreages: Implications for the Impact of Climate Change." Paper presented at AAEA (Agricultural and Applied Economics Association) annual meeting, Denver, Colorado. http://ageconsearch.umn.edu/handle/61527.

Jain, A., M. Khanna, M. Erickson, and H. Huang. 2010. "An Integrated Biogeochemical and Economic Analysis of Bioenergy Crops in the Midwestern United States." *Global Change Biology Bioenergy* 2:217–34.

Lemus, R., and D. L. Parrish. 2009. "Herbaceous Crops with Potential for Biofuel Production in the USA." *CABI Reviews: Perspectives in Agriculture, Veterinary Science, Nutrition and Natural Resources* 4:1–23.

Lewandowski, I., J. M. O. Scurlock, E. Lindvall, and M. Christou. 2003a. "The Development and Current Status of Perennial Rhizomatous Grasses as Energy Crops in the U.S. and Europe." *Biomass and Bioenergy* 25:335–61.

———. 2003b. "The Development and Current Status of Potential Rhizomatous Grasses as Energy Crops in the U.S. and Europe." *Biomass and Bioenergy* 25: 335–61.

Malcolm, S. 2008. "Weaning Off Corn: Crop Residues and the Transition to Cellulosic Ethanol." Paper presented at the conference on the Transition to a Bio-Economy: Environmental and Rural Development Impacts, St. Louis, Missouri.

McCarl, B. A., and U. A. Schneider. 2001. "Greenhouse Gas Mitigation in U.S. Agriculture and Forestry." *Science* 294:2481–82.

McCarl, B. A., and T. H. Spreen. 1980. "Price Endogenous Mathematical Programming as a Tool for Policy Analysis." *American Journal of Agricultural Economics* 62:87–102.

Mcphail, L. L., and B. A. Babcock. 2008. "Short-Run Price and Welfare Impacts of Federal Ethanol Policies." Working Paper no. 08-WP 468. Ames, IA: Center for Agricultural and Rural Development, Iowa State University.

Metcalf, G. E. 2008. "Using Tax Expenditures to Achieve Energy Policy Goals." *American Economic Review: Papers and Proceedings* 98:90–94.

National Research Council. 2002. *Effectiveness and Impact of Corporate Average Fuel Economy (CAFE) Standards.* Washington, DC: National Academies Press.

Önal, H., and B. A. McCarl. 1991. "Exact Aggregation in Mathematical Programming Sector Models." *Canadian Journal of Agricultural Economics* 39:319–34.

Perlack, R. D., L. L. Wright, A. F. Turhollow, R. L. Graham, B. J. Stokes, and D.C. Erbach. 2005. *Biomass as Feedstock for a Bioenergy and Bioproducts Industry: The Technical Feasibility of a Billion-Ton Annual Supply.* Oak Ridge, TN: Oak Ridge National Laboratory.

Pfuderer, S., G. Davies, and I. Mitchell. 2010. *The Role of Demand for Biofuel in the Agricultural Commodity Price Spikes of 2007/2008.* London: Food and Farming Analysis, Department for Environment Food and Rural Affairs (Defra).

Piggott, N. E., and M. K. Wohlgenant. 2002. "Price Elasticities, Joint Products, and International Trade." *Australian Journal of Agricultural and Resource Economics* 46:487–500.

Reilly, J., A. Gurgel, and S. Paltsev. 2009. "Biofuels and Land Use Change." In *Transition to a Bioeconomy: Environmental and Rural Development Impacts,* edited by M. Khanna, 1–18. St. Louis, MO: Farm Foundation.

Semere, T., and F. M. Slater. 2007. "Invertebrate Populations in Miscanthus (Miscanthus×Giganteus) and Reed Canary-Grass (Phalaris Arundinacea) Fields." *Biomass and Bioenergy* 31:30–9.

Sheehan, J., A. Aden, K. Paustian, K. Killian, J. Brenner, M. Walsh, and R. Nelson. 2003. "Energy and Environmental Aspects of Using Corn Stover for Fuel Ethanol." *Journal of Industrial Ecology* 7:117–46.

Takayama, T., and G. G. Judge. 1971. *Spatial and Temporal Price and Allocation Models.* Amsterdam: North-Holland.

Tokgoz, S., A. Elobeid, J. F. Fabiosa, D. Hayes, B. A. Babcock, T.-H. Yu, F. Dong, C. E. Hart, and J. C. Beghin. 2007. *Emerging Biofuels: Outlook of Effects on U.S. Grain, Oilseed, and Livestock Markets.* Ames: Center for Agricultural and Rural Development, Iowa State University.

Tyner, W., and F. Taheripour. 2008. "Policy Options for Integrated Energy and Agricultural Markets." *Review of Agricultural Economics* 30:387–96.

Ugarte, D. G., B. C. English, C. Hellwinckel, T. O. West, K. L. Jensen, C. D. Clark, and R. J. Menard. 2009. "Analysis of the Implications of Climate Change and Energy Legislation to the Agricultural Sector." Working Paper. Knoxville, TN: Department of Agricultural Economics, Institute of Agriculture, University of Tennessee.

Ugarte, D. G., M. E. Walsh, H. Shapouri, and S. P. Slinsky. 2003. *The Economic Impacts of Bioenergy Crop Production on U.S. Agriculture.* Washington, DC: U.S. Department of Agriculture.

U.S. Department of Agriculture (USDA). 2010. *Feed Grains Database.* Washington,

DC: U.S. Department of Agriculture. http://www.nass.usda.gov/ and http://www
.ers.usda.gov/Data/FeedGrains/.

U.S. Department of Agriculture/Economic Research Service (USDA/ERS) 1997.
General Crop Management Practices. Washington, DC: U.S. Department of Agriculture. http://www.ers.usda.gov/publications/sb969/sb969e.pdf.

———. 2009. *Commodity and Food Elasticities: Demand Elasticities from Literature.*
Washington, DC: U.S. Department of Agriculture. http://www.ers.usda.gov/Data/
Elasticities/query.aspx.

———. 2010. *Agricultural Outlook: Statistical Indicators.* Washington, DC: U.S.
Department of Agriculture. http://www.ers.usda.gov/Publications/AgOutlook/
AOTables/.

U.S. Department of Agriculture/National Agricultural Statistics Service
(USDA/NASS). 2009a. *Quick Stats.* Washington, DC: U.S. Department of Agriculture. http://quickstats.nass.usda.gov/.

———. 2009b. *U.S. & All States County Data—Crops.* Washington, DC: U.S.
Department of Agriculture. http://www.nass.usda.gov/QuickStats/Create_County
_All.jsp.

Van Den Wall Bake, J. D., M. Junginger, A. Faaij, T. Poot, and A. Walter. 2009.
"Explaining the Experience Curve: Cost Reductions of Brazilian Ethanol from
Sugarcane." *Biomass and Bioenergy* 33:644–58.

Wallace, R., K. Ibsen, A. Mcaloon, and W. Yee. 2005. *Feasibility Study for Co-
Locating and Integrating Ethanol Production Plants from Corn Starch and Ligno-
cellulogic Feedstocks.* Golden, CO: National Renewable Energy Laboratory.

Walsh, M. E., D. G. De La Torre Ugarte, H. Shouri, and S. P. Slinsky. 2003. "Bio-
energy Crop Production in the United States: Potential Quantities, Land Use
Changes, and Economic Impacts on the Agricultural Sector." *Environmental and
Resource Economics* 24 (4): 313–33.

Wilcke, W., and G. Wyatt. 2002. *Grain Storage Tips.* Twin Cities, MN: University of
Minnesota Extension Service, University of Minnesota. http://www.extension
.umn.edu/distribution/cropsystems/M1080-FS.pdf.

Wortmann, C. S., R. N. Klein, W. W. Wilhelm, and C. Shapiro. 2008. *Harvesting
Crop Residues.* Lincoln, NE: University of Nebraska.

Wullschlegera, S. D., E. B. Davisb, M. E. Borsukb, C. A. Gundersona, and L. R.
Lyndb. 2010. "Biomass Production in Switchgrass across the United States: Database Description and Determinants of Yield." *Agronomy Journal* 102:1158–68.

Land for Food and Fuel Production
The Role of Agricultural
Biotechnology

Steven Sexton and David Zilberman

8.1 Introduction

The global food crisis of 2008 ended three decades of declining food prices and highlighted a growing challenge for agriculture: to supply food and clean energy to a world population growing to nine billion by 2050. In roughly the last half of the twentieth century, agriculture accommodated a near doubling of the world population through intensification. Farm yields more than doubled with the use of high-yielding seed varieties, agricultural chemicals, irrigation, and mechanization. Per capita calorie production grew despite the rapid population growth and despite an exodus of land from production. Since the 1990s, however, yield growth in staple crops has been slowing and stalling as traditional sources of yield improvements are depleted. Absent intensification, demand growth will be met by extensification, which is unpalatable amid growing concern about climate change and biodiversity loss.

First-generation agricultural biotechnology has been promoted as a tool for improving the control of agricultural pests that diminish effective yields. To the extent adoption of the technology generates yield growth, it constitutes a mechanism for expanding farm output without expanding the

Steven Sexton is a PhD candidate in agricultural and resource economics at the University of California at Berkeley. David Zilberman is professor of agricultural and resource economics at the University of California at Berkeley and a member of the Giannini Foundation.

We thank the Energy Biosciences Institute for supporting this research. We are grateful to Graham Brookes for providing some of the data used in this analysis and to David Popp and seminar participants at the National Bureau of Economic Research (NBER) for helpful comments on an earlier draft.

area under cultivation.[1] A number of studies in a variety of countries have documented yield gains caused by the adoption of genetically engineered (GE) crops. The studies have been limited in size and scope, however, and have generated widely varying estimates of the yield gains from GE crop adoption. Absent agreement among empiricists on the magnitude of yield improvements, agricultural biotechnology remains controversial. Potential risks to human health and the environment are weighted heavily against the uncertain benefits. This chapter overcomes some of the limitations in earlier empirical work in order to assess the degree to which the technology has increased food supply on a global scale.

8.2   Background

8.2.1   Agricultural Biotechnology

Farmers around the world have rapidly adopted GE seeds since they were first commercialized in 1996. The GE seeds are intended to reduce pest damage and lower production costs. By 2008, 13.3 million farmers in twenty-five countries annually planted 8 percent of global cropland with transgenic crops. In 2009, U.S. farmers planted more than 80 percent of the sugar beet crop with transgenic varieties that had only been introduced one year earlier (James 2009). Despite the popularity of agricultural biotechnology on the farm, its introduction in the marketplace has met strong resistance from critics who advocate a precautionary approach to the technology because of potential risks to humans and the environment. Consequently, GE seeds and crops are banned in some countries and highly regulated in others, including those that lead in adoption. The European Union, for instance, imposed a de facto ban on GE seeds in 1998. The ban was lifted in 2008 amid pressure from the United States and the World Trade Organization. Consumer sentiment against GE foods has also constrained the market for GE seed. Products derived from GE seed have been relegated to feed and fiber uses only. Producers must segregate GE crop output throughout the supply chain in order to ensure the transgenic material is not comingled with

---

1. Yield improvements from exogenous technical changes can, in theory, induce cropland expansion by making farming more profitable. Yield gains increase output on existing land, which tends to reduce prices, but also lowers costs of production, potentially making expansion to more marginal lands profitable, as we note in section 8.3. Feng and Babcock (2010) provide analytical results that show yield improvements in maize induce cropland expansion under unregulated free markets. However, an extensive empirical body of research suggests the opposite is true: that yield improvements are associated with reductions in cropland expansion (e.g., Waggoner 1995; Matson et al. 1997; Balmford, Green, and Scharlemann 2005). Alston, Beddow, and Pardey (2009), for instance, document dramatic increases in agricultural productivity and only "slow growth" in the use of agricultural land. Barbier (2001) estimates a negative elasticity of crop yield with respect to land expansion in tropical forests. This point is also made in Zilberman et al. (1991), Mundlak (2001), and Mundlak (2011).

conventionally bred crop output. In early 2010, China was poised to approve the first use of a GE crop for human consumption.

The GE traits have been introduced to four principal crops: cotton, maize, rapeseed, and soybean. Rapeseed and soybean seeds have been engineered to tolerate broad-spectrum herbicides like glyphosates and gluphosinates, chemicals that target a host of weed species and are lethal to conventional crops. Adoption of such herbicide-tolerant (HT) varieties permits farmers to more effectively control weeds. Absent the HT trait, farmers are forced to apply more toxic and narrowly targeted chemicals in order to kill weeds and keep the crop safe. They also use mechanical control, like tilling operations, to control weeds. Because glyphosates have historically sold at prices below the targeted chemicals, adoption of HT varieties is likely to reduce damage control expenditures. Some cotton and maize varieties have also been engineered with the HT trait, while others are engineered to produce *Bacillus thuringiensis* (Bt), a naturally occurring toxin that is lethal if ingested by a number of common insect pests. These are referred to as Bt crops or insect-resistant (IR) crops. Some maize and cotton varieties are engineered to express both traits and are commonly referred to as "stacked" varieties. The HT traits have also been introduced into sugar beets and alfalfa, though both are planted on a relatively small scale. Crops with HT traits have always been the dominant GE crop, occupying 63 percent of total GE crop area in 2008, followed by "stacked" traits (22 percent) and IR traits (15 percent). The HT soybeans occupied the majority of total GE cropland (53 percent) and constituted 70 percent of the world soybean crop in 2008 (James 2009). The GE maize constituted 30 percent of all GE crop areas in 2008 and 24 percent of the world maize crop.

Adoption of GE crops has been rapid. By 2009, half of all U.S. cropland was planted with GE seed. Approximately 80 percent of the 2008 cotton, maize, and soybean crops in the United States were each produced from transgenic varieties. The United States has been a leader in adoption, planting more than half (62.5 million hectares) of all GE areas in 2008. But other countries have been similarly aggressive in their adoption. South Africa, Australia, and Argentina all planted more than 90 percent of their 2008 cotton crops with GE varieties, up from 1 to 2 percent a decade earlier. Canada planted virtually its entire maize crop with GE seed in 2008. Of the twenty-five countries that planted GE crops in 2008, fifteen were developed countries and ten were developing (James 2009). Figure 8.1 shows the annual area planted with GE crops from 1996 to 2008 by country type.

8.2.2   The Economics of Agricultural Biotechnology

There is a large and growing literature on the adoption and impact of GE crops. It is summarized in Qaim (2009) and National Research Council (2010). Much of the literature on GE crop adoption follows the threshold adoption framework of David (1969). This framework assumes that firms

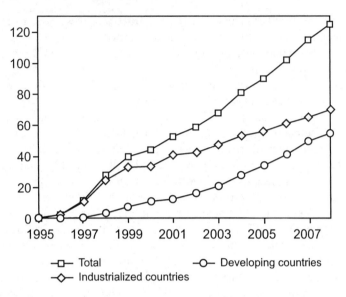

Fig. 8.1 **Genetically engineered crop adoption over time**

are heterogeneous, that they make choices that are consistent with an explicit economic decision-making criterion (e.g., profit maximization), and that the costs and benefits of technology adoption vary over time in response to changes in economic conditions and learning (Feder, Just, and Zilberman 1985). The threshold model is readily employed in applications with data on the behavior of individual agents by using discrete and discrete-continuous choice models.

Much of the literature on adoption of GE-crop technology estimated the factors that affect whether producers adopt the technology and the extent of adoption. These studies found that biophysical conditions (e.g., vulnerability to pest damage), economic conditions (e.g., output and input prices), and regulatory conditions affect adoption. The scale of operation and human capital are not major factors affecting adoption because GE-crop technology is simpler than alternative damage-control mechanisms and does not exhibit increasing returns to scale. Crost et al. (2007), however, did find evidence that farmers in India with higher human capital were more likely to adopt.

Another significant body of literature has investigated the impact of GE-crop technology. Most of this literature is surveyed in Qaim (2009) and National Research Council (2010). For the most part, these studies compared the performance of GE with non-GE crops under various conditions. Some conducted surveys of farmers to assess the reasons for adoption and the cause of yield changes post–GE crop adoption. Most existing studies were conducted in the early days of GE-crop adoption (from 1996 to 2003) or considered early data.

The potential gains associated with adoption of first-generation GE crops are several. They include reduced crop losses from insect pests; reduced expenditures on damage control inputs like herbicides, pesticides, and fuel; improved worker safety; greater flexibility in farm management; and lower risk of yield variability (National Research Council 2010). The magnitude of these benefits varies by location, crop, and time. Table 8.1, which is borrowed from Qaim (2009), summarizes existing empirical estimates of some of these benefits, including yield gains, gross margin impacts, and pesticide use. It demonstrates the heterogeneity of estimates in the extant literature.

There has been no rigorous assessment of the impact of adoption of GE technologies in aggregate even though there is a rich literature on the welfare implications of adoption based on stylized assumptions about shifts in supply. These studies, too, mostly cover the earlier period of adoption of GE crops. The National Research Council (2010) identified the lack of recent market impact assessments as one of the major gaps in the economic research on agricultural biotechnology. In this chapter, we employ data on acreage of major crops and the share of land for each crop that is allocated to biotechnology. We use analysis of variance to decompose yield per acre to different components. Our analysis applies an approach introduced by Just et al. (1990) to decompose variable input among crops. The approach is used to allocate output among crop types. We assume that at each time and

Table 8.1   **Farm-level effects of genetically engineered crops**

Country	Insecticide reduction (%)	Increase in effective yield (%)	Increase in gross margin (US$/ha)	Reference(s)
	Bacillus thuringiensis cotton			
Argentina	47	33	23	Qaim and de Janvry (2003, 2005)
Australia	48	0	66	Fitt (2003)
China	65	24	470	Pray et al. (2002)
India	41	37	135	Qaim et al. (2006), Sadashivappa and Qaim (2009)
Mexico	77	9	295	Traxler et al. (2003)
South Africa	33	22	91	Thirtle et al. (2003), Gouse et al. (2004)
United States	36	10	58	Falck-Zepeda et al. (2000b), Carpenter et al. (2002)
	Bacillus thuringiensis maize			
Argentina	0	9	20	Brookes and Barfoot (2005)
The Philippines	5	34	53	Brookes and Barfoot (2005), Yorobe and Quicoy (2006)
South Africa	10	11	42	Brookes and Barfoot (2005), Gouse et al. (2006)
Spain	63	6	70	Gómez-Barbero et al. (2008)
United States	8	5	12	Naseem and Pray (2004), Fernandez-Cornejo and Li (2005)

Source: Qaim (2009, 672).

location, the yield per acre of each crop with a given technology is fixed, but these yields per acre vary across crops, technologies, and time. This approach has been generalized by Lence and Miller (1998) and applied by Khanna and Zilberman (1999) to decompose aggregate data of energy generation and greenhouse gas (GHG) emissions in different locations. This rather simple approach allows us to rely upon a minimal amount of data to decompose yields. We use our estimate of the partial effect of GE adoption on yields of adopting farmers (a population averaged treatment effect on the treated) to estimate the change in food supply attributable to agricultural biotechnology and parameterize a model of the food market in 2008 in order to assess the effect of GE seeds on food prices during the food crisis.

8.3   Conceptual Model

In this section, we present a conceptual model that provides the theoretical foundation for the empirical analysis that follows in the next section. We adopt a modeling approach that follows Qaim and Zilberman (2003) and Ameden, Qaim, and Zilberman (2005) that employs the damage control framework of Lichtenberg and Zilberman (1986). This framework distinguishes between inputs that directly affect production, like capital and fertilizer, and inputs that indirectly affect production by reducing crop damage, such as pesticides and mechanical and biological control. Specifically, assume a constant-returns-to-scale agricultural production function. Let yield per acre, y, be the product of potential output, $f_j(z,a)$, and damage abatement, $g_i(x,N)$. Potential output is the output that would obtain if there were no pest damage. It is increasing in production inputs, z, like fertilizer, and a heterogeneity parameter, a, which characterizes farm quality and is a function of climate, human capital, and land quality. Potential output is also a function of seed variety, j, where $j = 0$ denotes a generic seed variety and $j = 1$ denotes a local seed variety. It is assumed that for all z and a, $f_1(z,a) \geq f_0(z,a)$. Damage abatement is the share of crop not lost due to pest damage. It is increasing at a decreasing rate in use of damage control inputs, x, like pesticides, and decreasing in effective pest pressure, n. Effective pest pressure is the product of a seed-technology parameter, δ_j, and initial pest pressure N, that is, $n = \delta_i N$, where $\delta_0 = 1$ denotes conventional seed technology, and $\delta_1 < 1$ denotes GE seed technology. Consequently, for all x and all positive N, $g_1(x,N) \geq g_0(z, N)$. Effective yield per acre under technology ij, then is given by:

(1) $$y_{ij} = g_i(x_{ij}, N)f_j(z_{ij},a).$$

With this specification, farmers face at most four distinct seed technology packages: generic-conventional ($i = 0$, $j = 0$), local-conventional ($i = 0$, $j = 1$), generic-GE ($i = 1$, $j = 0$), and local-GE ($i = 1$, $j = 1$).

The farmer's problem is:

(2) $\qquad \max_{z,x,i,j} \pi_{ij} = pg_i(x_{ij},N)f_j(z_{ij},a) - wz_{ij} - vx_{ij} - I_{ij},$

where p, w, and v are exogenously determined prices for output, production inputs, and damage control inputs, respectively, and where I_{ij} is a technology fee associated with technology ij. It is assumed $I_{00} < I_{01} < I_{10} < I_{11}$.

Farmers adopt the technology that yields the highest expected profits. We solve the farmer's problem recursively. First, conditional on seed technology choice and farm quality endowments, producers choose inputs to maximize profits. The profit maximizing quantity of inputs given technology ij are functions of prices and land quality, such that:

$$x_{ij}^* = x_{ij}^* (w,v,p,N)$$
$$z_{ij}^* = z_{ij}^* (w,v,p,N).$$

Maximum profits under each technology are obtained by substituting the optimal input demands into the profit function. Farmers select the technology that yields highest expected profits conditional on profits being nonnegative.

Analysis of these optimality conditions yields several results important for the subsequent empirical analysis. First, the adoption of GE crops increases damage abatement, which boosts effective yield under typical conditions. This is true so long as farmers face some pest pressure, and the adoption of GE crops does not require farmers to switch to a low-yield generic seed variety that would lower potential output. In theory, effective output may decline with adoption of GE crops either because a given farmer must switch from a local seed variety to a generic variety in order to adopt the GE technology or because the insertion of the GE trait into the seed germplasm causes an interaction that reduces potential output. In order for effective yield to decline with adoption, the percentage change in potential output must exceed the percentage change in damage abatement in absolute value. In practice, such reductions in effective output with GE adoption, termed "yield drag," have not been a significant problem (National Research Council 2010). Furthermore, the optimizing farmer would only choose to adopt GE seed that exhibited these yield drag effects and thereby reduced total output if the cost savings from reduced damage control expenditures exceeded the revenue loss from foregone yields.

Second, the damage-abatement gain is increasing in pest damage and the price of conventional damage control inputs like fertilizer. We can define the change in damage abatement due to GE crop adoption, assuming no change in the j-dimension, as:

(3) $\qquad \Delta g = g_{1j}(x,N) - g_{0j}(x,N).$

Then it can be shown that $d\Delta g/dN > 0$ and $d\Delta g/dw > 0$.

Third, GE crop adoption causes an increase in the use of production inputs like fertilizer. It boosts potential output as long as it does not require a switch from a local seed variety to a generic seed variety. As damage abatement increases, so, too, does the value of marginal product of production inputs increase, holding prices constant. Therefore, farmers employ more production inputs. The increase in production inputs raises potential output, which boosts effective output by more than the reduction in crop damage. Though we are unable to test impacts of GE crop adoption on input-use in the subsequent empirical analysis due to a lack of global data on input-use, this result suggests that the yield gain associated with GE crop adoption exceeds the "gene effect" estimated in much of the previous literature. Our empirical estimates of the yield gain associated with GE crop adoption incorporates this additional yield effect that operates through the potential yield function as opposed to the damage abatement function. This makes our yield estimates unique among the estimates of previous analyses.

Fourth, the change in yield due to GE crop adoption is increasing in farm quality, a, and pest pressure, N. We can decompose the total change in effective yield due to GE crop adoption as:

$$(4) \qquad \Delta y = y_{1j} - y_{0j} = f_{1j_0}\Delta g + \Delta f_z g_{1j_1} + \Delta f_j g_{1j_1},$$

where the first term on the right-hand side of the second equality is the damage abatement effect, the second term is the production input effect, and the third term is the yield drag effect, which can be negative but is typically zero (i.e., if $j_0 = j_1$ or if $j_0 = 0$ and $j_1 = 1$). It is easy to show, then, that $d\Delta y/da > 0$ and $d\Delta y/dN > 0$. We do not observe α and N in our data, so to the extent these theoretical predictions hold in practice, our empirical estimates of the yield gain associated with GE crop adoption may be biased. Failure to control for farm quality may induce an upward bias in the results. However, because the yield gains are expected to be greater with high pest pressure and because high pest pressure may be associated with low-quality farms, failure to control for pest pressure may induce an offsetting downward bias in our results.

8.4   Data and Methods

The empirical strategy of this chapter is motivated by the global pattern of GE seed adoption. By 2008, farmers in twenty-five countries had planted at least one of the four major GE crops. In most cases, the share of these crops planted to GE seed increased year over year in adopting countries from 1996 to 2008. In the United States, for instance, 12 percent of cotton was planted to GE seeds in 1996, but by 2007, the GE share had reached 87 percent. Some countries adopted multiple GE crops. Many others did not adopt any GE crops. Even some countries that are expected to experience significant benefits from adoption have not adopted because of political economy considerations. This was the case in European and African countries until 2010. Germany and Romania had deregulated GE technologies but then banned them for political reasons unrelated to their performance on the farm. Countries that did adopt GE crops continued to plant other crops exclusively to conventional seed either because GE alternatives did not exist or because regulation banned some GE crops.

The variation in GE adoption across countries and across time enables the econometrician to control for confounding factors at the country level by employing a panel fixed effects approach that relies on assumptions similar to, but weaker than, those required for estimation in triple differencing procedures. This procedure controls for endogeneity of adoption at the country level, that is, endogeneity of GE crop deregulation. However, estimation of a population average effect of GE crop adoption is subject to the biases described at the end of the preceding section, which stem from the endogeneity of adoption at the farm level, that is, selection on farm quality, which is unobservable in this data. These biases do not impede estimation of a population average effect of GE adoption among adopters, which is the critical coefficient for estimating the increase in food supply attributable to GE technologies.

Motivated by Just et al. (1990), we observe that total output of crop j in country i at time t, Q_{jit}, is the sum of output produced by each seed technology, k. Thus:

$$(5) \qquad Q_{jit} = \sum_{k=1}^{K} Q_{jitk},$$

where Q_{jitk} is the unobserved quantity of crop j produced by country i at time t using seed technology k. Define L_{jitk} as the amount of land planted to crop j with seed technology k in country i at time t. Then $q_{jitk} = Q_{jitk}/L_{jitk}$ is the output of crop j per unit of land using seed technology k in country i at time t. The deterministic component of the q_{jitk}, which is denoted q^*_{jitk}, can be decomposed into a crop-specific average seed-technology effect, β_{jk}, a crop specific time effect, γ_{jt}, and a country-specific crop effect, δ_{ji}. Then q^*_{jitk} is given by:

$$(6) \qquad q^*_{jitk} = \beta_{jk} + \gamma_{jt} + \delta_{ji}.$$

The β_{jk} are of interest and can be estimated by:

$$(7) \qquad Q_{jit} = \delta_j L_{jit} + \beta_{j1} L_{jit}^{GE} + \gamma_{jt} \mathbf{D}_{jt} + \varepsilon_{jit},$$

where L_{jit} is total land planted to crop j in country i at time t, L_{jit}^{GE} is the land planted to GE seed for crop j in country i at time t, $\mathbf{D}_{jt}$ is a crop-specific time dummy (the time dummy for the year 2008 is omitted), and ε_{jit} is a random deviation that is assumed normal and identically distributed. Equation (7) is estimated using fixed effects to control for country effects and secular trends. The fixed effects regression also controls for correlated random trends (Wooldridge 2005). Results are reported with White robust standard errors. The δ_j is the average yield on land that does not adopt GE seeds. The β_{j1} is the marginal effect on yield attributable to adoption of GE seeds ($k = 1$ denotes GE seed technology).

Data on total crop output are reported in tonnes and come from the Food and Agriculture Organization (FAO) of the United Nations. Total crop area is reported in hectares by FAO. The area of land planted to GM crops and specific traits was developed by Graham Brookes using data from the International Service for the Acquisition of Agri-Biotech Applications (ISAAA). The data cover the period 1990 to 2008. We include data on every country that adopted any GE crop from 1996 to 2008, as well as the top 100 gross producers of eight principal row crops during the period 1990 to 2008. For these 100 countries, we include observations on each of the four major GE crops (corn, cotton, soybean, and rapeseed) and each of four other principal row crops: wheat, rice, sorghum, and oats. These data comprise 10,717 annual country-level observations on crop output and GE seed area covering 627 country-crop groups. Because not all countries planted all eight crops in every year, the data constitute an unbalanced panel. Summary statistics are provided in tables 8.2 and 8.3.

8.5   Empirical Results

In the first econometric analysis of the global yield effects of GE seed adoption, we find that agricultural biotechnology generally produces significant yield improvements relative to non-GE seed on adopting farms. Table 8.4 reports results from estimation of equation (7).[2] In all cases, the coefficients of interest, the β_j, are statistically significant at the 99 percent level. Thus, the partial effect of GE seed adoption among adopters is positive and significant. Row (1) of table 8.5 reports the gain in yield from adoption of GE seed as a percent of total yield per acre.[3] The GE-seed effect on yields is greatest for crops with IR traits, that is, maize and cotton. Yield gains for GE cotton and maize—available in IR, HT, and stacked varieties—are estimated to be 65 percent and 45.6 percent, respectively. Yield gains for HT rapeseed and soybean are 25.4 percent and 12.4 percent, respectively. These estimates reflect the theoretical prediction that yield gains are larger for seeds expressing IR traits than for seeds expressing only HT traits because the HT trait largely permits substitution to cheaper and less-toxic chemicals. The primary effect of HT seed, then, is to reduce the cost of damage control and lessen the toxicity of chemicals applied to fields. As damage control becomes more cost-effective, however, increased damage control effort will be undertaken, which boosts effective yields and may boost potential yield as well.

2. Only coefficients of interest are reported. Full results are available from the authors by request.
3. Determined as $100 \cdot \delta_{ji}/\beta_{jk}$.

Table 8.2   **Summary statistics: Genetically engineered and trait shares**

	All	Developing	Developed	Adopters	Nonadopters
		Cotton			
Yield	15,521.02	14,155.02	27,981.82	19,070.02	14,492.22
	(9,278.30)	(7,954.58)	(11,074.55)	(10,174.24)	(8,741.64)
Seed share					
Genetically engineered	0.03	0.02	0.11	0.13	
	(0.14)	(0.11)	(0.26)	(0.27)	
Herbicide tolerant	0.01	0.01	0.08	0.06	NA
	(0.06)	(0.21)	(0.09)	(0.18)	
Insect resistant	0.02	0.02	0.08	0.11	NA
	(0.11)	(0.09)	(0.20)	(0.21)	
No. of observations	1,326	1,195	131	298	1,028
		Maize			
Yield	34,603.04	25,987.91	68,774.78	43,716.00	31,515.07
	(26,844.58)	(17,823.54)	(29,293.47)	(25,478.89)	(26,601.66)
Seed share					
Genetically engineered	0.01	0.01	0.03	0.05	NA
	(0.09)	(0.07)	(0.13)	(0.17)	
Herbicide tolerant	0.00	0.00	0.01	0.01	NA
	(0.03)	(0.01)	(0.07)	(0.07)	
Insect resistant	0.01	0.00	0.02	0.05	NA
	(0.07)	(0.06)	(0.09)	(0.14)	
No. of observations	1,778	1,420	358	450	1,328
		Rapeseed			
Yield	16,164.46	13,623.73	20,363.35	17,313.31	15,421.09
	(8,082.97)	(6,935.72)	(8,104.34)	(7,674.74)	(8,259.82)
Genetically engineered seed	0.02	0.01	0.05	0.05	NA
share	(0.11)	(0.07)	(0.18)	(0.18)	
No. of observations	756	471	285	297	459
		Soybean			
Yield	15,760.13	14,334.70	21,177.71	18,841.01	14,559.26
	(8,049.53)	(7,789.70)	(6,594.89)	(5,634.42)	(8,518.93)
Seed share					
Genetically engineered	0.03	0.01	0.04	0.12	NA
	(0.15)	(0.07)	(0.17)	(0.27)	
Herbicide tolerant	0.03	0.03	0.04	0.12	NA
	(0.16)	(0.15)	(0.17)	(0.27)	
No. of observations	1,469	1,163	306	412	1,119

Notes: Means with standard deviations in parentheses. NA = not applicable.

In order to test the theory that yield gains from GE crop adoption will be greatest in regions that suffer high pest pressure and have diminished access to chemical pest control agents, we estimate equation (7) separately for developed and developing countries. Because many developing countries effectively employ chemical pest control agents and because pest pressure is expected to be greatest in tropical regions, categorizing countries by eco-

Table 8.3   Summary statistics: Harvest, genetically engineered, and trait areas

	All	Developing	Developed	Adopters	Nonadopters
		Cotton			
Harvest area	474,349.90	428,056.00	896,649.60	1,379,338.00	212,009.10
	(36,980.19)	(37,104.66)	(155,609.40)	(145,291.90)	(14,420.28)
Area					
Genetically	68,553.91	40,843.28	321,334.10	305,041.90	NA
engineered	(13,715.36)	(11,320.57)	(90,135.19)	(59,087.57)	
Heat tolerant	14,809.95	794.37	142,662.00	65,899.31	NA
	(4,238.49)	(326.75)	(41,290.66)	(18,581.88)	
Insect resistant	45,593.07	39,889.99	97,617.34	202,873.90	NA
	(10,514.57)	(11,313.96)	(25,651.71)	(45,686.30)	
No. of observations	1,326	1,195	131	298	1,028
		Maize			
Harvest area	1,479,825.00	1,360,254.00	1,954,099.00	4,148,485.00	575,534.70
	(98,446.21)	(88,076.30)	(341,315.70)	(355,597.30)	(21,051.80)
Genetically	109,796.70	15,909.59	482,198.30	433,819.10	NA
engineered area	(30,228.59)	(4,282.14)	(147,695.10)	(118,219.00)	
Heat-tolerant area	48,679.08	2,454.09	232,029.60	192,336.50	NA
	(18,522.17)	(861.15)	(91,386.47)	(72,822.68)	
Insect resistant area	97,552.94	14,295.37	427,792.50	385,442.50	NA
	(29,210.24)	(3,861.43)	(143,092.30)	(114,434.10)	
No. of observations	1,778	1,420	358	450	1,328
		Rapeseed			
Harvest area	579,795.00	586,433.90	568,823.40	1,378,898.00	62,728.59
	(56,032.14)	(78,956.79)	(71,337.53)	(129,412.50)	(5,906.97)
Heat-tolerant area	56,013.80		148,584.00	142,580.60	NA
	(16,089.23)		(42,155.01)	(40,484.63)	
No. of observations	756	471	285	297	459
		Soybean			
Harvest area	955,104.90	729,134.40	1,813,940.00	3,208,778.00	76,662.81
	(100,410.50)	(78,176.62)	(376,048.00)	(333,191.70)	(5,633.53)
Heat-tolerant area	324,252.10	185,842.00	850,301.00	1,156,132.00	NA
	(62,136.70)	(42,322.81)	(249,257.40)	(216,403.60)	
No. of observations	1,469	1,163	306	412	1,119

Note: See table 8.2 notes.

nomic status is admittedly crude. The development literature has struggled, however, to develop appropriate country classifications according to agro-ecological factors and doing so is beyond the scope of this chapter. Nevertheless, estimated yield effects from the separate regressions of the developed and developing country samples does support the theory from section 8.3. The separate estimation of GE-seed effects for developed and developing countries are reported in tables 8.6 and 8.7, respectively. The magnitudes of these effects relative to conventional seed effects are summarized in rows (2) and (3) of table 8.5. The estimated yield gains associated with GE seed

Table 8.4   **Genetically engineered seed adoption effects**

Crop	Total area (1)	Genetically engineered area (2)
Cotton	1.313***	0.854***
	(0.220)	(0.130)
Maize	6.363***	2.902***
	(0.548)	(0.419)
Rapeseed	1.499***	0.382***
	(0.128)	(0.107)
Soybean	2.461***	0.307***
	(0.203)	(0.112)
Oats	1.202***	
	(0.0917)	
Rice	5.094***	
	(0.545)	
Sorghum	1.236***	
	(0.194)	
Wheat	2.257***	
	(0.254)	
Constant	−366,994	
	(239,633)	
No. of observations	10,717	
No. of groups	627	
R^2	0.728	

Note: Robust standard errors in parentheses.
***Significant at the 1 percent level.

Table 8.5   **Yield gain from genetically engineered seed as percent of yield**

Variable	Cotton (1)	Maize (2)	Rapeseed (3)	Soybean (4)
All countries	65.042	45.607	25.484	12.475
Developed countries	22.886	15.193	24.057	7.040
Developing countries	109.510	56.403	NA	30.189

Note: NA = not applicable.

are greater in developing countries than in developed countries for each GE crop. These differences are statistically significant at the 95 percent level.

We further estimate equation (7) with the addition of GE and non-GE time trends. These results are reported in table 8.8. We find a positive and significant trend associated with non-GE crop yields for cotton, maize, rapeseed, rice, and wheat. These correspond to 1.37 percent, 0.99 percent, 2.17 percent, 0.65 percent, and 1.16 percent annual growth from 1990 to 2008 for each of these crops, respectively. The GE cotton, rapeseed, and soybean exhibited statistically significant positive yield growth over the same time period, suggesting that learning by doing and learning

Table 8.6   Genetically engineered seed adoption effects in developed countries

Crop	Total area (1)	Genetically engineered area (2)
Cotton	1.407***	0.322***
	(0.267)	(0.105)
Maize	12.440***	1.890***
	(2.867)	(0.485)
Rapeseed	1.538***	0.370***
	(0.126)	(0.099)
Soybean	2.784***	0.196
	(0.624)	(0.164)
Oats	2.149***	
	(0.115)	
Rice	5.381***	
	(1.154)	
Sorghum	4.572***	
	(0.366)	
Wheat	2.189***	
	(0.222)	
Constant	−453,968*	
	(262,868)	
No. of observations	2,208	
No. of groups	150	
R^2	0.848	

Note: Robust standard errors in parentheses.
***Significant at the 1 percent level.
*Significant at the 10 percent level.

by using have fueled yield growth that dominates declines caused by the pattern of adoption (i.e., expansion of GE seed to farms that benefit less) and development of resistance to complementary chemicals. When the GE-seed trends are introduced, however, significance of the average GE-seed effect is lost except in maize.

The foregoing results demonstrate that GE crop adoption generally has statistically and economically significant effects on yields. As the threshold adoption model introduced in section 8.3 demonstrates, farmers select to adopt GE technologies based on their expected gain. These gains are expected to increase in pest pressure and farm quality. Our estimates do not control for the selection at the farm level. To the extent that GE crops are adopted on farms of higher quality, these estimates will be upwardly biased estimates of the population average treatment effect (PATE). However, they represent unbiased estimates of the population-average treatment effect of the treated (Imbens and Wooldridge 2009). These estimates of yield gains among adopters are not inconsistent with some estimates in the existing literature based on field trials that control for the farmer selection problem.

Table 8.7   **Genetically engineered seed adoption effects in developing countries**

Crop	Total area (1)	Genetically engineered area (2)
Cotton	1.062***	1.163***
	(0.239)	(0.219)
Maize	5.404***	3.048***
	(0.508)	(0.409)
Rapeseed	1.476***	
	(0.210)	
Soybean	2.120***	0.640***
	(0.273)	(0.191)
Oats	1.123***	
	(0.091)	
Rice	5.058***	
	(0.549)	
Sorghum	0.966***	
	(0.124)	
Wheat	2.250***	
	(0.390)	
Constant	−453,968*	
	(262,868)	
No. of observations	8,509	
No. of groups	477	
R^2	0.650	

Note: Robust standard errors in parentheses.
***Significant at the 1 percent level.
*Significant at the 10 percent level.

Furthermore, unlike studies based on field trials, we have not endeavored to estimate a "gene" effect, but rather the "GE-adoption" effect, which incorporates behavioral responses to GE adoption, including the adoption of other technologies and farming practices and changes in production input-use (e.g., fertilizer-use) that theory predicts will boost potential output. The GE-adoption effect that we estimate should dominate the gene effect estimated in the extant literature.

While the potential for upward bias of a PATE estimate is real, it should also be noted that the upward bias traditionally associated with the endogeneity of technology adoption should be somewhat minimized in this case for several reasons. First, the technology under consideration serves to reduce the complexity of farming, suggesting that farmers with less human capital may benefit the most from adoption. Second, while theory predicts the gains increase in land quality, it also suggests the benefits of adoption will be greater where pest pressure is higher. It is not clear this land will be of higher quality than land with less pest pressure. It is quite possible that pest pressure is negatively correlated with land quality such that the positive

Table 8.8   Genetically modified and conventional seed yield trends

Crop	Total area (1)	Genetically engineered area (2)	Conventional trend (3)	Genetically engineered trend (4)
Cotton	1.240***	−0.164	0.017**	0.077***
	(0.294)	(0.297)	(0.009)	(0.026)
Maize	5.055***	2.586***	0.050**	−0.033
	(1.610)	(0.515)	(0.024)	(0.030)
Rapeseed	1.262***	−0.049	0.027***	0.016***
	(0.101)	(0.092)	(0.009)	(0.005)
Soybean	2.374***	0.005	0.008	0.026**
	(0.158)	(0.122)	(0.015)	(0.012)
Oats	1.336***		0.015	
	(0.092)		(0.012)	
Rice	5.267***		0.034***	
	(0.545)		(0.002)	
Sorghum	1.250***		0.002	
	(0.194)		(0.007)	
Wheat	2.584***		0.030***	
	(0.254)		(0.007)	

Note: Robust standard errors in parentheses.
***Significant at the 1 percent level.
*Significant at the 5 percent level.

selection bias will be muted. Depending on the distribution of pest pressure and quality, the selection bias could be negative. Third, GE seed is adopted on marginal land that was not profitably farmed before the introduction of the technology. This land expansion effect further diminishes the likelihood that the quality of farms that adopt GE crops far exceeds the quality of farms that do not adopt.

8.6   Simulating Impacts during the 2008 Food Crisis

In 2008, a global food crisis induced hunger and starvation in poor regions of the world as prices for grains rose dramatically and major food producing countries slashed exports to protect domestic markets. Food prices reached near-record levels in 2008, with some commodity prices nearly doubling in just a few years and food indexes climbing 56 percent in one year. The dramatic run-up in food prices in 2008 coincided with record biofuel production, so much of the blame for food insecurity was leveled at the diversion of harvest from food to fuel uses.

Without the increased food supply afforded by agricultural biotechnology adoption, prices would have climbed even higher. Using partial equilibrium analysis, it is possible to consider what would have happened to food markets in 2008 if observed levels of biofuel production had prevailed and the

Table 8.9   **Simulation scenarios**

	Scenario 1	Scenario 2	Scenario 3
Own-price elasticity of demand	−0.300	−0.500	−0.300
Own-price elasticity of supply	0.300	0.300	0.300
Cross-price elasticities of demand	0.050	0.050	0.050
Cross-price elasticities of supply	−0.100	−0.100	−0.075

additional output attributable to GE seed adoption had not. To this end, we employ a multimarket framework to model the impacts of 2008 biofuel production on soybean, maize, wheat, and rapeseed. We assume a global market for commodities and simulate three separate assumptions on own and cross-price elasticities of demand and supply. These scenarios are summarized in table 8.9. Scenario 1 is characterized by reasonable elasticity assumptions based on estimated elasticities in the literature. Scenario 2 is characterized by more elastic demand, and Scenario 3 incorporates greater substitutability among crop supply. The supply attributable to GE crop adoption is determined by multiplying the estimated GE yield gain by the area planted to GE crops for each crop.[4] We further parametrize the model based on observed prices and quantities in 2008. We then consider the price effect of biofuel production by subtracting biofuel demand and finding the new equilibrium price.

Global biofuel production in 2008 recruited 86 million tons (10 percent) of global maize production and 8.6 million tons of global vegetable oil, which we assume was equally drawn from soybean and rapeseed production to constitute 7 percent of the global rapeseed harvest and 2 percent of the global soybean harvest. This increased demand for maize, soybean, and rapeseed increased prices 67 percent, 40 percent, 36 percent, and 57 percent for maize, soybean, wheat, and rapeseed, respectively. As reported in table 8.10, world prices for these four commodities would have been between 26 percent and 40 percent lower without biofuel demand given the assumptions of Scenario 1. Without the yield gains of global biotechnology production, 2008 prices would have been considerably higher. Corn prices would have been 35 percent higher, soybean prices 43 percent higher, wheat prices 27 percent higher, and rapeseed prices 33 percent higher.[5] As is also shown in table 8.10, even under the assumptions of more elastic demand (Scenario 2) and supply substitutability (Scenario 3), GE crop adoption in 2008 alone

4. We employ the developing and developed-country estimates in the simulations.
5. An estimate of the global production gains attributable to biotechnology adoption was determined for each maize, soybean, and rapeseed by multiplying observed country-level production in 2008 by the country-appropriate estimate of the GE-induced percentage increase in yield and the country-crop-year-specific GE-crop share. These estimates determined GE-induced output gains to constitute 5 percent, 11 percent, and 4 percent of total output for maize, soybean, and rapeseed, respectively.

Table 8.10   Simulating food price effects of biofuel with and without biotechnology

				Percent change	
	2008 price	No biofuel	No biotech	No biofuel	No biotech
Scenario 1: Base					
Corn	223.13	133.28	300.24	−40.27	34.56
Soybean	474.74	337.96	676.55	−28.81	42.51
Wheat	268.59	197.87	342.25	−26.33	27.42
Rapeseed	604.92	385.70	802.32	−36.24	32.63
Scenario 2: Elastic demand					
Corn	223.13	178.70	256.40	−19.91	14.91
Soybean	474.74	337.96	575.33	−28.81	21.18
Wheat	268.59	197.87	293.51	−26.33	9.27
Rapeseed	604.92	385.70	685.91	−36.24	13.38
Scenario 3: Increased substitutability					
Corn	223.13	157.19	274.76	−29.55	23.14
Soybean	474.74	390.71	623.64	−17.70	31.36
Wheat	268.59	227.95	310.92	−15.13	15.76
Rapeseed	604.92	451.37	732.85	−25.38	21.15

significantly reduced food prices. The cumulative effect of GE yield gains over the past fourteen years is likely greater still, as inventories carried into 2008 would have been larger, serving to dampen upward pressure on prices. Given the degree of suffering that near-record-high commodity prices in 2008 induced among poor populations, it is likely that agricultural biotechnology adoption helped to avert starvation and death. A more complete characterization of the welfare effects of biofuel and biotechnology adoption is the subject of ongoing research.

8.7   Discussion and Conclusions

In 2008, food riots and the doubling of commodity prices in some regions served as a reminder that with slowing agricultural productivity growth and growing demand for farm output, the victory over hunger could only be ephemeral. Agricultural production must grow in order to feed and fuel a global population that is at once increasing in size and wealth. Because of growing concern about climate change and biodiversity loss, production may need to grow without expanding into natural lands. This chapter provides new econometric analysis of aggregate farm yields that suggests that among adopting farms, agricultural biotechnology boosts yields of the four main crops in which it has been introduced. Consistent with the theory developed in this chapter, we find that the yield gains are greatest in developing countries, which are generally characterized by high pest pressure and limited access to insecticides. We also show that the yield effect of

GE crop adoption is growing over time, suggesting that learning effects have dominated the effects of expansion into less suitable applications and the development of resistance. This analysis, which points to the capacity for agricultural biotechnology to drive productivity growth, is constrained by data limitations that preclude controls for farm-level endogeneity of adoption. Consequently, our estimates can conservatively be interpreted as a population average treatment effect on the treated.

Simulation analysis based on the econometric estimation shows that, at the height of the 2008 global food crisis, the additional output generated by GE-crop yield gains mitigated price increases, perhaps saving lives in poor countries. Absent the intensification permitted by agricultural biotechnology, an additional twenty million hectares of land—an area equal in size to the state of Utah—would have been required to produce the 2008 harvest of staple crops. Such expansion of farmland would come at a cost in terms of GHG emissions (from land conversion) and risk to biodiversity, especially if forests were cleared to accommodate the additional crops. This analysis suggests that agricultural biotechnology constitutes a tool to overcome challenges posed by macro trends at the outset of the twenty-first century. First-generation GE crops permit the intensification of agriculture, which effectively frees land for production of biofuel, or at least diminishes the demand for new cropland induced by rising food and fuel needs. In future research, we intend to investigate the capacity for yield improvements associated with increased adoption of agricultural biotechnology and to explore empirically the degree to which agricultural biotechnology adoption is land-saving.

References

Alston, J. M., J. M. Beddow, and P. G. Pardey. 2009. "Agricultural Research, Productivity, and Food Prices in the Long Run." *Science* 325 (5945): 1209–10.

Ameden, H., M. Qaim, and D. Zilberman. 2005. "Adoption of Biotechnology in Developing Countries." In *Agricultural Biodiversity and Biotechnology in Economic Development,* edited by Joseph Cooper, Leslie Marie Lipper, and David Zilberman, 329–57. New York: Springer Science+Business Media.

Balmford, A., R. E. Green, and J. P. W. Scharlemann. 2005. "Sparing Land for Nature: Exploring the Potential Impact of Changes in Agricultural Yield on the Area Needed for Crop Production." *Global Change Biology* 11 (10): 1594–1605.

Barbier, E. B. 2001. "The Economics of Tropical Deforestation and Land Use: An Introduction to the Special Issue." *Land Economics* 77 (2): 155–71.

Crost, B., B. Shankar, R. Bennett, and S. Morse. 2007. "Bias from Farmer Self-Selection in Genetically Modified Crop Productivity Estimates: Evidence from Indian Data." *Journal of Agricultural Economics* 58 (1): 24–36.

David, P. A. 1969. *A Contribution to the Theory of Diffusion.* Stanford, CA: Stanford University, Research Center in Economic Growth.

Feder, G., R. E. Just, and D. Zilberman. 1985. "Adoption of Agricultural Innovations in Developing Countries: A Survey." *Economic Development and Cultural Change* 33 (2): 255–98.

Feng, H., and B. A. Babcock. 2010. "Impacts of Ethanol on Planted Acreage in Market Equilibrium." *American Journal of Agricultural Economics* 92 (3): 789–802.

Imbens, G. W., and J. M. Wooldridge. 2009. "Recent Developments in the Econometrics of Program Evaluation." *Journal of Economic Literature* 47 (1): 5–86.

James, C. 2009. "Global Status of Commercialized Biotech/GM Crops: 2008." ISAAA Brief no. 41. Ithaca, NY: International Service for the Acquisition of Agri-biotech Applications.

Just, R. E., D. Zilberman, E. Hochman, and Z. Bar-Shira. 1990. "Input Allocation in Multicrop Systems." *American Journal of Agricultural Economics* 72 (1): 200–209.

Khanna, M., and D. Zilberman. 1999. "Barriers to Energy-Efficiency in Electricity Generation in India." *Energy Journal* 20 (1): 25–42.

Lence, S. H., and D. J. Miller. 1998. "Estimation of Multi-Output Production Functions with Incomplete Data: A Generalised Maximum Entropy Approach." *European Review of Agricultural Economics* 25 (2): 188–209.

Lichtenberg, E., and D. Zilberman. 1986. "The Econometrics of Damage Control: Why Specification Matters." *American Journal of Agricultural Economics* 68 (2): 261–73.

Matson, P. A., W. J. Parton, A. G. Power, and M. J. Swift. 1997. "Agricultural Intensification and Ecosystem Properties." *Science* 277 (5325): 504–9.

Mundlak, Y. 2001. "Production and Supply." In *Handbook of Agricultural Economics.* 1st ed., Vol. 1, edited by B. L. Gardner and G. C. Rausser, 3–85. Oxford, UK: Elsevier.

———. 2011. "Plowing through the Data." *Annual Review of Resource Economics,* forthcoming.

National Research Council. 2010. *The Impact of Genetically Engineered Crops on Farm Sustainability in the United States.* Washington, DC: National Academy of Sciences.

Qaim, M. 2009. "The Economics of Genetically Modified Crops." *Annual Review of Resource Economics* 1:665–94. doi:10.1146/annurev.resource.050708.144203.

Qaim, M., and D. Zilberman. 2003. "Yield Effects of Genetically Modified Crops in Developing Countries. *Science* 299 (5608): 900–902.

Waggoner, P. E. 1995. "How Much Land Can Ten Billion People Spare for Nature? Does Technology Make a Difference?" *Technology in Society* 17 (1): 17–34.

Wooldridge, J. M. 2005. "Fixed-Effects and Related Estimators for Correlated Random-Coefficient and Treatment-Effect Panel Data Models." *Review of Economics and Statistics* 87 (2): 385–90.

Zilberman, D., A. Schmitz, G. Casterline, E. Lichtenberg, and J. B. Siebert. 1991. "The Economics of Pesticide Use and Regulation." *Science* 253 (5019): 518–22.

Contributors

Bruce A. Babcock
Department of Economics
Iowa State University
578F Heady Hall
Ames, IA 50011

Jayson Beckman
Economic Research Service/U.S.
  Department of Agriculture
1800 M Street NW
Room S4178
Washington, DC 20036-5831

Xiaoguang Chen
Institute for Genomic Biology
University of Illinois at Urbana-
  Champaign
1206 West Gregory Drive
Urbana, IL 61801

Rachael E. Goodhue
Department of Agricultural and
  Resource Economics
University of California, Davis
One Shields Avenue
Davis, CA 95616-8512

Barry K. Goodwin
Agricultural and Resource Economics
North Carolina State University
Box 8109
Raleigh, NC 27695-8109

Joshua S. Graff Zivin
School of International Relations and
  Pacific Studies
University of California, San Diego
9500 Gilman Drive
MC 0519
La Jolla, CA 92093-0519

Thomas W. Hertel
Department of Agricultural
  Economics
and Center for Global Trade Analysis
Purdue University
403 West State Street
West Lafayette, IN 47907

Haixiao Huang
Institute for Genomic Biology
University of Illinois at Urbana-
  Champaign
1206 West Gregory Drive
Urbana, IL 61801

Madhu Khanna
Department of Agricultural and
  Consumer Economics
University of Illinois at Urbana-
  Champaign
1301 West Gregory Drive
Urbana, IL 61801

Jeffrey LaFrance
School of Economic Sciences
Washington State University
PO Box 646210
Pullman, WA 99164-6210

Ethan Ligon
Department of Agricultural and
    Resource Economics
University of California, Berkeley
207 Giannini Hall
Berkeley, CA 94720-3310

Ashok K. Mishra
Department of Agricultural
    Economics and Agribusiness
LSU AgCenter, Louisiana State
    University
211 Martin D. Woodin Hall
Baton Rouge, LA 70803

Hayri Önal
Department of Agricultural and
    Consumer Economics
University of Illinois at Urbana-
    Champaign
1301 Gregory Drive
Urbana, IL 61801

François Ortalo-Magné
School of Business
University of Wisconsin
975 University Avenue
Madison, WI 53706

Jeffrey M. Perloff
Department of Agricultural and
    Resource Economics
University of California, Berkeley
207 Giannini Hall, MC 3310
Berkeley, CA 94720-3310

Rulon Pope
Department of Economics
Brigham Young University
154 Faculty Office Building
Provo, UT 84602

Carlo Russo
Faculty of Economics
University of Cassino
via S. Angelo—Località "Folcara"
03043 Cassino (FR), Italy

Steven Sexton
Department of Agricultural and
    Resource Economics
207 Giannini Hall, MC 3310
University of California, Berkeley
Berkeley, CA 94720

Jesse Tack
Department of Agricultural
    Economics
Mississippi State University
Box 5187
Mississippi State, MS 39762

David Zilberman
Department of Agricultural and
    Resource Economics
207 Giannini Hall
University of California, Berkeley
Berkeley, CA 94720

Author Index

Subject Index